Water Science and Technology Library

Volume 73

The aim of the Water Science and Technology Library is to provide a forum for dissemination of the state-of-the-art of topics of current interest in the area of water science and technology. This is accomplished through publication of reference books and monographs, authored or edited. Occasionally also proceedings volumes are accepted for publication in the series.

Water Science and Technology Library encompasses a wide range of topics dealing with science as well as socio-economic aspects of water, environment, and ecology. Both the water quantity and quality issues are relevant and are embraced by Water Science and Technology Library. The emphasis may be on either the scientific content, or techniques of solution, or both. There is increasing emphasis these days on processes and Water Science and Technology Library is committed to promoting this emphasis by publishing books emphasizing scientific discussions of physical, chemical, and/or biological aspects of water resources. Likewise, current or emerging solution techniques receive high priority. Interdisciplinary coverage is encouraged. Case studies contributing to our knowledge of water science and technology are also embraced by the series. Innovative ideas and novel techniques are of particular interest.

Comments or suggestions for future volumes are welcomed.

Vijay P. Singh, Department of Biological and Agricultural Engineering & Zachry Department of Civil Engineering, Texas A and M University, USA
Email: vsingh@tamu.edu

More information about this series at http://www.springer.com/series/6689

Arup K. Sarma · Vijay P. Singh
Suresh A. Kartha · Rajib K. Bhattacharjya
Editors

Urban Hydrology, Watershed Management and Socio-Economic Aspects

Springer

Editors
Arup K. Sarma
Department of Civil Engineering
Indian Institute of Technology Guwahati
Guwahati, Assam
India

Suresh A. Kartha
Department of Civil Engineering
Indian Institute of Technology Guwahati
Guwahati, Assam
India

Vijay P. Singh
Department of Biological and Agricultural Engineering and Zachry Department of Civil Engineering
Texas A and M University
College Station, TX
USA

Rajib K. Bhattacharjya
Department of Civil Engineering
Indian Institute of Technology Guwahati
Guwahati, Assam
India

ISSN 0921-092X ISSN 1872-4663 (electronic)
Water Science and Technology Library
ISBN 978-3-319-40194-2 ISBN 978-3-319-40195-9 (eBook)
DOI 10.1007/978-3-319-40195-9

Library of Congress Control Number: 2016941318

Printed on acid-free paper

This Springer imprint is published by Springer Nature
The registered company is Springer International Publishing AG Switzerland

Preface

The twentieth century witnessed massive urbanization at a rapid pace in developing countries, and this pace is continuing in the twenty-first century. Currently, nearly half of the world's population lives in urban areas. The rapid urbanization is a direct result of economic growth and people's urge toward higher comfort and access to education, health, and travel facilities. Urbanization, more often, causes strain on natural and man-made resources that are meant for the well-being of people. One direct and most perceptible consequence of urbanization is the change in land use and land cover, which, in turn, impacts the hydrological resources and the hydrological cycle. This impact is witnessed through changes in surface and groundwater levels, surface runoff patterns, diurnal variations in temperature, humidity, cloud cover, smog, pollution, radiation, and wind. Therefore, scientific studies on urban hydrology are vital for appropriate design of urban landscapes and civil infrastructure works.

A center of excellence for "Integrated Landuse Planning and Water Resources Management" sponsored by Ministry of Urban Development, Government of India, was established at the Indian Institute of Technology Guwahati in the year 2010 to undertake studies on urban hydrology of the northeast region of India. As part of the mandate, the center conducted an international conference ENSURE 2012 and invited authors to submit articles related to urban hydrology. A number of authors presented highly thoughtful research papers on the topic, and seeing their value, it was decided to publish these conference articles in a book form. Consequently, we revisited the conference themes and articles and selected some of them for inclusion in this volume.

The subject matter of the book is, therefore, divided into six parts. One major impact of urbanization is the rapid increase in floods and storm waters. Part I, comprising of four chapters, deals with the impact of urbanization and mitigation measures to reduce the negative impacts of urbanization. The impact on hydrological processes, such as infiltration and surface runoff of a hilly terrain, due to changing slopes brought out by urban development, is analyzed using laboratory experiments in one of the chapters. These experiments highlighted the impacts of the presence and absence of vegetation on surface runoff. The role of imperviousness on surface runoff and infiltration due to urbanization is analyzed in chapter "Impact of Total and Effective Imperviousness on Runoff Prediction" by comparing the outputs of hydrological models of a region using total and effective impervious areas. Chapter "Issues of Urban Drainage—Present Status and the Way Forward" discusses on issues related to urban drainage and the way to move forward. The advantages of optimal allocation of ecological management practices in a hilly urban watershed to alleviate the impacts of urban floods were discussed in the final chapter in this part.

Remote sensing and GIS are widely used in various applications of natural resources, and hydrology is one of the most benefitted sciences from the use of remote sensing. GIS, remote sensing, and digital elevation models are very much important in urban hydrological modeling, and these tools are efficient in assessing the quantity of water resources for sustainable management. Therefore, Part II deals with remote sensing and GIS applications especially in urban hydrology. It is comprised of three chapters. One of the most adopted forms to quantify the losses due to infiltration is the use of curve number technique. The first chapter of this part describes the use of GIS and satellite images in the Soil Conservation Services (SCS) [now called Natural Resources Conservation Service (NRCS)] curve number method to estimate surface runoff, infiltration, and groundwater recharge. The next chapter discusses the effect of rainfall, and land use and land cover change on surface runoff as well as on surface soil erosion, mainly due to the change in vegetation area, of an urban region using remote sensing and GIS. The final chapter in this part describes the use of digital elevation models to characterize flood plains of a river passing through urban locations.

Any hydrological study will be incomplete without dealing with groundwater and subsurface water of the region. Special emphasis is given to subsurface hydrology in this book. Therefore, groundwater and subsurface water hydrology of urban areas is the theme of Part III that consists of five chapters. Some ideas in this part emphasize the need and importance of groundwater management and rainwater harvesting in arid and semiarid areas. The authors of some chapters have critically reviewed the anthropogenic causes of saltwater intrusion in urban coastal regions due to overexploitation of groundwater. Applications of groundwater models, like MODFLOW, to predict and forecast water levels and describe the aquifer stresses are also important studies in this part. There are chapters on the need of groundwater management due to rapid urbanization and linked optimization for groundwater contamination monitoring networks. Also discussed in this part is the conjunctive use of rainwater harvesting methods as tools for artificial recharge of groundwater to aid urban water supply systems. The velocities and patterns of flow of subsurface water through porous media depend on the type of soil, degree of water saturation, and the soil–water characteristics. A useful study on soil–water characteristics for urban hill soils of northeast India is presented in one of the chapters in this part.

Considering the various expenditures on field and laboratory exercises to study hydrological phenomena, computational modeling mechanisms allow for gaining an understanding of these phenomena at low cost. Therefore, Part IV is on applications of computational and numerical models to urban hydrological problems. It contains five chapters. The hydraulic models for urban drainage systems often need to consider erosion and sedimentation. Some research studies, in this part, highlight the need for two-dimensional computational models for urban drainage. These computational models determine the critical sections of urban drains that are quite important in drainage design. Urban floods can be modeled using computational software, such as HEC-HMS and HEC-RAS. Simulation results from these computational models can assess the risk associated with these floods. Some chapters in this part deal with risk assessment due to urban floods. An interesting chapter in this part is the use of well water-level fluctuations to develop seismographs for highly sensitive seismic regions. In yet another chapter in this part, the authors present the use of momentum transfer methods, in their computational models, to evaluate flow in compound open channels. There are chapters that discuss optimal reservoir operation policies, including environmental flows, for hydropower projects of northeast India.

Soft computing techniques in urban hydrology are dealt with in Part V that contains five chapters. There are many soft computing tools used in hydrological analysis—artificial neural network, fuzzy logic, fuzzy sets, wavelets, etc. The first chapter in this part deals with the application of fuzzy sets in the design of a water distribution network. The uncertainties, if any, in the inputs for urban networks are shown to be overcome by the use of fuzzy sets. The use of wavelet transforms in hydrology is discussed in the next chapter. Artificial intelligence methods, such as artificial neural networks and fuzzy logic, are used in urban hydrology with large success. Soft computing tools can also be used independently or in combination like wavelet–neural networks and adaptive neurofuzzy. The use of such soft computing tools can also be extended to predict pipeline leakages in urban centers or river stages adjacent to urban locations or prediction of surface runoff, etc. Soft computing models depend extensively on time series data, and an article describes that the preprocessing of these data can enhance the performance of such models.

The concluding part, i.e., Part VI, is on socio-economic aspects and the role of society in overall urbanization impacts. It comprises six discourses. The importance of socio-economic studies to understand population, land requirement, economic status, etc., of the people residing in urban hilly areas to assess the carrying capacity of hills is discussed in the first write-up of this part. The effect of urbanization is critically remarked in a chapter that describes the performance of students in the course curriculum. The rapid population growth in the university affects the overall well-being of the campus community, and several environmental parameters are compromised, as a result. Another chapter discusses the role and importance of community in river basin management. Some chapters discuss the role of local bodies in municipal solid waste management, role of government in sustainable urban development, etc. Another important aspect of urbanization is the migration of large population from rural and semiurban areas to urban areas. Such events cause a huge strain on natural and man-made resources, and there are chapters in this part that discuss such issues. The final study in this part discusses the benefits of using urban and natural water bodies as means of transport and communication.

Overall, the book comprising six parts covers a wide range of topics on urban hydrology, urban watershed management, and the socio-economic aspects of urbanization in a single volume. As this book consolidates most of the water and environmental issues related to urbanization, the reader will be benefitted in gathering useful information at one go. Moreover, the parts are selected on thematic aspects, and the reader can directly go to the respective parts for his or her topic of interest.

The authors admit that there are limitations of this book and it is not complete. However, best efforts have been made to consolidate all the related themes in a single book. The authors express their gratitude to the organizing committee of ENSURE 2012 in providing the contents of this book and acknowledge the efforts of students of Department of Civil Engineering, IIT Guwahati, in assisting in the development of the book.

Guwahati, Assam — Arup K. Sarma
College Station, USA — Vijay P. Singh
Guwahati, Assam — Suresh A. Kartha
Guwahati, Assam — Rajib K. Bhattacharjya

Contents

About the Editors

Prof. Arup K. Sarma a professor and former head of Civil Engineering Department, Indian Institute of Technology Guwahati, is presently holding the prestigious *B.P. Chaliha Chair Professor* position given by Ministry of Water Resources, Government of India. Professor Sarma has also served as visiting professor in the Asian Institute of Technology, Bangkok, Thailand during 2015. The NPTEL video course on *hydraulic engineering* developed by Prof. Sarma is receiving wide appreciation from different parts of the globe and has entered into the top 5 most visited course. Till date, he has completed 22 sponsored research project and 44 consultancy projects from India and abroad. Reviewer of several reputed international journals, Prof. Sarma, has also served as member and adviser of various national and international scientific committees. He has also devoted himself for the promotion of scientific temperament in the society through music and drama. As an approved lyricist of All India Radio, Prof. Sarma has composed several songs, drama, and musical features for All India Radio and Television to bring scientific awareness among masses.

Prof. Vijay P. Singh is a University Distinguished Professor and Caroline and William N. Lehrer Distinguished Chair in Water Engineering at Texas A & M University. He received his B.S., M.S., Ph.D. and D.Sc. degrees in engineering. He is a registered professional engineer, a registered professional hydrologist, and an Honorary diplomate of ASCE-AAWRE. He has published more than 870 journal articles; 25 textbooks; 59 edited reference books; 100 book chapters; and 308 conference papers in the area of hydrology and water resources. He has received more than 75 national and international awards, including Arid Lands Hydraulic Engineering Award, Ven Te Chow Award, Richard R. Torrens Award, Norman Medal, and EWRI Lifetime Achievement Award, given by American Society of Civil Engineers; Ray K. Linsley Award and Founder's Award, given by American Institute of Hydrology; Crystal Drop Award, given by Onternational Water Resources Association, and three honorary doctorates. He is a Distinguished Member of ASCE, and a fellow EWRI, AWRA, IWRS, ISAE, IASWC and IE. He is a member of 10 international engineering academies. He has served as President of the American Institute of Hydrology (AIH). He is editor-in-chief of three journals and two book series and serves on editorial boards of more than 25 journals and three book series.

Dr. Suresh A. Kartha is presently an associate professor in the Department of Civil Engineering, IIT Guwahati. He joined IIT Guwahati in the year 2007 as a Senior Lecturer in the specialization Water Resources Engineering and Management in the Department of Civil Engineering. As a faculty, Dr. Kartha has taught courses on numerical methods, hydrology, subsurface hydrology, fluid mechanics, etc., and his research focus is in the area of flow and transport processes in porous media. His Internet Web profile can be accessed through the link http://www.iitg.ac.in/kartha/homepage/index.html.

Dr. Rajib K. Bhattacharjya is a professor in the Department of Civil Engineering, Indian Institute of Technology Guwahati, India. His current research interests include groundwater management and pollution control, management of saltwater intrusion in coastal aquifers, classical and non-classical optimization methods, and artificial neural networks. He has more than 15 years of teaching and research experience and has authored more than 50 peer-reviewed scientific publications in various reputed international journals and conference proceedings. As a faculty at Indian Institute of Technology Guwahati, he has taught courses on subsurface hydrology, hydrology, optimization methods, water resources planning and management, etc. More about his research and academic activities can be found at http://www.iitg.ac.in/rkbc.

Part I
Impacts of Urbanization on Hydrological Processes and Mitigation Measures

Impact of Slope and Vegetation on Hydrological Processes

Achintyamugdha S. Sharma, Dipankar Das, Kumari Koustuvee, Bhupali Dutta, Ruchika Agarwala, Sagar Sen, Dhrubajyoti Thakuria and Arup K. Sarma

Abstract At the face of urbanization, the assessment of the behavior of precipitated water with terrain of varying slopes and varying soil types with or without vegetative cover has gained utmost importance. The fast growing cities and urban hubs clamor for a complete utilization of precipitation thus preventing the extreme situations of incessant floods or drought. This in turn creates an urge for making a study of the behavior of hydrological parameters such as infiltration and runoff with varying slopes and vegetation. These hydrological parameters may also be dependent upon the soil type. In this study, a series of experiments have been performed on an experimental setup known as Rainfall Apparatus and various

A.S. Sharma (✉)
Construction Management and Engineering Department/Department of Civil and Environmental Engineering, North Dakota State University, Fargo, USA
e-mail: achintyamugdha.sharm@ndsu.edu

D. Das
NEEPCO Ltd, Shillong, India
e-mail: dipankar424@gmail.com

K. Koustuvee
Assam Engineering College, Guwahati, India
e-mail: kumarikoustuvee@gmail.com

B. Dutta · R. Agarwala
Department of Civil Engineering, Indian Institute of Technology, Kanpur, India
e-mail: bhupali08@gmail.com

R. Agarwala
e-mail: ruchikaagarwal03@gmail.com

S. Sen · D. Thakuria
Department of Civil Engineering, National Institute of Technology, Silchar, India
e-mail: sagarh1734@gmail.com

D. Thakuria
e-mail: djt2588@gmail.com

A.K. Sarma
Department of Civil Engineering, Indian Institute of Technology, Guwahati, India
e-mail: aks@iitg.ernet.in

A.K. Sarma et al. (eds.), *Urban Hydrology, Watershed Management and Socio-Economic Aspects*, Water Science and Technology Library 73,
DOI 10.1007/978-3-319-40195-9_1

hydrographs of infiltration flow-rate versus time and runoff flow-rate versus time are plotted from the obtained data thereby drawing certain specific and generalized conclusions. Regression equations with a coefficient of correlation of more than 97 % have been used to define the relationship between infiltration and other parameters.

Keywords Infiltration · Runoff · Infiltration capacity · Flow-rate · Flow-rate versus time curves · Catchment with or without vegetation

1 Introduction

Water exists in the earth in three states, viz., solid, liquid, and vapor. Water from the large water reservoirs such as seas, oceans, and rivers gets evaporated by the action of the sun and the formation of cloud follows. Soon, a part of the water vapor in the cloud condenses and falls back upon the ocean or sea while the remaining part of the cloud is transported by the wind to other parts. There the cloud condenses and precipitates in the form of rain. A part of this precipitation is absorbed by the vegetation and given back to the atmosphere by the process of evapotranspiration. Another part of the water enters the soil through pores, if present, and travels down further until it reaches the phreatic surface, which is the free ground water surface in an unconfined aquifer. This process is known as *Infiltration* (Balasubramanium 1990). At some point of time when all the pores or voids in the soil are filled or saturated, the water cannot infiltrate into the soil any further and therefore the excess water flows along the ground surface along a downwards gradient following the laws of gravity. This is known as *Runoff* (Garg 1998). The water in the form of runoff over the ground or infiltrated water though cracks and fissures ultimately reaches the sea from where it is again evaporated and the whole process is repeated. This repetitive process is known as *Hydrological cycle.*

Hydrological processes such as infiltration and runoff are highly variable in space and time. Earlier researchers have determined that the knowledge of the sources and patterns of variation in these processes and their controlling factors is crucial for understanding and modeling the hydrological functioning of semiarid ecosystems (Mayor et al. 2009). Previous researchers have also demonstrated that sloped surfaces behave as a mosaic of runoff generation and infiltration patches which in turn depend strongly upon morphometric characteristics of the slopes, the lithology, the land use, and the different development of soils and their cover (Yair and Lavee 1985; Abrahams and Parsons 1991; Solé-Benet et al. 1997; Cantón et al. 2002; Calvo-Cases et al. 2005; Yair and Lavee 1985). Although it had been indicated in past literature that slope and vegetation can influence hydrological processes such as infiltration and runoff, their interactions are still somewhat unclear (Seeger 2007; Mayor et al. 2009; Li et al. 2011). Li et al. (2011) demonstrated that vegetation cover led to a reduction in runoff. Again, infiltration was found to be reduced with increasing slope angle (Sharma et al. 1983; Morbidelli et al. 2015).

In this work, laboratory scale experiments have been conducted to study the impacts of slope and vegetation upon hydrological processes such as infiltration and runoff and their interactions. Regression equations were also used to define a relationship between infiltration and other accompanying parameters and runoff conditions.

2 Background

The definitions and related theory of some of the terms used in this paper have been thrown light upon in this section.

2.1 *Infiltration*

When rain falls upon the ground, it first wets the vegetation or the bare soil. When the surface cover is completely wet, subsequent rain must either penetrate the surface layers if the surface is permeable, or runoff the surface toward a stream channel if the surface is impermeable.

If the surface layers are porous and have minute passages available for the passage of water droplets, the water infiltrates into the subsurface soil. This movement of water through a porous surface into the subsurface soil is called infiltration. Different types of soil allow water to infiltrate at different rates, measured in mm/h or inches/h. Moreover, various other factors like rainfall intensity, slope, degree of saturation, climatic conditions, etc., also play a vital role in the rate of infiltration.

2.2 *Runoff*

Runoff means the draining or flowing off of precipitation from a catchment area through a surface channel. It thus represents the output from the catchment in a given unit of time.

For a given precipitation, the evapotranspiration, initial loss, infiltration, and detention storage requirements will have to be first satisfied before the commencement of runoff. When these are satisfied, the excess precipitation moves over the land surfaces to reach smaller channels. This portion of the runoff is called *overland flow* and involves building up of storage over the surface and draining off of the same. Usually the lengths and depths of overland flow are small and the flow is in the laminar regime. Flows from several small channels join bigger channels and flows from these in turn combine to form a larger stream and so on, till the flow reaches the catchment outlet. The flow in this mode, where it travels all the time over the surface as overland flow and through the channels as open channel flow and reaches the catchment outlet is called *surface runoff*.

A part of the groundwater that infiltrates moves laterally through upper crusts of the soil and returns to the surface at some location away from the point of entry into the soil. This component of runoff is known variously as *interflow, through flow, storm seepage, subsurface storm flow* or *quick return flow*. Depending upon the time delay between the infiltration and the outflow, the interflow is sometimes classified into *prompt interflow*, i.e., the interflow with the least time lag and *delayed interflow*.

Another route for the infiltrated water is to undergo deep percolation and reach the groundwater storage in the soil. The groundwater follows a complicated and long path of travel and ultimately reaches the surface. The time lag, i.e., the difference in time between the entry into the soil and outflows from it is very large, being of the order of months and years. This part of runoff is called *groundwater runoff* or *groundwater flow*. Groundwater flow provides the dry-weather flow in perennial streams.

2.3 Permeability

Permeability is an engineering property of a soil mass which allows the water to flow through its interconnected voids. The resistance to flow of water is greater when the pores are smaller in size and pore channels are irregular. A material is said to be highly pervious or highly permeable when the water flows through its interconnected voids easily. A completely impervious soil does not permit the water to flow through it. If the volume of void is less the soil is said to be semi-permeable or semi-pervious.

In soils, the interconnected pores provide passage for water. A large number of such flow paths act together and the average rate of flow of water is termed as *Coefficient of Permeability*. It is the measure of ease that the soil provides to the water to flow through its pores.

2.4 Catchment Area

By catchment area is meant the whole of the land and water surface contributing to the discharge at a particular stream or river cross-section. There are many properties of the catchment area which affect the rate and quantity of discharge from it. The area as defined above is usually, but not necessarily, bounded by topographic water divide. The true boundary of the catchment area is indeterminate.

2.5 *Watershed with Slopes*

In most cases, catchment areas contain a mixed topography of steep and plain slopes. A watershed created by a system of rivers and tributaries can have significant behavioral dependence on the topography of the watershed. An example of this topography can be observed in the city of Guwahati in India. The river Basistha and its tributaries flow through the city (Fig. 1). It can be seen from Figs. 2 and 3 that the city of Guwahati consists of a topography with slopes of varying degree surrounding plain areas. Now, it may be anticipated that these topographical features can be potent driving factors for the behavior of hydrological processes such as infiltration and runoff. This lab scale study can pave the way for more field scale studies to understand the relationships of slope and hydrological processes of watersheds with slopes and plains such as the Basistha watershed in Guwahati city.

3 Modes and Methods

3.1 *Experimental Study*

The project work has been carried out at the Basic Rainfall Apparatus (Fig. 4) at the Hydraulics and Water Resources Laboratory of IIT Guwahati. The catchment area of the apparatus is rectangular ($L \times B = 197$ cm $\times$ 90.5 cm). The apparatus

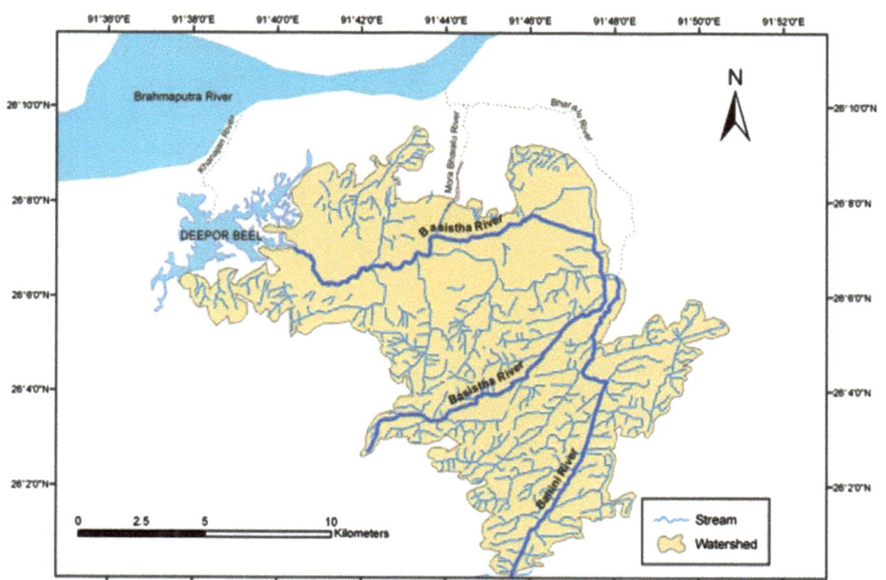

Fig. 1 The Basistha watershed

Fig. 2 3-D view of the hilly part of Basistha watershed

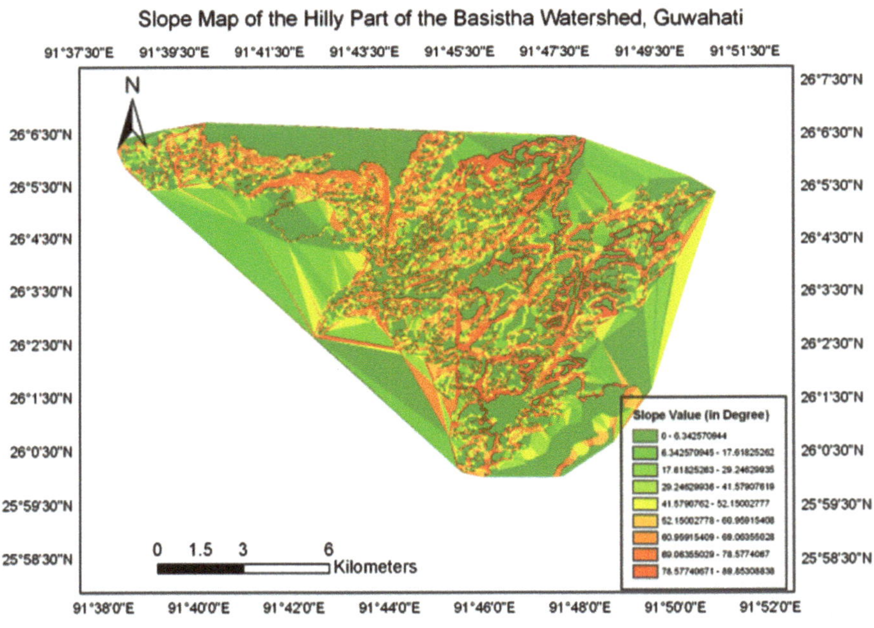

Fig. 3 Slope map of the hilly part of Basistha watershed

consists of two sets of sprinklers each containing four nozzles, by which artificial rainfall is created. The two sets of sprinklers are regulated by two flow valves. The bottom surface of the catchment area is provided with 20 pipes equally spaced, connected to a piezometer (Fig. 5) which shows the level of ground water at the respective positions. On the catchment area, at one-third from either ends, two wells —well 1 and well 2, are present through which water percolates down through the pipes. This water can either be collected to measure the infiltration or can be sent back to the main tank in case of overflow by two knobs.

The apparatus (Fig. 4) consists of two side tanks provided on either side of the catchment where the runoff is collected. Runoff measurement is regulated by two

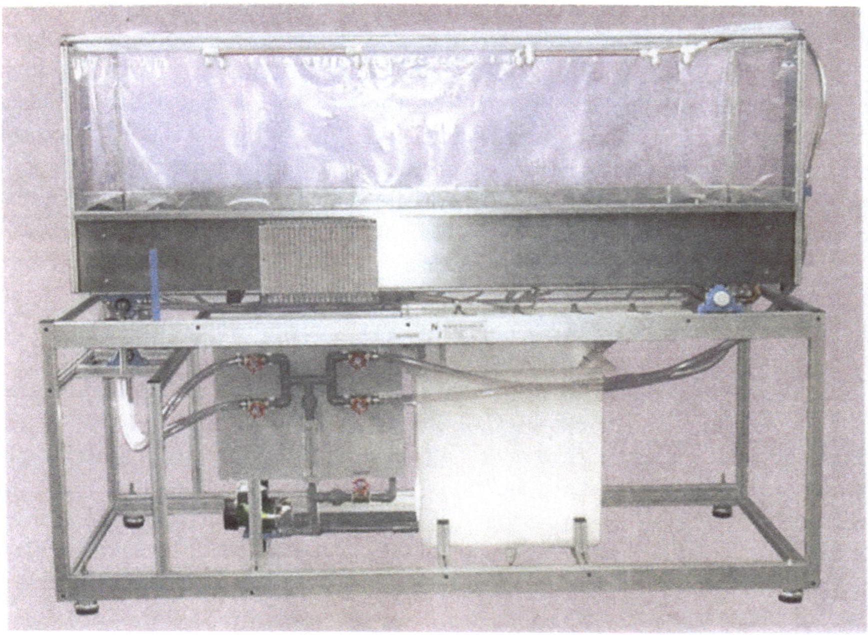

Fig. 4 Rainfall apparatus

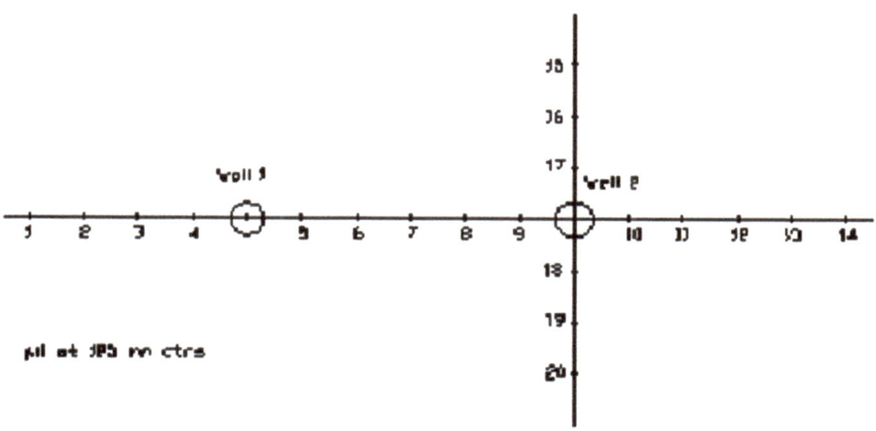

Fig. 5 Plan of piezometer tapings and well position (Basic Rainfall Apparatus Instruction Manual, 2000)

knobs present to the left and right of the knobs controlling the infiltration. The water in the two tanks is regulated by two valves present above the flow valves (Fig. 5).

The intensity of the artificial rainfall can be regulated as per the experimental requirements and can be observed from the flow meter.

There is also provision for slope adjustment of the catchment area which can be varied from a minimum of 0 mm to a maximum of 55 mm vertical slope as per the required conditions.

The circuit is arranged for flow-bypass regulation for maintaining the pump.

All the experiments have been performed under varying slope conditions with or without allowing runoff subject to rainfall in a catchment area with or without vegetation. Initially, the catchment area of the rainfall apparatus is filled with a bed of sand consisting of irregular sand particles. Having filled up the sand into the catchment of the apparatus power is supplied to the apparatus and the motor is switched on and the bypass is opened to full flow condition for about 5 min so that no heating occurs and the coil is saved from damage. After that the bypass valve is partially closed and either or both of the sprinkler valves are opened for about 1 min to create artificial rainfall of required intensity. Then the sprinklers are closed and the water is allowed to percolate, following which the knobs regulating infiltration are adjusted to the flow meter position and the infiltrated water is collected in a measuring cylinder of 500 ml capacity. This process is continued after an interval of every 2 min, for a period of 20–22 min for a particular slope. The above mentioned operations are carried out for a total of 5 slope positions (0, 5, 10, 15 and 20 mm vertical) to measure infiltration flow-rate (with or without vegetation) without allowing runoff. However, in order to allow runoff while measuring infiltration, the plate on the downstream side of the catchment area is opened to collect the runoff in the side tank. It is worth mentioning that the catchment area in this case is reduced to half.

4 Results and Discussion

4.1 Particle Size Gradation Curve

First, the gradation and quality of the sand used in the catchment is assessed by performing sieving operation. The operation of sieving is performed by taking a sample of the sand used in the experiment in order to ascertain the gradation of different sizes of particles present in the same.

The percentage finer is calculated and plotted against particle size to obtain the curve as shown in Fig. 6. This curve is prepared by plotting the values of particle size and percentage finer in the *X*- and *Y*-axes, respectively. It may be mentioned here that the *Y*-axis is taken in a logarithmic scale.

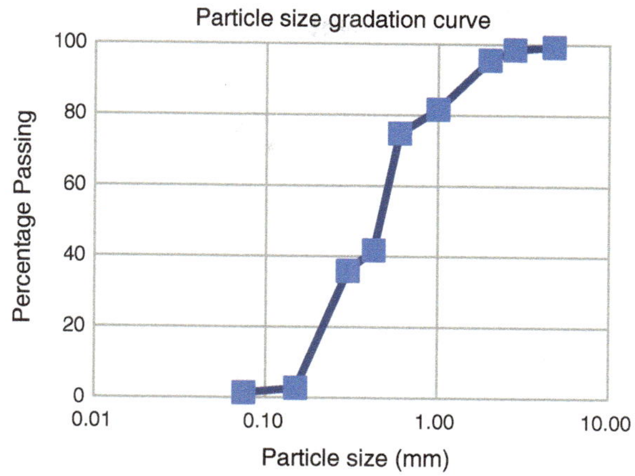

Fig. 6 Particle size versus percentage finer curve

4.1.1 Calculation of Coefficient of Uniformity and Coefficient of Curvature

From the particle size gradation curve obtained the *Coefficient of Uniformity* (C_u) is calculated as under

$$C_u = D_{60}/D_{10} = 1.722/0.287 = 6$$

Thus, the sand can be classified as *well graded*.

The *Coefficient of Curvature* (C_c) is calculated as under

$$C_c = D_{30}^2/(D_{10} \times D_{60}) = 0.902^2/(0.287 \times 1.722) = 1.646$$

Thus, the soil can be classified as *well graded*, as it is between 1 and 3.

4.2 Infiltration by Restricting Runoff Without Vegetation

While keeping the plate on the downstream side of the catchment area closed so as to restrict the water coming from the artificial rainfall of the water sprinkler nozzles to flow out as runoff into the side tanks, the knobs regulating infiltration are brought to the flow meter position and the discharge from infiltrated water is collected in a measuring cylinder for an interval of 2 min. Considering the whole of the catchment area (1,782,850 mm^2) without applying any form of vegetation, the flow-rate is calculated and then plotted against time. Five curves are obtained for the five slopes in that setup and are shown in Fig. 7.

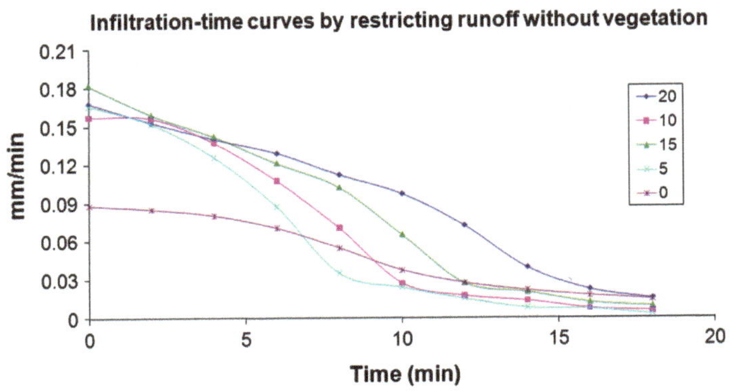

Fig. 7 Infiltration-time curves by restricting runoff without vegetation

4.2.1 Conclusions Drawn from the Curves

- Infiltration decreases with time for all the slopes nonlinearly. It is observed from the graph that the infiltrated water discharges out in high quantities initially and abruptly trails off with time until reaching constancy.
- With the increase in slope the infiltration flow-rate for a particular instant of time increases due to enhanced infiltration through the downstream well due to a pondage created on the downstream side of the apparatus and a resultant water head.
- Due to the presence of the two wells considerable amount of water from the artificial rain reach the bottom of the sand bed earlier than that traveling through the sand bed and it takes considerable amount of time for the latter to discharge out completely. In other words, the combined discharge from the twenty pipes underneath the infiltrating sand layer and the two wells forms the initial higher flow-rate regions of the curves, while the discharge from the twenty pipes alone after the early exhaustion of the water flowing through the wells forms the drastic fall and later constancy of the flow-rate versus time curves.

4.3 Infiltration Without Considering Runoff with Vegetation

While keeping the plate on the downstream side of the catchment area closed so as to keep the water from flowing into the side tanks, this time a vegetative cover is applied into the catchment area in the form garden grass. The five flow-rate versus time graphs obtained are shown in Fig. 8.

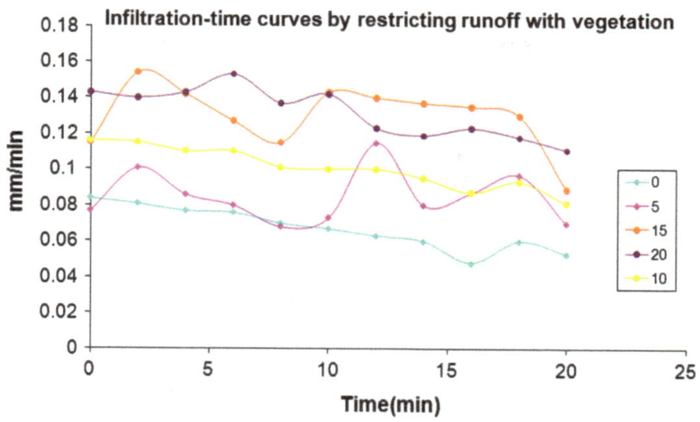

Fig. 8 Infiltration-time curves by restricting runoff with vegetation

4.3.1 Conclusions Drawn from the Curves

- The infiltration is high for 20 and 10 mm slopes rather than for the experiment conducted with horizontal slope.
- All the curves have shown some well-defined peaks. The first peak has been observed due to infiltration from both the wells.
- The second peak is due to infiltration through the second well only when the slope is changed and the bed is no longer horizontal. When the bed is made inclined, water cannot stand near the first well and percolate down through it. All the water then concentrates around the second well and moves down through it.
- With time the sand becomes saturated and hence the infiltration rate decreases. Due to vegetation some of the rainfall water is retained by the grass blades and infiltrates later on giving rise to the third peak of the curves.

4.4 Infiltration by Allowing Runoff Without Vegetation

For the sake of considering the runoff during the measurement of infiltration flow-rate, half the catchment area (891,425 mm^2) is considered and the plate on the downstream side of the catchment is opened so as to allow the water to flow as runoff into the side tanks. It is worth mentioning that the catchment area is devoid of any vegetation whatsoever. The graphs obtained by plotting infiltration flow-rate against time for the five slope positions are as shown in Fig. 9.

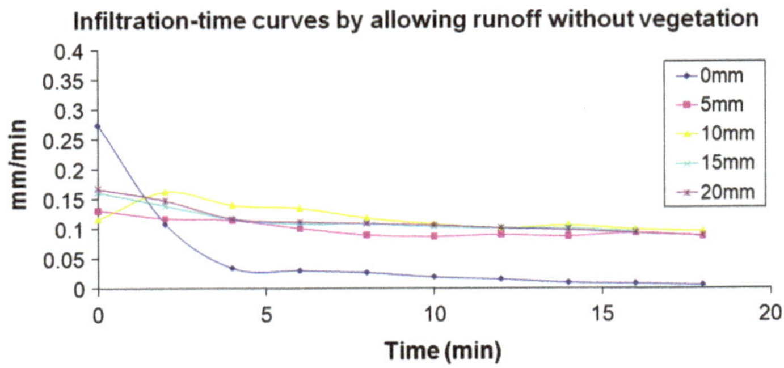

Fig. 9 Infiltration-time curves by allowing runoff without vegetation

4.4.1 Conclusions Drawn from the Curves

- When the catchment is in the horizontal position, it is observed that the flow-rate versus time curve exhibits a pattern similar to the case of infiltration by restricting runoff. This is because most of the water artificially sprinkled upon the catchment is passing through the sand bed as infiltration before flowing off as runoff. The discharge from the one well and that from the twenty pipes underneath the sand layer amount to the initial high rate of discharge, while the discharge from only the pipes after the exhaustion of water flowing through the wells imparts the drastic fall in discharge rate and thus flow-rate with time.
- In general, it is observed that at inclined slope positions of the apparatus the flow-rate versus time curves do not show much fluctuation. This may be due to fast draining off of most of the rain water as runoff due to effect of gravity and the infiltration of the remaining water and discharge through the pipes at a relatively constant rate. Moreover, due to the water flowing through the second well (the first well is out of consideration as only half the catchment area is subject to artificial rainfall) results in the formation of an initial faint peak in the flow-rate versus time curve. In fact, this peak is more prominently formed in the graph plotted for the 10 mm vertical setup.
- It is also observed that the infiltration flow-rate for a particular time in general decreases with a decrease in slope. It may be mentioned that the experiments had been performed in a reverse order from 20 mm vertical slope to horizontal setup of the apparatus. Thus, it can be inferred upon that due to repeated application of water from the sprinkler system into the catchment the voids in the sand layer have become saturated due to which the rate of infiltration is considerably reduced for the subsequent experimental setups.

4.5 *Infiltration by Allowing Runoff with Vegetation*

While keeping the plate on the downstream side of the catchment open so as to allow the water to runoff into the side tanks, one half of the catchment area is subjected to a vegetative cover of garden grass. The hydrographs obtained are shown in Fig. 10.

4.5.1 Conclusions Drawn from the Curves

- It is observed that the infiltration-time curves fall off drastically initially and then maintain relative constancy with occasional minor peaks. This may be due to the fact that initially the grass blades retain some of the water before flowing off as runoff and which gets a chance to infiltrate. Eventually, the water gets exhausted out of the catchment area as runoff and only a little amount is left for further infiltration. The minor peaks may be due to the rapid flow through the well.
- It is seen that the infiltration flow-rate for the 5 and 20 mm vertical slopes are less than that of the horizontal setup of the apparatus. This may be explained by the fact that with the increase in slope of the catchment the effect of gravity becomes more predominant and thus more amount of water flows off as runoff earlier than before, thereby reducing the water infiltrating through the sand bed.
- Again, the infiltration flow-rate for the 10 and 15 mm slopes is more than that of the horizontal setup of the apparatus. This fact quite contradicts the earlier theory about the predominance of gravity effect with increasing slope. This may have resulted due to the repeated application of water from the sprinkler system into the catchment and the subsequent saturation of the voids in the sand layer and thereby reduction of the infiltration rates (the experiments being performed in a reverse order from 20 mm vertical to the horizontal setup). Moreover, various anomalies in observation such as higher intensity of rainfall than the

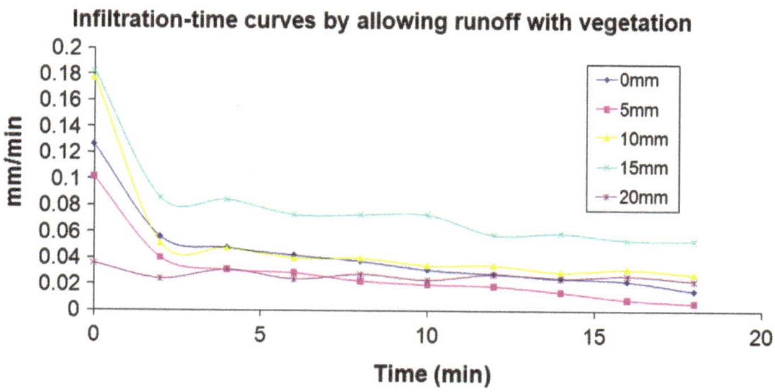

Fig. 10 Infiltration-time curves by allowing runoff with vegetation

other slope setups or waiting for a longer time after the application of artificial rainfall into the catchment and before the commencement of taking observations of discharge may also serve as reasons for this behavior.

4.6 Runoff Without Vegetation

The plate on the downstream side of the catchment is kept open so as allow the water to flow as runoff into the side tanks. The runoff and infiltration are collected separately by two pipes and measured in two different measuring cylinders, each of 500 ml capacity for every 2 min. This process is continued for the next 20–22 min for a particular slope. In this setup only half of the catchment area is considered which is devoid of any vegetation. The five runoff flow-rate versus time curves obtained for this setup are shown in Fig. 11.

4.6.1 Conclusions Drawn from the Curves

- The runoff-time curve initially decreases drastically maintaining a constant value thereafter tending to become zero. This may be due to the fact that water when flowing as runoff over the catchment travels at a higher velocity than that flowing as infiltration through the sand layer, and thus much of the water from the sprinkler system after infiltration is discharged out as runoff within a short time period after the commencement of taking observations and the near exhausted remaining water comes out for the latter observations at a constant rate almost in negligible quantities.
- It is observed that the curves for 5 and 10 mm vertical have their starting points showing higher magnitudes than that of the curve for the horizontal setup. This can be explained by the fact that with an increase in slope the component of the

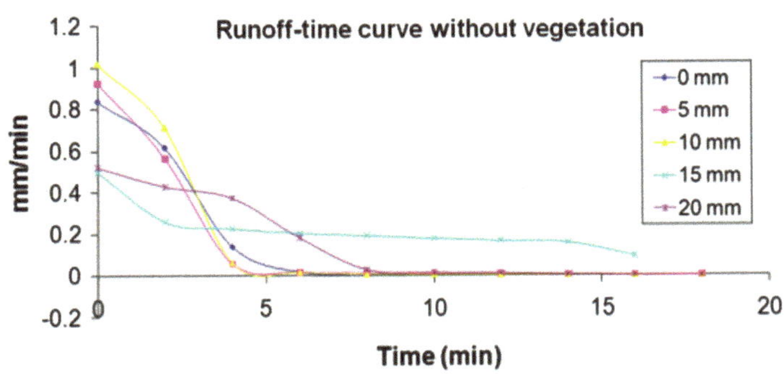

Fig. 11 Runoff-time curve without vegetation

force of gravity increases and thus runoff becomes more rapid and higher discharge is received initially.

- The curve for 15 mm vertical slope shows a drastic initial fall and thereby maintaining a constancy which is high above zero for the remaining observations. This may be attributed to the fact that due to repeated conduct of the experiments over the catchment the sand bed got saturated and the voids are filled with water thereby making it more difficult for infiltration to occur and thus higher amounts of water goes off as runoff.
- It is observed that the curve for the 20 mm vertical slope continues at a level higher than the 15 mm vertical slope curve well justifying the earlier theory of saturated sand and lack of infiltration, and then falls to a level almost equal to zero thus maintaining constancy. This may be because at this stage the effect of gravity becomes so predominant that the water gets exhausted and almost reaches the zero mark before the time required for the 15 mm vertical slope curve to reach the same.
- It is also worth mentioning that the initial points of the curves for 15 and 20 mm vertical slopes are having a lower magnitude than the horizontal slope curve and those for 5 and 10 mm vertical slopes thereby contradicting the earlier theory of increasing effect of gravity with increasing slope. This may be due to possible observation errors like continuation of rainfall for an interval of less than 1 min or waiting for a longer time than 2 min before commencement of observations.

4.7 *Runoff with Vegetation*

This time one half of the catchment area is filled with vegetative cover (garden grass) and runoff is calculated. The five runoff flow-rate versus time curves are shown in Fig. 12.

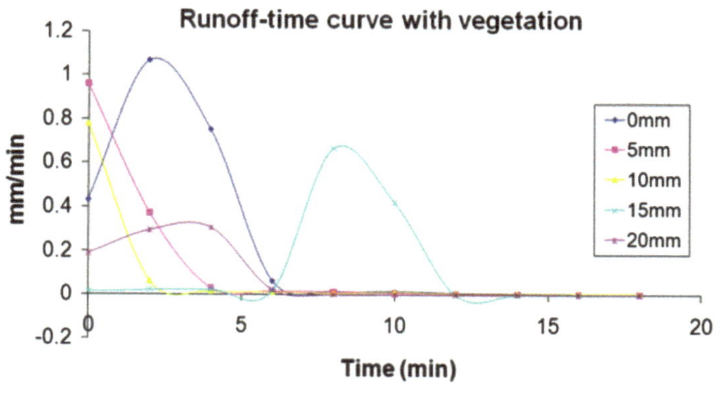

Fig. 12 Runoff-time curve with vegetation

4.7.1 Conclusions Drawn from the Curves

- The runoff-time curves have a tendency of showing peaks which can be speculated to be formed due to deferred reaching of surface water to the runoff zone due to hindrance of vegetation.

4.8 Ideal Conditions for Conducting the Experiment in Rainfall Apparatus

Some of the basic conditions which should be given due consideration for conducting the experiment in an ideal manner can be summarized as follows:

- The sand particles of the sand bed used in the catchment should be uniformly graded.
- The sand bed should be made completely dry before conducting any experiment.
- The artificial rainfall should be of equal intensity and should be allowed to fall on the catchment for a fixed period of time for each setup.
- The nozzles of the sprinklers should be arranged in such a way so that the whole of the catchment area receives equal and uniform amount of rainfall.
- The surface of the sand bed or the sand bed covered with vegetation should be more or less perfectly leveled with very little or no surface slope.

The circulated water should be changed periodically and the pipes should be cleaned.

4.9 Data Analysis

All experiments conducted in this work were randomized complying with statistical significance. Based on the results obtained from all the experiments measuring infiltration under different conditions, the following regression equations were obtained:

$$
\begin{aligned}
&\text{For } x_4 = 0, x_5 = 0,\\
&\ln(y) = -4.809 + 0.0481x_1 + 0.02840x_2 - 0.0230x_3\\
&\qquad - 0.00421x_1^2 - 0.000115x_2^2 + 0.00282x_3^2 - 0.000160x_1x_2\\
&\qquad - 0.001114x_1x_3 - 0.000100x_3^3 + 0.000019x_1^2x_2 + 0.000084x_1x_3^2
\end{aligned}
$$

For $x_4 = 0,\ x_5 = 1,$

$$\ln(y) = -3.913 - 0.0529x_1 + 0.02717x^2 - 0.0230x_3 - 0.001333x_1^2 - 0.000115x_2^2 + 0.00282x_3^2 + 0.000603x_1x_2 - 0.001114x_1x_3 - 0.000100x_3^3 + 0.000019x_1^2x_2 + 0.000084x_1x_3^2$$

For $x_4 = 1,\ x_5 = 0,$

$$\ln(y) = -4.909 + 0.0754x_1 + 0.02943x_2 - 0.0230x_3 - 0.00360x_1^2 - 0.000115x_2^2 + 0.00282x_3^2 - 0.000391x_1x_2 - 0.001114x_1x_3 - 0.000100x_3^3 + 0.000019x_1^2x_2 + 0.000084x_1x_4^2$$

For $x_4 = 1,\ x_5 = 1,$

$$\ln(y) = -4.0132 - 0.0256x_1 + 0.02820x_2 - 0.0230x_3 - 0.000722x_1^2 - 0.000115x_2^2 + 0.00282x_3^2 + 0.000372x_1x_2 - 0.001114x_1x_3 + 0.000100x_3^3 + 0.000019x_1^2x_2 + 0.000084x_1x_3^2$$

where

y Flow-rate (mm/min)
x_1 Time (min)
x_2 Discharge (mm^3/min)
x_3 Slope (mm vertical)
x_4 Vegetation (with = 1, without = 0)
x_5 Runoff (with = 1, without = 0)

The coefficient of correlation associated with these equations is 97.75 %.

4.10 Scope of Further Study

The experiments performed during this project work may be conducted on field under natural conditions of slope, vegetation, and rainfall.

The type of soil of the catchment area has a major role to play in the infiltration and runoff rates. It may be mentioned here that during this project work mainly a layer of well graded sandy soil has been used. But the same set of experiments if conducted by using clayey, silty or any other type of soil under the same set of conditions will yield results different from the one obtained in this project. However, on field a particular area does not necessarily consist of only single type of soil, due to which mixed results may be obtained.

During this project only grass cover has been used as vegetation. However, on field, mixed type of vegetation like dense forests and scanty grass cover may be present, which will yield different results.

4.11 Practical Application

In case of many cities in developing countries like India, it has been observed that during incessant rains the streets are inundated with runoff water from the surrounding hill slopes, which exceeds the design discharge of the present drains. As a result, flash floods occur on the said streets. In lieu of the above, it can be inferred upon that if the runoff can somehow be effectively minimized then the problem of flash floods can be tackled to considerable extent. In order to achieve the same the following setup has been suggested:

A barrier of small height will be constructed at the foothills of the slope along its breadth which will restrict flow of water into the street drain till water level exceeds the barrier height. This will create a pondage on upstream of the barrier and will promote infiltration at the foothill. Now, if flood wells are provided at the downstream portion of the slope at regular intervals resembling the second well (well-2) of the experimental setup presented in this paper, significant amount of runoff will infiltrate and will result in flood moderation. This system will provide benefit for both deforested and forested slope as seen in the experimental set up. The barrier will also help in restricting eroded soil to reach the drain. In case of infiltration by restricting runoff with vegetation, as apparent from the curves, some amount of water is retained in the grass blades which results in a delayed infiltration in the downstream well. Therefore, a set of well may also be placed at upstream in case of forested area if rainfall intensity is very high. The well to be placed on the downstream of the slope also serves the purpose of ground water recharge as a part of rain water harvesting, a study of which was carried by Sarma et al. (1983). However, while achieving the above, the stability of the slope due to the pondage created should be taken note of. In general, the saturation of the soil at the downstream parts of the slope may affect the stability of the slope. Therefore, various remedial measures to check slope instability such as flattening of slope, soil nailing, soil reinforcement, etc., may be adopted after proper research on the same. It may be mentioned here that the provision of well on the downstream of the slope reduces the runoff to considerable extent which in turn helps in providing a check upon soil erosion which itself serves as a means to stabilize the slope.

The above arrangement of restricting runoff can also be achieved by providing a pond having a higher elevation on the downstream side than that on the upstream side. While providing the same, the water going through induced infiltration is subjected to filtration through various layers before actually reaching the ground water table, thereby putting a check on possible contamination of ground water.

5 Conclusion

This project has been performed with a motive of preparing a comparative study of infiltration and runoff characteristics of sandy soil with and without vegetation. The quantities of infiltration and runoff are found to be dependent upon the intensity of rainfall incident on the surface, surface characteristics and varying slope conditions. However, due to unavailability of ideal conditions and the constraint of time utmost accuracy could not be maintained while performing the experiments. Nevertheless, this project has paved the way for future and more fruitful research on the Rainfall Apparatus thereby working for the betterment of mankind.

Acknowledgements The authors would like to acknowledge Civil Engineering Department of Indian Institute of Technology, Guwahati for all the support. The authors would also like to thank Dr. Diganta Goswami, Professor, Assam Engineering College, India, Mr. Anupam Das of Gammon India Limited and Mr. Navaratnam Leelaruban of North Dakota State University, USA for their invaluable suggestions and support.

References

Abrahams AD, Parsons AJ (1991) Relation between infiltration and stone cover on a semiarid hillslope, southern Arizona. J Hydrol 122:49–59

Balasubramanium K (1990) Engineering hydrology. Tata McGraw-Hill Publishing Company Limited

Basic Rainfall Apparatus Instruction Manual (2000) Norwood Instruments Limited, New Mill Road, Honley, Huddersfield, Great Britain

Calvo-Cases A, Boix-Fayos C, Arnau-Rosalén E (2005) Patterns and thresholds for runoff generation and sediment transport on some Mediterranean hillslopes. In: García C, Batalla R (eds) Catchment dynamics and river processes: Mediterranean and other climate regions. Elsevier, Amsterdam, pp 31–51

Cantón Y, Domingo F, Solé-Benet A, Puigdefábregas J (2002) Influence of soil surface types on the overall runoff of the Tabernas badlands (SE Spain). Field data and model approaches. Hydrol Process 16:2621–2643

Garg SK (1998) Soil mechanics and foundation engineering. Khanna Publishers

Li X, Contreras S, Solé-Benet A, Cantón Y, Domingo F, Lázaro R, Lin H, Wesemael BV, Puigdefábregas J (2011) Controls of infiltration–runoff processes in Mediterranean karst rangelands in SE Spain. Catena 86:98–109

Mayor AG, Bautista S, Bellot J (2009) Factors and interactions controlling infiltration, runoff, and soil loss at the microscale in a patchy Mediterranean semiarid landscape. Earth Surf Proc Land 34:1702–1711

Morbidelli R, Saltalippi C, Flammini A, Cifrodelli M, Corradini C, Govindaraju RS (2015) Infiltration on sloping surfaces: laboratory experimental evidence and implications for infiltration modeling. J Hydrol 523:79–85

Sarma AK, Giraud G, Baishya MD (2006) Rainwater harvesting for urban flood peak reduction, My Green Earth. J Soc Socio Econ Awareness Environ Prot SSEAEP 3(2):14–21

Seeger M (2007) Uncertainty of factors determining runoff and erosion processes as quantified by rainfall simulations. Catena 71:56–67

Sharma K, Singh H, Pareek O (1983) Rain water infiltration into a bar loamy sand. Hydrol Sci J 28:417–424

Solé-Benet A, Calvo A, Cerdà A, Lázaro R, Pini R, Barbero J (1997) Influences of micro-relief patterns and plant cover on runoff related processes in badlands from Tabernas (SE Spain). Catena 31:23–38

Yair A, Lavee H (1985) Runoff generation in arid and semi-arid zones. In: Anderson MG, Burt TP (eds) Hydrological forecasting. Wiley, New York, pp 183–220

Impact of Total and Effective Imperviousness on Runoff Prediction

Sahoo Sanat Nalini and P. Sreeja

Abstract Urbanization in a broad sense means increase in imperviousness and thus putting more obstruction to free flow of water into the ground surface. The most commonly used measure of imperviousness is total impervious area (TIA), which is a measure of the area that resists the rain water to infiltrate the down soil, whereas effective impervious area (EIA) is that fraction of TIA that has a direct hydraulic connection to the downstream drainage. Imperviousness being an important parameter in most of the hydrologic models has brought considerable attention in recent years among the researchers community. Hence, the need arises to evaluate the degree of accuracy of using TIA or EIA in hydrologic modeling. Toward this end, the current study presents a comparison between the predicted runoff of an Indian catchment using TIA and EIA in a hydrologic model and hence the better predictor has been found out.

Keywords Urbanization · Total impervious area · Effective impervious area · Hydrologic modeling

1 Introduction

Imperviousness being a direct measure of urbanization (Snyder et al. 2005) information, it is very much essential for urban hydrology and watershed management. Imperviousness may be of two types, directly connected (effective impervious area i.e. EIA) and indirectly connected (non effective) impervious areas. Together they constitute total impervious area (TIA). EIA is that portion of TIA that has a direct hydraulic connection to the downstream drainage network. Many times due to lack of proper and accurate data TIA has been used as EIA in hydrologic models. Many

---

S.S. Nalini (✉) · P. Sreeja
Department of Civil Engineering, IIT Guwahati, Assam, India
e-mail: sanat@iitg.ernet.in

P. Sreeja
e-mail: sreeja@iitg.ernet.in

A.K. Sarma et al. (eds.), *Urban Hydrology, Watershed Management and Socio-Economic Aspects*, Water Science and Technology Library 73,
DOI 10.1007/978-3-319-40195-9_2

studies (Han and Burian 2009; Ravagnani et al. 2009) have been reported that emphasize the effect of imperviousness on changing hydrological characteristics of a watershed. However, studies need to be addressed to access the capability of either type of imperviousness to predict the runoff accurately. Toward this end, the current study compares the TIA-predicted runoff and EIA-predicted runoff for an Indian catchment. In thistudy, TIA has been estimated by a combined approach of remote sensing and GIS. EIA has been estimated (Alley and Veenhuis 1983) using an empirical equation. The estimated imperviousnesses are further utilized to predict the runoff of the watershed and thus a comparison between both are presented.

2 Study Area

The study area, Guwahati as depicted in Fig. 1 is a part of Kamrup District in Assam (North East India), and is situated between 26° 4′ 45″ and 26° 13′ 25″ North Latitude and between 91° 34′ 25″ and 91° 52′ 00″ East Longitude. Located on the Bank of River Brahmaputra, it is the largest commercial, industrial, and educational centre of the north east India and a rapidly expanding urban city. The city is situated on an undulating plane of varying altitude of 47.0 to 55.5 m above Mean Sea Level (MSL). The southern and eastern sides of the city are surrounded by hillocks. Apart from the hilly tracts, swampy/marshy lands and water bodies cover a considerable portion of the city. In this study, the city area has been divided into seven watersheds. The methodology used in this study is demonstrated with respect to one watershed (*Silsako* watershed).

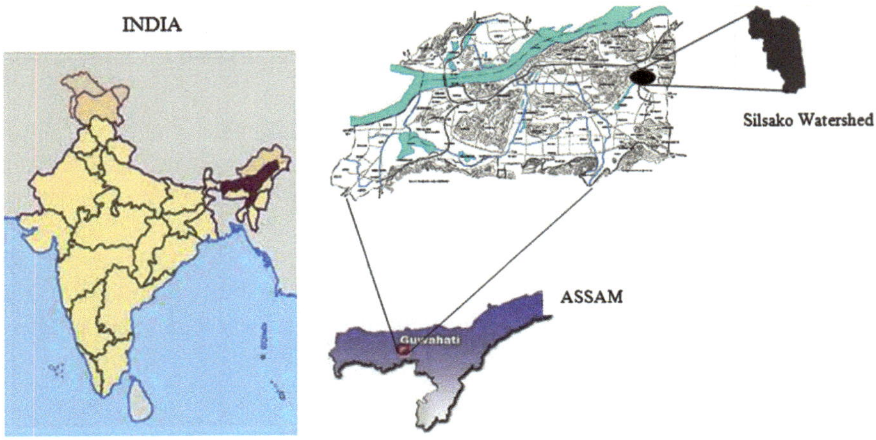

Fig. 1 Location of *Silsako* watershed of Guwahati city

3 Methodology

In the present study, a methodology is followed which combines the GIS and remote sensing technologies (Ravagnani et al. 2009) to obtain the imperviousness of the study site. The image (LISS4 image of Guwahati for 2006) has been classified into seven basic classes namely built-up area, water body, forest, scrub land, agricultural land, swampy/marshy land, and grass land. For classification process, training areas having good separability values (Han and Burian 2009) were selected at various locations of the image. The built-up area estimated from the classification of the imageries that includes the building rooftops and the transportation networks which are collectively called as TIA. TIA includes both effective impervious area (EIA) and noneffective impervious area (NEIA). The most commonly used measure of imperviousness is TIA which is a measure of the area that prevents water infiltration into the soil. Since direct determination of EIA is a data and time-demanding process, an indirect methodology has been followed here to calculate EIA. The TIA estimated from the classified image was then converted to EIA by an empirical relationship given in Eq. (1) (Alley and Veenhuis 1983).

$$\text{EIA} = 0.15\,\text{TIA}^{1.41} \tag{1}$$

4 Runoff Simulation of Watersheds

In this study, SWMM is used as the hydrologic model because of its success in modeling the urban watersheds (Delleur 2003; Spry and Zhang 2006). SWMM developed by EPA (Environmental Protection Agency, USA) (Rossman 2005) is a dynamic rainfall runoff simulation model that computes runoff originating primarily from urban areas from single or continuous events (Huber and Dickinson 1988). It consists of several different blocks (Wang and Altunkaynak 2011) to be simulated separately. Blocks used in the current study are the runoff block for runoff estimation and transport block for routing of the estimated runoff. Infiltration has been modeled by Green–Ampt method. The physical characteristics such as area, width, and slope of the sub-watersheds are determined from ArcGIS 9.3.1. The output from runoff block is used as the input for the subsequent transport block, which models the drainage channels as a series of geometrical hydraulic "elements", that are nodes or conduits. All the conduit properties and the node properties were derived from the available drainage network details. To best capture the spatial variability of the catchment parameters, the study site is further segregated into 22 sub-watersheds. Runoff is simulated at the outlet of each of the sub-watersheds using the above said methodology and then dynamic wave routing is employed to route the flow in the drains to the common outlet (Silsakolake) of the catchment. A continuous simulation has been performed here due to lack of short interval rainfall data.

5 Analysis and Results

Due to lack of short interval rainfall data, a continuous hydrologic modeling has been performed here. Continuous hydrologic modeling also synthesizes hydrologic processes and phenomena (i.e., synthetic responses of the basin to a number of rain events and their cumulative effects) over a longer time period that includes both wet and dry conditions (Chu and Steinman 2009).

Figure 2 shows the variation of runoff in monsoon seasons from the month of June to September for the year 2006 due to TIA of the study area. It can be noticed that comparatively large flood events are found to hit the study area more frequently. The peak runoff in 2006 was observed to be 36.3 m^3/s on 11th June from a rainfall intensity of 4.24 mm/h.

Figure 3 shows the simulated runoff in monsoon seasons for the year 2006 due to EIA. The peak runoff in 2006 was 26.15 m^3/s on 11th June from a rainfall intensity of 4.24 mm/h.

Figure 4 shows a comparison between the runoff predicted by TIA and EIA of the study area. It is clear from the figure that the TIA predicts 38.9 % more runoff in the catchment than it is predicted by EIA. However, it can be mentioned here that the time to peak remains the same for both the cases and only the peak is magnified. However, the results need to be refined using any direct methodology of EIA determination that can be employed to predict runoff.

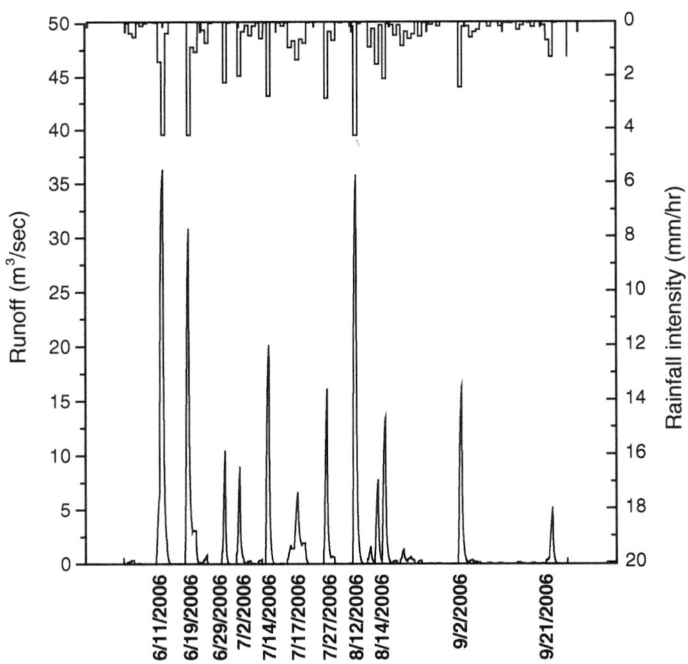

Fig. 2 Runoff simulated in monsoon season taking TIA as imperviousness

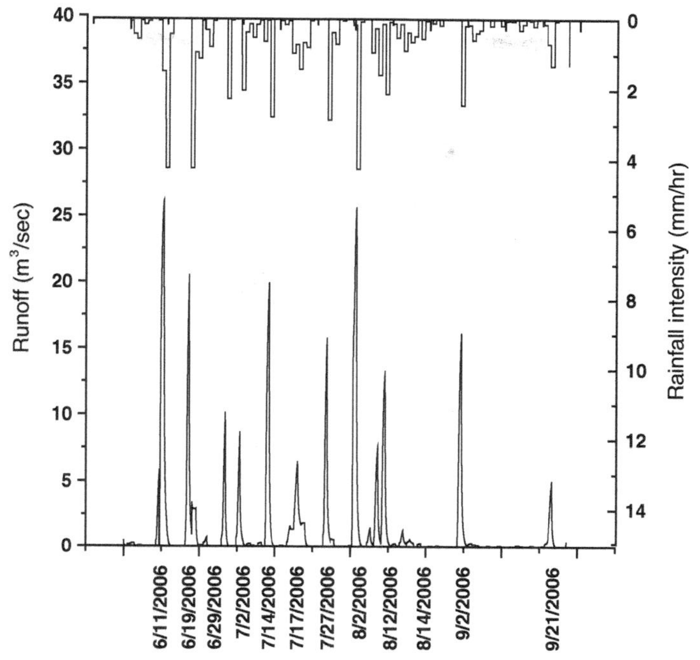

Fig. 3 Runoff simulated in monsoon season taking EIA as imperviousness

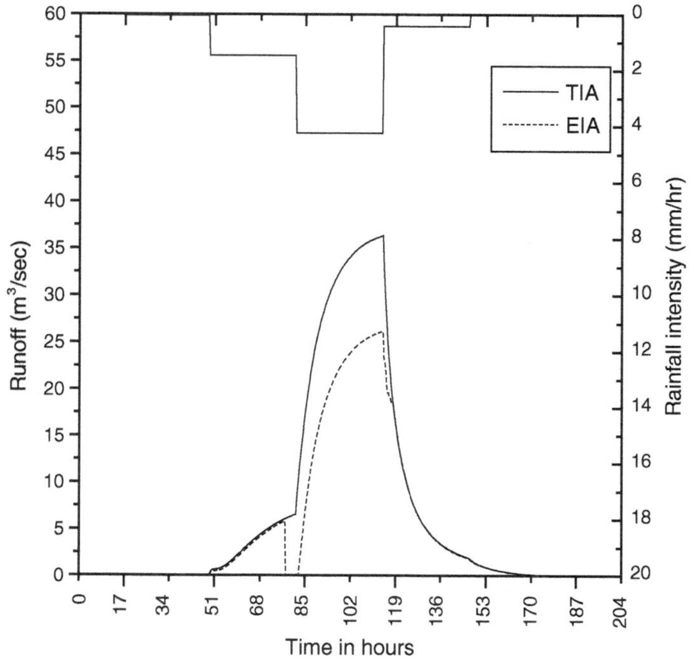

Fig. 4 Runoff predicted due to TIA and EIA

6 Conclusion

Being the discharge data unavailable to check the accuracy of the model, a comparison has been done for simulated runoffs by TIAs and EIAs. Peak runoff in 2006 is overestimated by 38.9 % if TIA is used in the model instead of EIA. However, the result can be further modified by feeding short duration rainfall data as input file.

References

Alley WM, Veenhuis JE (1983) Effective impervious area in urban runoff modelling. J Hydraul Eng ASCE 109(2):313–319

Delleur JW (2003) The evolution of urban hydrology: Past, present, and future. J Hydraul Eng 129 (8):563–573

Han W, Burian SJ (2009) Determining effective impervious area for urban hydrologic modelling. J Hydrol Eng ASCE 14(2):111–120

Ravagnani F, Pellegrinelli A, Franchini M (2009) Estimation of urban impervious fraction from satellite images and its impact on peak discharge entering a storm sewer system. Water Resour Manage 23:1893–1915

Rossman LA (2005) Storm water management model-users manual version 5.0, EPA-United States, (http//www.epa.gov/ednnrmrl/models/swmm5/manual.pdf)

Snyder MN, Goetz SJ, Wright RK (2005) Stream health rankings predicted by satellite derived land cover metrics. J Am Water Resour Assoc 41(3):659–677

Spry R, Zhang S (2006) Modelling of drainage systems and overland flow paths at catchment's scales. In: Proceedings of urban drainage modelling and water sensitive urban design, Monash University, Melbourne

Wang KH, Altunkaynak A (2011) A comparative case study of rainfall-runoff modeling between SWMM and fuzzy logic approach. J Hydrol Eng ASCE (posted ahead of print, April-2011)

Issues of Urban Drainage—Present Status and the Way Forward

Kapil Gupta

Abstract The 2007 Fourth Assessment Report by the Intergovernmental Panel on Climate Change (IPCC) has indicated that the global surface temperature is likely to rise a further 1.1–6.4 °C during the twenty-first century, while studies by NASA indicate that the urban heat island effect over cities has resulted in an increase in rainfall over several urban areas. A major concern is the likely effects of changes in the frequency and intensity of extreme weather events, especially droughts and floods—"an increased risk of drought" while "precipitation is projected to be concentrated into more intense events, with longer periods of little precipitation in between." Cities/towns located on the coast, on river banks, upstream/downstream of dams, inland cities, and in hilly areas can all be affected. There has been an increasing trend of urban flood disasters in India over the past several years whereby major cities in India have been severely affected. The most notable amongst them are Hyderabad in 2000, Ahmedabad in 2001, Delhi in 2002 and 2003, Chennai in 2004, Mumbai in 2005, Surat in 2006, Kolkata in 2007, Jamshedpur in 2008, Delhi 2009, and Guwahati and Delhi in 2010 and again in September 2011. There is therefore an urgent need for adaptation and mitigation options for tackling the impacts of climate change on water resources, especially flooding from extreme rainfall. This paper describes the measures being taken at the local, national, and international level to make our cities flood resilient.

Keywords Urban drainage · JNNURM · NDMA guidelines · Drainage design · Climate change

K. Gupta (✉)
Department of Civil Engineering, Indian Institute of Technology, Mumbai, India
e-mail: k.gupta@civil.iitb.ac.in

A.K. Sarma et al. (eds.), *Urban Hydrology, Watershed Management and Socio-Economic Aspects*, Water Science and Technology Library 73,
DOI 10.1007/978-3-319-40195-9_3

1 Introduction

The 2007 Fourth Assessment Report by the Intergovernmental Panel on Climate Change (IPCC) has indicated that the global surface temperature is likely to rise a further 1.1–6.4 °C during the twenty-first century while studies by NASA indicate that the urban heat island effect over cities has resulted in an increase in rainfall over several urban areas. A major concern is the likely effects of changes in the frequency and intensity of extreme weather events, especially droughts and floods—"an increased risk of drought" while "precipitation is projected to be concentrated into more intense events, with longer periods of little precipitation in between." Cities/towns located on the coast, on river banks, upstream/downstream of dams, inland cities, and in hilly areas can all be affected. There has been an increasing trend of urban flood disasters in India over the past several years whereby major cities in India have been severely affected. The most notable amongst them are Hyderabad in 2000, Ahmedabad in 2001, Delhi in 2002 and 2003, Chennai in 2004, Mumbai in 2005, Surat in 2006, Kolkata in 2007, Jamshedpur in 2008, Delhi 2009, and Guwahati and Delhi in 2010 and again in September 2011.

There is therefore an urgent need for adaptation and mitigation options for tackling the impacts of climate change on water resources, especially flooding from extreme rainfall.

2 Adaptation and Mitigation Measures in Mumbai

The Mumbai floods of July 2005 turned out to be an eye-opener not only for Mumbai but also for India. On July 26, 2005, Mumbai suffered severe flooding due to 944 mm rainfall in 24 h recorded at Santa Cruz observatory at Mumbai airport. According to the Government of Maharashtra, over 60 % of Mumbai was inundated to various degrees. At that time, there was no reliable real-time rainfall forecast mechanism and IMD was unable to issue advance warnings due to the lack of state-of-the-art equipment like Doppler weather radar and tipping bucket rain gauges. Thus, disaster managers had no means of knowing the spatial or temporal variation of rainfall in real time. To improve the response and determine the spatial and temporal variation of rainfall in real time, a network of 35 weather stations with tipping bucket rain gauges has been setup in the city by the Municipal Corporation of Greater Mumbai (MCGM) and Indian Institute of Technology Bombay in June 2006. Majority of them are installed on the roof of the fire station control rooms. These rain gauges have been programmed to give rainfall intensity in real time (every 15 min) to the emergency control room at MCGM headquarters through Internet. The average rain gauge density is 1 per 16 km^2 and interstation distances ranges from 0.68 to 4.56 km. This network has enabled monitoring of rainfall in real time and has been of immense benefit to disaster managers for mobilizing rescue and relief to the flood affected areas during heavy rainfall since 2006. An automatic Doppler flow gauge has also been set up in the upstream reaches of Mithi River to measure the flow levels and issue early warnings for downstream areas. Under an international European Union funded project Collaborative Research on Flood Resilience in Urban areas (CORFU), work is presently ongoing to mitigate flooding by issuing advance warnings using real-time rainfall data and improve forecasted flood levels along the Mithi River using hydraulic flow modeling software in real time. The Municipal Corporation of Mumbai is also finalizing a comprehensive Disaster Risk Management Master Plan (DRMMP) for Mumbai which includes other disasters like earthquakes, floods, epidemics, oil fires, transportation escape routes, etc. before July 2011.

Future strategies should recognize that sea-level rises worldwide cannot be reversed. The only alternative is to have increased investment in flood defenses. For example, the MCGM is now in the process of installing floodgates in combination with high-discharge pumps at eight of the hitherto un-gated sea outlets.

3 Adaptation and Mitigation Measures in India

Realizing that the causes of urban flooding are different and so also are the strategies to deal with them, the National Disaster Management Authority, Government of India has addressed urban flooding as a separate disaster and has released the Urban Flood Guidelines in September 2010.

Urban flooding is significantly different from rural flooding as urbanization leads to developed catchments which increases the flood peaks from 1.8 to 8 times and flood volumes by up to 6 times. Consequently, flooding occurs very quickly due to faster flow times, sometimes in a matter of minutes. Problems associated with urban floods range from relatively localized incidents to major incidents, resulting in cities being inundated from a few hours to several days. Therefore, the impact can also be widespread, including temporary relocation of people, damage to civic amenities, deterioration of water quality and risk of epidemics.

Gist of Some of the Key Action Points of the NDMA Flood Guidelines

1. Ministry of Urban Development will be the Nodal Ministry for Urban Flooding;
2. Establishing Urban Flood Early Warning System;
3. Establishment of Local Network of Automatic Rainfall Gauges for Real time Monitoring with a density of 1 in every 4 km^2 in all 2325 Class I, II, and III cities and towns;
4. Strategic Expansion of Doppler Weather Radar Network in the country to cover all Urban Areas for enhanced Local-Scale Forecasting Capabilities with maximum possible Lead-time;

5. India Meteorological Department will develop a Protocol for Sub-Division of Urban Areas on the basis of Watershed and issue Rainfall Forecast on the Watershed-basis;
6. Catchment will be the basis for Design of Stormwater Drainage System;
7. Watershed will be the basis for all Urban Flooding Disaster Management Actions;
8. All 2325 Class I, II, and III cities and towns will be mapped on the GIS platform;
9. Contour Mapping will be prepared at 0.2–0.5 m contour interval;
10. Inventory of the existing stormwater drainage system will be prepared on a GIS platform;
11. Pre-Monsoon De-silting of Drains should be completed before March 31 every year;
12. Involve the Residents' Welfare Associations and Community-Based Organizations in monitoring this and in all Urban Flood Disaster Management actions;
13. Every building shall have Rainwater Harvesting as an integral component of the building utility;
14. Better Compliance of the Techno-legal Regime will be ensured;
15. Establish the Incident Response System for Coordinated Response Actions;
16. Capacity Development at the Community and Institutional level to enhance UFDM capabilities;
17. Massive Public Awareness programmes covering Solid Waste Disposal, problems of Encroachments, relevance of Techno-legal Regime besides all other important aspects; and
18. Involve elected Public Representatives in Awareness Generation.

4 National Mission on Sustainable Habitat: Climate Change Adaptation and Urban Drainage Parameters

The Prime Minister released India's first National Action Plan on Climate Change (NAPCC) outlining existing and future policies and programs addressing climate mitigation and adaptation in June, 2008. The NAPCC has set out eight "National Missions" as the way forward in implementing the Government's strategy and achieving the National Action Plan's objective. The focus of these missions is on "promoting understanding of climate change, adaptation and mitigation, energy efficiency and natural resource conservation." The National Mission on Sustainable Habitat is one of the eight missions.

The National Mission on Sustainable Habitat aims to address sustainability concerns related to habitats, primarily in urban areas through improved management of solid and liquid waste is to improve the ability of habitats to adapt to climate change by improving resilience of infrastructure, measures for improving advance warning systems for extreme weather events and conservation through appropriate changes in legal and regulatory framework. The development of parameters is essential for developing legal frame work/regulations to improve urban planning in respect of storm water drainage. These parameters/indicators are generally in the form of indices, for systematic and scientific assessment of situation, progress and deficit. In February 2011, 20 Sustainable Habitat parameters on Urban Stormwater Management have been formulated by the Ministry of Urban Development. Parameters such as Climate Change Stress Index, Preparedness Index/Early Warning Index, Rainfall Intensity Index, etc., have been formulated.

5 Summary and Conclusions

Some of the aspects related to climate change, adaptation, and mitigation have been presented. Ignoring the change will not make the problem go away, but we can make use for science and technology to improve the warning, response, and mitigation to reduce our losses.

References

Report of Sub-committee for Development of 'National Sustainable Habitat Parameters' on 'Urban Stormwater Management', CPHEEO, Ministry of Urban Development, New Delhi. http://www.mpurban.gov.in/pdf/UrbanStormWaterMgmt.pdf

National Disaster Management Guidelines: Management of Urban Flooding (2010) A publication of the National Disaster Management Authority, Government of India, September 2010, New Delhi. ISBN: 978-93-80440-09-5

Precipitation Extremes, section 10.3.6.1 in IPCC Fourth Assessment Report: Climate Change 2007

Optimal Allocation of Ecological Management Practices in a Hilly Urban Watershed

Banasri Sarma and Arup K. Sarma

Abstract Urbanization is increasing at a rapid rate and in many places expanding into the hilly areas, thereby inducing significant alteration in the hydrological response of watershed. In the developing world, the process of urbanization is more often unplanned and disorganized, which results in higher yield of sediment and surface runoff, which manifested itself in the form of hazards like flash flood and landslide. Washing off of pollutants from the urbanized impermeable upper catchment is also causing downstream water quality declination. Therefore, urban developments in hilly watersheds require application of efficient management practices that can handle adverse consequences of urban developments in an ecologically sound and sustainable manner. Such eco-friendly sustainable management practices can be termed as Ecological Management Practices (EMPs). However, the cost, efficiencies, and applicability of EMPs vary widely from place to place depending upon the site condition. Therefore, it is necessary to determine the optimal combination of EMPs that satisfies all requirements at minimum cost. In this study, allocation of EMPs for managing sediment and water yield from hilly urban watershed has been done through an optimization model OPTEMP-LM (OPTimal EMP model with Linear programming for Multiple ownership). The model allocates EMPs in such a way that the undesirable hydrological consequences of urban development can be alleviated in a sustainable manner at minimum possible cost while addressing various other constraints imposed by topography and owner's choice. The model was applied to a micro watershed of Guwahati, Assam, India with three EMPs, namely: grass, garden and detention pond and was found suitable for the proposed application.

Keywords Ecological management practices · Optimization · Urban watershed

B. Sarma (✉)
CoE(ILPWRM), Department of Civil Engineering, Indian Institute of Technology, Guwahati, India
e-mail: banasris@gmail.com

A.K. Sarma
Department of Civil Engineering, Indian Institute of Technology, Guwahati, India
e-mail: aks@iitg.ernet.in

A.K. Sarma et al. (eds.), *Urban Hydrology, Watershed Management and Socio-Economic Aspects*, Water Science and Technology Library 73,
DOI 10.1007/978-3-319-40195-9_4

1 Introduction

Release and transport of nonpoint source pollutants from the catchment can be controlled to a great extent in a sustainable way through land management practices. While emphasis is increasing on structural measures like pond, retaining wall, etc. non-structural measures can also be used for controlling nonpoint source pollutants and water yield. A judicious combination of structural and non-structural measures can provide a sustainable solution to the problems that arise due to increase in the release of runoff and nonpoint source pollutants from the upper catchment.

Based on this idea, the concept of Ecological Management Practices (EMPs) has been developed, which can be defined as eco-friendly sustainable management practices used for maintaining and enhancing land-uses in a natural way. In other words, EMPs are basically sustainable measures that consist of combination of different structural and nonstructural land management practices adopted for controlling runoff volume and yield of sediments and other pollutants from an area and transport of the same by the runoff.

EMPs like contour terracing, mulching, grass, shrubs, detention/retention pond, buffer zone with vegetation and tree, sediment trap, rainwater harvesting systems, and vegetated waterways are commonly used for controlling sediment yield and runoff volume from land surface. One of the natural land covers, wetlands, and swamps were observed to be a good nonpoint pollutant trap (Robb 1992; Smith 2001). Knowing the effectiveness of wetlands to control nonpoint source pollution, construction of wetlands as a pollution control measure was also suggested by Wang et al. (2006) and Moreno et al. (2007). Besides, some studies have also reported effectiveness of detention pond in controlling nonpoint pollution like sediment and nutrient. Hsieh and Yang (2007) used optimization method and an annual average reservoir water quality model to develop optimal Best Management Practice (BMP) strategy for controlling nonpoint pollution in the Fei-Tsui Reservoir watershed; they suggest installation of several detention ponds in the watershed to attain oligotrophic conditions in the reservoir. The riparian buffer zones with grass and forest were also reported to be effective in controlling nonpoint source pollution in some of the recent studies by Anbumozhi et al. (2005), Shiono et al. (2007) Maillard and Santos (2008) and Parajuli et al. (2008). Besides the traditional practices, some new ideas on landuse management practices like wood filters (Boving and Neary 2007) and biofilter (Hatt et al. 2009) for controlling nonpoint pollution have also emerged in recent years.

While many investigators have reported successful application of land management practices for controlling nonpoint source pollution, study toward application of such practices in urban sector is limited. So far, there is no well-established EMP recognized as suitable for urban residential developments. However, various traditional land management practices as described in the previous section are available for controlling sediment yield and runoff volume from agricultural areas. These practices, if suitably modified considering the necessity of an urban area, can also be used for land management of urban residential areas. Some of the possible EMPs for urban residential areas are:

(i) Grass land: Grass reduces the velocity of surface runoff, minimizes the impact of rain on soil, and its root system helps in increasing infiltration. In urban area, grass land can also serve as open land that is needed according to the municipal rules.
(ii) Forest land: Tree canopy reduce the direct impact of rain on soil. Besides, forest land, covered with falling leaves, also reduces the surface runoff velocity and increases infiltration. Falling leaves and decaying branches act as mulches and thus tree cover can control sediment yield and runoff volume.
(iii) Covering rain impacted areas with pebble, vegetation or wood chips: Erosive power of rain drop depends on size of rain drop and its falling height. Size of rain drop that falls from inclined roof of a house became quite large and thus strikes the ground with very high erosive power. Thus, the portion of ground lying below the line of the roof edges is prone to more erosion as the accumulated rain over the roof falls with a high velocity and with larger drop size. Such drop line areas of water around the house can be covered with pebble or wood chips or erosion resisting vegetation can be allowed to grow, which protect the soil from the direct impact of rain drop and also allow more infiltration.
(iv) Detention drain and Retention pond: To capture the excess surface runoff, detention drains can be constructed across the slope and retention ponds can be constructed in a suitable location. This can minimize downstream erosion and flooding.
(v) Vegetated waterways: If the paths (or channels) of accumulated surface runoff are covered with vegetation, then the vegetation provide an obstruction to the flowing water. It reduces velocity and hence the erosive power of the flowing water. This reduces erosion of channel bed and bank and prevents gully formation. Root systems of a vegetated waterway not only increase bondage of soil and make it resistant to water erosion, but also promote infiltration. Depending on the status of degradation of the waterway, different types of vegetation can be suggested.
(vi) Rainwater harvesting system: Rainwater harvesting is a technique of collection and storage of rain water in surface (storage tanks) or subsurface aquifer before it is lost as surface runoff. Rainwater harvesting system helps in reducing the peak runoff and also recharges ground water. The collected rainwater through the rainwater harvesting system can also be used during the period of water shortage.

Scope of using optimization technique for landuse planning for achieving sustainable solutions to wide range of environmental problems has drawn attention of researchers in recent years (Seppelt and Voinov 2002; Gabriel et al. 2006; Holzkamper and Seppelt 2007; Riveira et al. 2008; Lin et al. 2009). In case of urban development in hilly terrain, soil erosion is the major cause of hazard like flood, water quality degradation and slope instability. In this study, allocation of EMPs at minimum possible cost for managing sediment and water yield from hilly urban watershed is done through a watershed-based optimization models.

A LPP model that considers different EMPs in different plots as separate variables and allocation is done with constraints for their upper limit, lower limit, etc., with respect to their respective plots while limiting the sediment and water yield at the outlet of the entire watershed. This model is called as OPTEMP-LM (**OPT**imal **EMP** model with **L**inear programming for a watershed having **M**ultiple ownership).

2 Mathematical Formulation

Objective function: Minimization of construction and maintenance cost for the EMPs in p number of plots

$$\text{Minimize } Z = \sum_{k=1}^{p}\sum_{i=1}^{n}(\mathrm{Cc}_{ik} + \mathrm{Cm}_{ik})a_{ik} \tag{1}$$

Constraints:

Sediment yield constraints:

$$\mathrm{St}_{\min} \le \sum_{k=1}^{p}\left[R_K K_K (\mathrm{LS})_k \left\{\sum_{i=1}^{n} C_{ik}a_{ik} + \sum_{j=1}^{m}\left(A_{jk} - \sum_{i=1}^{n} a_{ik}\right)C_{ik} + \mathrm{Cc}_k \mathrm{Ac}_k\right\}\mathrm{P}_k\right] \le \mathrm{St}_{\max} \tag{2}$$

Peak discharge constraints:

$$Q_{k_{\min}} \le \left[\left\{\sum_{i=1}^{n} R_k C_{ik} a_{ik} + \left(\sum_{j=1}^{m} A_{jk} - \sum_{i=1}^{n} a_i\right)\mathrm{Rc}_{jk} + (\mathrm{Rc}_k \mathrm{Ac}_k)\right\} \div A_k\right]\mathrm{IA} \le Q_{k_{\max}} \tag{3}$$

Available EMP area constraints:

$$\sum_{i=1}^{n} a_{ik} \le \mathrm{AL}_k = A_k - A_k \mathrm{C}_{k_{\max}} \tag{4}$$

Suitable EMP area constraints:

$$a_{ki} \leq \text{as}_{ki} \tag{5}$$

Owner's Choice constraints:

$$(a_{\max})_{ki} \geq a_{ki} \geq (a_{\min})_{ki} \tag{6}$$

Non negativity constraints:

$$a_{ki} \geq 0 \tag{7}$$

where,

j — 1, 2, …, m, where m is the number of different land cover in the plot except the EMPs and coverage area

$(\text{Cc})_{ki}$ — Capital cost of the ith EMP in the kth plot (Rs)

$(\text{Cm})_{ki}$ — Maintenance cost of the ith EMP in the kth plot (Rs)

a_{ki} — Area of the ith EMP in the kth plot (m^2)

$\text{St}_{\min}$ — Minimum sediment yield required from the watershed (tones/year)

$\text{St}_{\max}$ — Maximum sediment yield allowed from the watershed (tones/year)

A_k — Area of the kth plot in the watershed (m^2)

R_k — Rainfall and runoff erosivity index of the kth plot (100 ft tonf in./acre/h/year)

K_k — Soil erodibility factor of the kth plot (tones/acre per unit of R)

$(\text{LS})_k$ — LS factor of the kth plot (dimensionless)

C_{ik} — cover factor for the ith EMP in the kth plot (dimensionless)

A_{jk} — Area of the jth land cover in the kth plot (m^2)

C_{jk} — Cover factor for the jth land cover in the kth plot (dimensionless)

$Q_{k_{\max}}$ — Maximum allowable peak rate of runoff from the kth plot (cumec)

$Q_{k_{\min}}$ — Minimum allowable peak rate of runoff at downstream from the kth plot, (cumec)

RC_{ki} — Runoff coefficient for the ith EMP in the kth plot (dimensionless)

RC_{kj} — Runoff coefficient for the jth landcover in the kth plot (dimensionless)

Rc_k — Runoff coefficient of the coverage area in the kth plot (dimensionless)

Ac_k — Coverage area of the plot (m^2) in the kth plot

I_k — Maximum intensity of rainfall for the time of concentration of the selected design storm for the kth plot (mm/h)

$(\text{as})_{ki}$ — Suitable area available for ith EMP in the kth plot (m^2)

$(\text{a}_{\min})_{ki}$ — Minimum area required for ith EMP in the kth plot (m^2)

AL_k — Available EMP area in the kth plot (m^2)

$C_{k_{\max}}$ — Maximum coverage allowed in the kth plot (%)

$(a_{\min})_{ki}$ — is the minimum area kept for ith EMP in the plot according to owner's choice in the kth plot (m^2)

$(\text{a}_{\max})_{ki}$ — is the maximum area kept for ith EMP in the plot according to owner's choice in the kth plot (m^2)

3 Case Study

A micro watershed of Guwahati (0.17 km^2) is considered for application of the optimization model to select optimal EMP combinations for the watershed from a set of feasible EMPs. The study watershed is located near the Games village area of Guwahati (Fig. 1), a site having potential for residential development in its hilly parts as well. The watershed delineation was carried out initially by using ASTER DEM data in ArcSwat and the selected watershed was refined following the drainage line of Survey of India (SoI) toposheets. The TIN model for the area was developed by using 20 m contour interval obtained from SoI toposheets (1:500,000 scale). From the TIN, the slope and aspect map of the area was developed. The slope ranges from 0 to 40 % and elevation values of this area ranges from 11 to 200 m. The major soil type of this watershed is fine and coarse loamy type as obtained from soil map of Assam Remote Sensing Application Centre (ARSAC). The watershed area was considered to be consisting of four different plots under different ownership (Fig. 1).

3.1 EMP Consideration

We considered three types of EMPs in this study-Grass, Garden (grass, flower, vegetable bushy vegetation) and Detention pond. Grass is traditionally practiced for controlling sediment yield, grazing purposes and as lawns as they are easy to grow and maintain. Therefore, grass is observed to have significant potential as one of the EMPs for hilly areas of Guwahati. The main reason of selecting garden as EMP is that many owners prefer gardening (kitchen gardens/flower) as their hobby or for economic benefit, which has the capability of controlling sediment yield and runoff generation. Detention pond that can be used to capture rainwater and runoff is observed to have highest potential of controlling sediment and water yield from a watershed area and thus is an important EMP for areas with high sediment yield.

Cost per unit area of the considered EMPs are estimated based on the prevailing market rate of the study area and are tabulated in Table 1.

3.2 Parameters of the RUSLE Model

Available literatures, as given below, helped in considering the values of the model parameters.

(i) Rainfall erodibility factor (*R*): A study conducted by Sarma et al. (2005) found the *R* value to be 544 for Guwahati City. In this study, the *R* value was taken as 544.

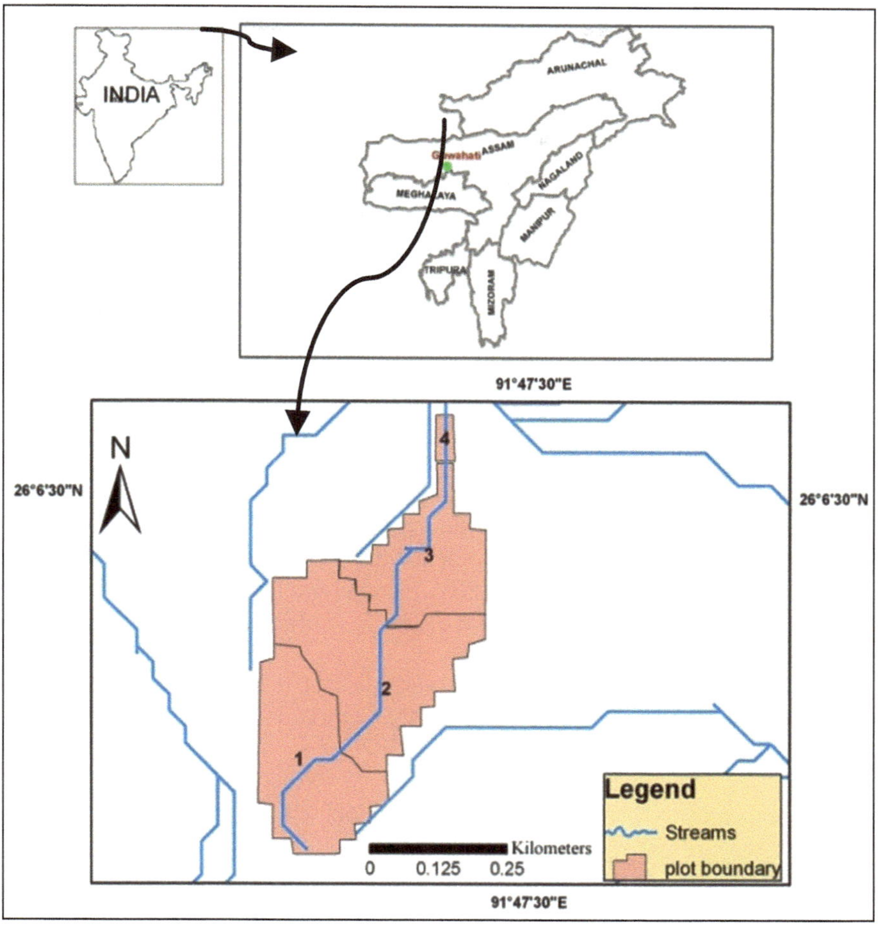

Fig. 1 The study watershed

Table 1 Cost of EMPs considered for the study

EMP No	EMP name	Material cost Rs/m^2	Construction cost Rs/m^2	Maintenance cost (year) Rs/m^2
EMP1	Grass	100	20	240
EMP2	Garden having grass and bushy vegetation/vegetable garden	150	20	120
EMP3	Detention pond	0	200	100

(ii) Soil Erodibility factor (K): For calculation of K value, the soil map for the area was referred and K value for each soil type was taken from the USLE table. The average K value for the entire plot is commutated by weighting over entire area.

(iii) Slope-Length factor (LS): Slope map generated in ArcGIS was used to calculate the LS factor. From the slope map, the slope value for each pixel was taken and its corresponding length was considered as length for that slope. Slope-Length factor for each of the pixel was then determined using Eq. 8.

$$\mathrm{LS} = \left[0.065 + 0.0456(\mathrm{slope}) + 0.00654(\mathrm{slope})^2 + (\mathrm{slopelength}) \div 72.5\right]^{\mathrm{NN}} \tag{8}$$

where

Slope	Slope steepness (%),
Slope length	Length of slope (ft),
NN	Slope steepness factor, ranging from 0.2 to 0.5.

LS value for the entire plot was determined by taking arithmetic average of LS value computed for each of these pixels.

(iv) Cover factor (*C*): The cover factor for the various land cover is considered based on available literature (USDA Agricultural Handbook 1978; Toy et al. 1998; Sarma et al. 2005; Jabbar et al. 2005; Wall et al. 2002) and are given in Table 2.

(v) Practice factor (*P*): The *P* factor was considered equal to 1, as impact of different EMPs are introduced indirectly by modifying the *C* factor of USLE.

3.3 Parameters for the Rational Method

(i) Runoff coefficient (RC): The runoff coefficients for various land cover were considered based on available literatures (Sarma et al. 2005; Iowa Storm Water Management Manual 2008) and are given in Table 2.

(ii) Rainfall intensity (*I*): To determine the design rainfall intensity, time of concentration for the area was first determined by using the Kirpich formula. The intensity of the designed rainfall is determined by using the intensity duration curve developed by Sarma and Goswami (2004) for the area under study for duration equal to time of concentration.

$$I = 51.307e^{0.2179D} \tag{9}$$

where

I Intensity (mm per h);

D duration (h)

Table 2 The USLE cover factor and the Rational method's runoff coefficient for the various land covers

Land cover	USLE cover factor	Rational method runoff coefficient
Built up area	0	1
Built up area with rainwater harvesting system	0	0.8
Forest	0.1	0.2
Barren land	1	0.5
Grass	0.01	0.2
Garden having grass and bushy vegetation/shrub/vegetable garden	0.01	0.3
Detention pond	0	0

We obtained the intensities of the design rainfall for the area as 50 mm/h.

(i) Area of the plot: The area was directly read from the ArcGIS shapefile.

3.4 *Other Consideration for the Optimization Model*

(i) Coverage in the plots: As per the guidelines followed in the State of Assam, India; we took 60 % coverage area as allowable maximum coverage which includes the area required for infrastructural facilities for that area.
(ii) Suitable area for EMPs: Suitability conditions for the EMPs are considered as given in Table 3.

By superimposing the soil map, slope map, and DEM of the plots in ArcGIS, we found that all the three EMPs are suitable for the area.

(iii) Total maximum EMP area: The EMPs will not be allowed within the coverage area. Therefore, we considered 40 % of the total area as the total maximum EMP area (Table 4).
(iv) Owner's choice for EMPs: The owner of the plot may have special choice for some EMP and no choice for some other. A hypothetical owner's choice was considered for the study and is presented in Table 5.

Table 3 Suitability conditions considered for EMPs in the study

EMP	EMP1	EMP2	EMP3
Suitable soil type	Loamy	Loamy	All, except rock
Suitable slope range	All	0–60 %	0–45 %
Elevation	All	All	<800 m

Table 4 Summary of the model parameters considered OPTEMP-LM model

Plot no	Area (m^2)	Soil erodibility factor	Slope-length factor
1	55,665	0.3	3.44
2	73,971	0.13	4.86
3	42,154	0.13	3.72
4	2741	0.13	0.24

(v) Maximum and minimum sediment yield and peak discharge: Allowable sediment yield from the plot was estimated by computing sediment yield from the plot for its natural land cover condition using RUSLE method. This value was considered as the permissible sediment yield when 60 % of the area is used for constructing building and rest for EMPs. The value of maximum allowable sediment yield from the area was thus obtained as 2000 tones/year. As the studied watershed is small, sediment delivery ratio of the watershed is taken as 1.

For model implementation, it is considered that there is no sediment requirement from the watershed area for the required norms of water quality at the outlet point. Therefore, minimum sediment requirement ($S_{\min}$) was considered as zero.

The peak discharge from the area with and without built up area was calculated using the Rational method. The value of peak discharge obtained with natural cover was considered as the lower limit of peak discharge ($Q_{\min}$). Maximum allowable peak discharge ($Q_{\max}$) was estimated based on the safe carrying capacity of the drainage system at downstream. In this study, we considered $Q_{\max}$ as 2 times of the $Q_{\min}$. Thus we got 0.5–1.5 cumec as the peak discharge range for the area.

We found that for some of the plots, restricting peak discharge within the maximum permissible limit becomes impossible even after applying lowest water yielding EMPs in the entire available EMP area. Thus, solution of the optimization model becomes infeasible. Such situations left us with two options:

(i) Application of EMPs within the coverage area in the form of rainwater harvesting or other water retention structures.
(ii) Reduction of coverage area percentage from the otherwise permissible limit.

Thus in case of infeasible solution option (i) and (ii) are tried in sequence to obtain a feasible solution. A study conducted by Sarma et al. (2006) revealed that the runoff coefficient of building area can be reduced from 1 to 0.8 by adopting rooftop rainwater harvesting. This value was considered in this study.

In this case, allowable peak discharges are considered as given in Table 6.

4 Results

Results of the OPTEMP-LM considering four plots in the watershed are presented in Tables 7 and 8.

Table 5 Owner's choice for EMPs

Plot no	Owner's choice for EMP
1	EMP1: min = 2 % of AL, max = AL
	EMP2: min = 5 % of AL, max = 75 % of AL
	EMP3: min = 20 % of AL, max = 50 % of AL
2	EMP1: min = 60 % of AL, max = 90 % of AL
	EMP2: min = 5 % of AL, max = 75 % of AL
	EMP3: min = 5 % of AL, max = AL
3	EMP1: min = 0, max AL
	EMP2: min = 0, max = AL
	EMP3: min = 0, max = AL
4	EMP1: min = 2 % of AL, max = AL
	EMP2: min = 5 % of AL, max = 75 % of AL
	EMP3: min = 0, max = 0

Table 6 The allowable peak discharge values for different plots considered in the OPTEMP-LM model

Plot no	Allowable peak discharge (cumec)
1	$Q_{min} = 0.15$, $Q_{max} = 0.46$
2	$Q_{min} = 0.21$, $Q_{max} = 0.62$
3	$Q_{min} = 0.12$, $Q_{max} = 0.35$
4	$Q_{min} = 0.01$, $Q_{max} = 0.02$

Table 7 Optimal EMP combinations as obtained from OPTEMP-LM model for the four plots of the watershed

Plot no	EMP1 (grass) area (m^2)	EMP2 (garden) area (m^2)	EMP3 (detention pond) area (m^2)
1	3901	11,133	5566
2	17,753	1479	3687
3	8016	0	4214
4	0	0	0
Watershed	29,670	12,612	13,467

Table 8 Optimal cost, sediment yield, and peak discharge considering the EMPs in the four plots of the watershed

Plot no	Optimal costs (Rs)	Sediment yield in plots (tones/year)	Peak discharge in plots (cumec)
1	6,303,035	454	0.46
2	7,926,349	1055	0.41
3	4,150,080	485	0.23
4	0	6	0.02
Watershed	18,379,464	2000	1.12

5 Conclusions

In this study, a scientific way of developing hilly watersheds with optimal EMPs is presented considering their hydrological behavior along with the relevant implementation constraints imposed by the plot owners and topographical conditions. Therefore, an optimization model was developed with an objective to minimize the total implementation and maintenance cost of EMPs for a hilly watershed subject to the constraints of desired sediment yield and peak discharge, maximum allowable EMP area, owner's choice and suitability conditions of EMPs.

In this study, the watershed was considered to be consisting of several plots in sequence from upstream to downstream under different ownership and different EMPs in different plots of the watershed was taken as separate variables and allocation was done with constraints for their upper limit, lower limit, etc., with respect to their respective plots while limiting the sediment and water yield at the outlet of the entire watershed. The models described in the study are application of optimization techniques with hydrological models and GIS, which are capable of providing economic solution in selecting management practices to tackle two independent but interconnected problems—soil erosion and high runoff generations from hilly urban watersheds. Judicious selection of EMPs can reduce their expenditure to a great extent, therefore cost effective and economic selection of EMPs is very important in view of a country's economy. Besides, selection of management practices by considering their hydrological behavior can enhance their sustainability in long run. Therefore, for urban developments in hilly watersheds, the model can assist in suggesting modification of its landscape in an ecologically and economically sustainable way.

References

Anbumozhi V, Radhakrishnan J, Yamaji E (2005) Impact of riparian buffer zones on water quality and associated management considerations. Ecol Eng 24:517–523

Boving TB, Neary K (2007) Attenuation of polycyclic aromatic hydrocarbons from urban stormwater runoff by wood filters. J Contam Hydrol 91:43–57

Gabriel SA, Faria JA, Moglen GE (2006) Multiobjective optimization approach to smart growth in land development. Socio Econ Plann Sci 40:212–248

Hatt BE, Fletcher TD, Deletic A (2009) Hydrologic and pollutant removal performance of stormwaterbiofiltration systems at the field scale. J Hydrol 365(3–4):310–321

Holzkamper A, Seppelt RF (2007) A generic tool for optimising land-use patterns and landscape structures. Environ Model Softw 22:1801–1804

Hsieh CD, Yang WF (2007) Optimal non point source pollution control strategies for a reservoir watershed in Taiwan. J Environ Manage. doi:10.1016/j.jenvman.2006.10.022

Iowa Storm Water Manual (2008) Technical Report Version 2. 2C-4 Rational Method. www.ctre.iastate.edu/pubs/stormwater/documents/2C-4RationalMethod.pdf

Lin YP, Verburg PH, Chang CR, Chen HY, Chen MH (2009) Developing and comparing optimal and empirical land-use models for the development of an urbanized watershed forest in Taiwan. Landscape Urban Plann 92(3–4):242–254

Maillard P, Santos NAP (2008) A spatial-statistical approach for modeling the effect of non-point source pollution on different water quality parameters in the Velhas river watershed—Brazil. J Environ Manage 86:158–170
Moreno D, Pedrocchi C, Comın FA, Garcıa M, Cabezas A (2007) Creating wetlands for the improvement of water quality and landscape restoration in semi-arid zones degraded by intensive agricultural use. Ecol Eng 30:103–111
Parajuli PB, Mankin KR, Barnes PL (2008) Applicability of targeting vegetative filter strips to abate fecal bacteria and sediment yield using SWAT. Agric Water Manag 95:1189–1200
Riveira IS, Marcos BM, Rafael CM, David MB (2008) Algorithm based on simulated annealing for land-use allocation. Comput Geosci 34:259–268
Robb DM (1992) The role of wetland water quality standards in nonpoint source pollution control strategies. Ecol Eng 1:143–148
Sarma AK, Goswami P (2004) Developing intensity duration curve with limited rainfall data. In: Predictions in ungauged basins for sustainable water resource planning and management. Jain Brothers, New Delhi, pp 187–194. ISBN:81-8360-044-1
Sarma AK, Chandramouli V, Singh B, Goswami P, Rajbongshi N (2005) Urban flood hazard mitigation of Guwahati city by silt monitoring and watershed modeling. Report submitted to Ministry of Human Resources Department (MHRD) by Department of Civil Engineering, IIT Guwahati
Sarma AK, Giraud G, Baishya MD (2006) Rainwater harvesting for urban flood peak reduction, my Green Earth. J Soc Socio Econ Awareness Environ Prot (SSEAEP) 3(2):14–21
Seppelt R, Voinov A (2002) Optimization methodology for land use patterns using spatially explicit landscape models. Ecol Model 151:125–142
Shiono T, Yamamoto N, Haraguchi N, Yoshinaga A (2007) Performance of grass strips for sediment control in Okinawa, Jarq-Japan. Agric Res Q 41(4):291–297
Smith E (2001) Pollutant concentrations of stormwater and captured sediment in flood control sumps draining an urban watershed. Water Resour 35(13):3117–3126
Toy TJ, Foster GR, Galetovic JR (1998) Guidelines for the use of the Revised Universal Soil Loss Equation (RUSLE). Version 1.06 Technical Report
USDA Agricultural Handbook (1978) Predicting rainfall erosion losses-a guide to conservative planning (No 537). Technical Report
Wall GJ, Coote DR, Pringle EA, Shelton IJ (2002) RUSLEFAC—revised universal soil loss equation for application in Canada: a handbook for estimating soil loss from water erosion in Canada. Research Branch, Agriculture and Agri-Food Canada, Ottawa, Contribution No. AAFC/AAC2244E. Technical Report, 117 pp
Wang L, Wang WD, Gong ZG, Liu YL, Zhang JJ (2006) Integrated management of water and ecology in the urban area of Laoshan district, Qingdao, China. Ecol Eng 27:79–83

Part II
Remote Sensing and GIS Applications in Urban Hydrology

Surface Runoff Depth by SCS Curve Number Method Integrated with Satellite Image and GIS Techniques

Thiyam Tamphasana Devi and Yashwant Baskar Katpatal

Abstract The existing scenario of groundwater recharge potential around the corner is one of the most talks about topic in today's world. And in the case if groundwater recharge scenario is primarily concerned, surface runoff is a prime factor to be quantified since it affects to the potential of groundwater recharge. Surface runoff also supports to the information required for groundwater recharging efficiency by providing the behavior of surface water flows for certain geographical terrain. The study area, Bhandara district (Maharastra, India), semi-arid region having geographical area of about 4090 Km^2 (24 watersheds)normally receives average annual rainfall of 1300 mm and is very much concerned about the existing groundwater recharge scenarios. So, in this study an attempt has been made to quantify the annual average surface runoff depth of each watershed so that the potential groundwater recharge could be revealed indirectly. Soil Conservation Services (SCS) curve number method has been employed for the estimation of surface runoff depth integrated with remote sensing (RS) and Geographical Information System (GIS) techniques. SCS curve number method is basically defined by the curve number (CN) derived from landuse/landcover and hydrological soil groups of the particular geographical area. Landuse/landcover and hydrological soil group have been derived from satellite image (LISS III) using GIS techniques. The watershed-wise results of surface runoff depth were classified into six classes (very low, low, medium low, medium high, high and very high) in the range of 700–1000 mm with 50 mm difference between the classes. The study area is mostly covered by low to medium low runoff depth indicating for medium groundwater recharge potential. These results were re-examined with the average

T.T. Devi (✉)
Department of Civil Engineering, National Institute of Technology, Manipur 795001, Manipur, India
e-mail: thiyam85@gmail.com

Y.B. Katpatal
Department of Civil Engineering, Visveswaraya National Institute of Technology, Nagpur 440010, Maharastra, India
e-mail: ybkatpatal@rediffmail.com

A.K. Sarma et al. (eds.), *Urban Hydrology, Watershed Management and Socio-Economic Aspects*, Water Science and Technology Library 73,
DOI 10.1007/978-3-319-40195-9_5

groundwater level fluctuation. Both the results were shown similar conclusion of having moderate groundwater recharge potential in this area.

Keywords SCS curve number method · RS · GIS · Runoff depth · Groundwater recharge

1 Introduction

India with 2.3 % of the world's supports more than 16 % of the world's population with only 4 % of world's freshwater resource (Patil et al. 2007). The per capita land and water availability is decreasing day by day, which in turn is posing a challenging task for the agricultural scientists to sustain the required level of food grain production. In the coming decades, the accelerating population growth, surface water pollution, and climate change together may produce a drastic decline in fresh water supply. Groundwater is considered to be the purest form of water available on earth. Keeping above factors in view, not only the quantification and conservation of surface water resources for the need of the hour for ensuring sustainable livelihood but also estimation of surface runoff is essential for the assessment of groundwater yield potential of the watershed, recharging the ground water zones and sustainable management of groundwater in future. As we know infiltration is indirectly proportional with runoff and hence estimation of runoff and its analysis is very important. In India, the availability of accurate information on runoff is scarcely available and that too in a few selected sites where recording and automatic hydrologic gauging stations are installed. Thus, there is an urgent need to generate information on basin runoff and management programmes. In India, most of the agricultural watersheds are ungauged, having no past record whatsoever of rainfall-runoff processes (Sarangi et al. 2005a, b).

Non-availability of continuous rainfall and runoff records in majority of Indian watersheds has led to the adoption of alternate techniques such as Soil Conservation Service Curve Number (SCS-CN) method for estimation of surface runoff from un-gauged basins (Chattopadhyay and Choudhury 2006). SCS-CN method (SCS 1956, 1972, 1973) since its inception has been widely accepted by scientists, hydrologists (Boughton 1989; Ruslin 2011; Mahboubeh et al. 2012; Sindhu et al. 2013), water resources planners, agriculturists, foresters, and engineers (Bosznay 1989) for estimation of surface runoff. It is revealed from the published literature that the CN based methods have been widely used in most of the hydrologic and water quality (H/WQ) models due to its robustness and ease of application under different watershed conditions.

The watershed hydrologic responses leading to generation of surface runoff are governed by the interaction of precipitation with the topological, land use and soil physical properties of the land surface. Therefore, use of Geographic Information System (GIS) is preferred over the traditional techniques in proper quantification of surface runoff by generating landuse/landcover without field investigation. The estimation process becomes more efficient, interactive and less cumbersome when the GIS is used for storing, interpreting and displaying the data required in curve number (CN) based runoff estimation techniques (Ahmed et al. 2015). As such GIS and Remote Sensing (RS) tools have been extensively applied by the researchers to estimate the watershed hydrological responses.

The integrated approach of SCS-CN method with GIS techniques in the estimation of surface runoff is an effective practice. Jasrotia et al. (2002) used a mathematical model to estimate rainfall, runoff in conjunction with RS data and GIS using SCS-CN method and runoff potential map. Amutha and Porchelvan (2009), Somashekar et al. (2011), Swetha et al. (2015) showed that estimation of runoff by SCS-CN method integrated with GIS can be used in watershed management effectively. The application of modified SCS-CN method (Mishra et al. 2003; Viji et al. 2015) in a particular circumstance (in Indian context) in order to achieve the appropriate result is also practiced by several researchers. Raina et al. (2009) estimates and compares the moisture storage values of treated and untreated Shivalik micro watersheds by modifying the initial abstraction value used in original SCS-CN method. Pandey and Dabral (2004) estimated the runoff from SCS-CN model modified for Indian condition by conventional data base and GIS for Dikrong river basin in Arunachal Pradesh. The most critical assumption of the SCS method is that the ratio of actual retention to potential retention is the same as the ratio of actual runoff to potential runoff; however, this assumption has not been empirically validated (Yu 2011). The lack of physical reality in the formulation of the method is an inherent limitation to any further development is another limitation of SCS-CN method. A major weakness is the sensitivity of estimated runoff to errors in the selection of the CN. Changes of about 15–20 % in the CN doubles or halves the total estimated runoff.

In this study using SCS-CN method integrated with RS and GIS techniques, the surface runoff of the Bhandara district is being estimated. The watershed wise results of surface runoff is again verified using the experimental data of average groundwater level fluctuation (AGLF) of wells exist in this study area.

2 Study Area

Bhandara district, Maharastra has been selected for the study for several results as it is a semi arid region which required adequate availability water for drinking and other purposes. Groundwater is considered to be the effective means for the fulfillment of this demand. And this region is famous for its geological complexities all over India is also, called as “District of Lakes” in Maharashtra State. Bhandara district lies in the northeastern part of the state of Maharashtra. The geographical area of the district is about 4090 km^2. It lies between North latitude 20° 35°–21° 05° and East longitude 79° 30°–80° 10°. Nagpur district is located on its western side and Gadchiroli district on its southern side. To the north and east of it, lies

Fig. 1 Location map of study area (Bhandara district)

Madhya Pradesh. The maximum part of the district falls in the survey if India topo-sheets 55 O/12, 55 P/9, 55 O/11, 55 O/15, 55 O/16, 55 P/13, 65 C/4. The location of the study area has been presented in Fig. 1.

3 Methodology

The current study has been focused on the estimation of runoff through RS and GIS techniques. The digitally processed IRS LISS III (Indian Remote Sensing satellite) satellite data and aerial photographs were used for the generation of thematic map on landuse/landcover. Satellite data were mainly used to generate land use and land cover information. Such land use and land cover detail, hydrological soil group and storm rainfall data were used for the calculation of surface runoff through the SCS-CN method.

The overall methodology adopted in the study is given as follows:

(i) Selection of study area
(ii) Finalization of method to be adopted
(iii) Data acquisition: RS satellite data, district map, watershed map, rainfall data, soil map and average groundwater fluctuation level map
(iv) GIS techniques: Demarcation of district boundary, generation of watershed map, landuse/landcover map and hydrological soil group through GIS software, generation of curve number
(v) Estimation of surface runoff depth and infiltration by SCS-CN method
(vi) Comparison of the results from SCS-CN method with average groundwater fluctuation level

3.1 *Data Acquisition and GIS Techniques*

The data used in this study and acquired from different authorities are given in detail as follows:

(i) LISS III (November, 2009), which is sensor of IRS 1C series of satellite, is having 23.5 m spatial resolution (Fig. 2). The spectral resolution of the sensor is 4, where as its radiometric resolution is 8, and the satellite covers the whole globe once in 24 days which is its temporal resolution.
(ii) District map, Soil map, watershed map and average groundwater fluctuation level of wells exist throughout the study area are acquired from the Journal of Land Resource atlas Bhandara District, published by National Bureau of Soil Services, publ. 22 (December 1994).
(iii) Rainfall data (2009) has been collected from Groundwater Survey and Development Agency Maharashtra.

The main process involved in GIS technique comprises of image registration with suitable projection system, digitization after the image is properly registered, attributes adding and analysis and visualization of final generated maps and later a significant conclusion is made. The projection system used in this study for Projected Co-ordinate System is Lambert Conformal Conic; for Geographic Co-ordinate System, it is GCS Clarke 1866; and for Datum, it is Clarke 1866.

The application of GIS techniques for the estimation of surface runoff of study area (Bhandara district) begins with digitization of its district boundary to hydrological soil group followed by generation of landuse/landcover thematic map. These results are required to be shown in watershed wise and hence, the numbers of watersheds classified in this study area were also digitized in GIS software. In the following section, a brief discussion of this process is being carried out.

Fig. 2 LISS III (April 2009)

3.1.1 Demarcation of District Boundary and Watershed Delineation

After the raw satellite image (LISS III) is properly registered with suitable projection system, demarcation of district boundary of Bhandara district was performed in GIS software using district map obtained from Journal of Land Resource atlas Bhandara district. Watersheds are generally delineated from drainage map by demarcating the enveloping surfaces of the least order drainages. There are total 25 watersheds in Bhandara district. Area Spread by each watershed lies between 5 and 325 km^2. A code name is assign to each watershed.

3.1.2 Landuse/Landcover Classification

The term land use refers to the human activity associated with a specific piece of land where as the term land cover related to the type of feature present on the surface of earth. Land use and land cover are dynamic. Land use changes associated with urbanization are invariably reflected in the stream flow regime. This is mainly as a result of changes to the characteristics of the surface runoff hydrograph. The quality and quantity impacts of urbanization have been well documented (ASCE 1975; Codner et al. 1988; Mein and Goyen 1988). The land use/land cover is an important characteristic of the runoff process that affects the infiltration, erosion and evapotranspiration and runoff vary from one land cover to another. The runoff yield is increased gradually from forest cover, grassland, farmland, barren land and urban built-up land (Yannian 1990). In this study, supervised classification has been used for the classification of landuse/landcover in vector format. Vector supervised classification was done in ArcGIS. In this study, map of land use and land cover are prepared for this study area from registered LISS-III satellite image. In this Level-II, nine categories of land use land cover are identified in this study area. Supervised approach of image classification was adopted. Based on the ground truth, training sets were identified on the imageries. The FCC (False Colour Composite) image with RGB format with band number 3, 2 and 1 is used for identification of various training sets. But sometimes RBG with band number 4, 2, 1 is also used because the spectral of water body was seen more clearly on band number 4.

3.1.3 Generation of Hydrological Soil Group

Soil map data collected from National Bureau of Soil Survey (NBSS) of study area were registered and digitized using ArcGIS tools and later according attributes are added to the final map. Hydrological soil group are mainly classified as A, B, C and D based on the rate of infiltration possess by the soils. Details of this hydrological soil group are discussed in the next Sect. (3.3).

3.1.4 Derivation of the Curve Number (*CN*)

Soil Conservation Service method is based on CN and is widely used methodology for understanding the response of the basin to rainfall. The CN values ranging from 40 to 100 reflect the land condition as a function of soil type, land use and soil moisture condition of the basin. SCS guidelines for estimation of runoff are given in detail by Chow et al. (1988), where CN values are given for AMC-II condition. If the area is consists of patches of landuse and soil group then the composite CN for the watershed is obtained by weighing them in proportion of the area.

Watersheds in this study area consists of the patches of different landuse and soil group, hence union of the watershed boundary along with landuse map and union of the watershed boundary along with soil map is done. After that percentage of the area corresponds to each watershed as per landuse classes is calculated and the composite CN by weighing the landuse with area is estimated. The same procedure is done for soil type by calculating the percentage of area for each group. Then average of this two is taken as the final CN for every watershed.

3.2 *Estimation of Runoff by Using USDA SCS-CN Method*

SCS-CN method was developed by Ogrosky and Mockus (1957) for determining peak rate of runoff from small watershed. A runoff CN is developed through field studied by measuring runoff from different soils at various locations. The antecedent moisture condition and physical characteristic of watershed are correlated to give hydrologic soil groups. Soils of any watershed can be classified into four hydrologic groups as A, B, C and D and are explained in this section. Runoff CNs are estimated by using landuse, soil and antecedent moisture condition (AMC). Land use is information obtained topo-sheets and digital analysis of IRS-1C LISS III image date. A GIS based methodology is used to get the input parameter at pixel level. The volume and the rate runoff depend on both meteorological and watershed characteristics and the estimation of runoff require an index to represent those two factors. The precipitation volume is probably the single most important meteorological characteristics and the estimation of volume of runoff. The soil type land use and hydrological condition of cover are the watershed factors. The antecedent soil moisture is also an important determinant of runoff volume.

Group A: A Low runoff potential group with very high infiltration rate. From such soils, even under wet condition, the runoff expectations are low. Infiltration rate is 8–12 mm/h. Transmission rate is very high for such soils.
Group B: Moderately low runoff potential soil groups with moderate rate of water transmission. Soil textures vary from fine to moderately course. Final infiltration rate is 4-8 mm/h.
Group C: Moderately high runoff potential with low infiltration rates with moderately good to well drained soils. Texture is moderately fine to moderately course with slow rate of water transmission. Final rate of infiltration is 1–4 mm/h.
Group D: Very slow infiltration rate when thoroughly wet. Clay soils form such groups. The final infiltration rate for such soils varies from 0 to 1 mm/h. Such soils have very low rate of transmission.

The above classification of soils is based on effective depth, average clay content, infiltration and probability. The AMC conditions are given as:

AMC I: Lowest runoff potential. The watershed soils are dried enough for satisfactory cultivation to take place.
AMC II: Average condition.
AMC III: Highest runoff potential. The watershed is practically saturated from antecedent rain. The AMC group is determined using five day antecedent rainfall.

If the area consists of patches of land used then the composite CN for the watershed can be obtained by weighing them in proportion of area. The value of CN is lower for soils with high infiltration than for soils with low infiltration. Calculation of potential maximum retention S in cm,

$$\mathrm{CN} = \frac{2540}{(25.4 + S)} \tag{1}$$

Calculation of runoff depth for all soils in India except Black soil region of AMC II & III, where initial losses (considering interception depression storage and infiltration) i.e. I_a equal to $0.2S$

$$R = \frac{(P - 0.2S)^2}{(P + 0.8S)} \tag{2}$$

But for Indian condition of AMC I & II type R is given as

$$R = \frac{(P - 0.3S)^2}{(P + 0.7S)} \tag{3}$$

where
P Mean rainfall in mm,
R Runoff depth in mm.

4 Results and Discussion

In this section, all the result comes from this study will be presented and accordingly the necessary discussion will also be reported.

4.1 Watershed Map

The generated watershed map of study area is shown in Fig. 3. There are total 25 watersheds classified in this district according to its drainage outlet pattern.

4.2 Landuse/Landcover Map

In this study area, landuse/landcover (Fig. 4) is classified (level II) as crop land, dense forest, fallow land, habitation, land with scrub, land without scrub, open

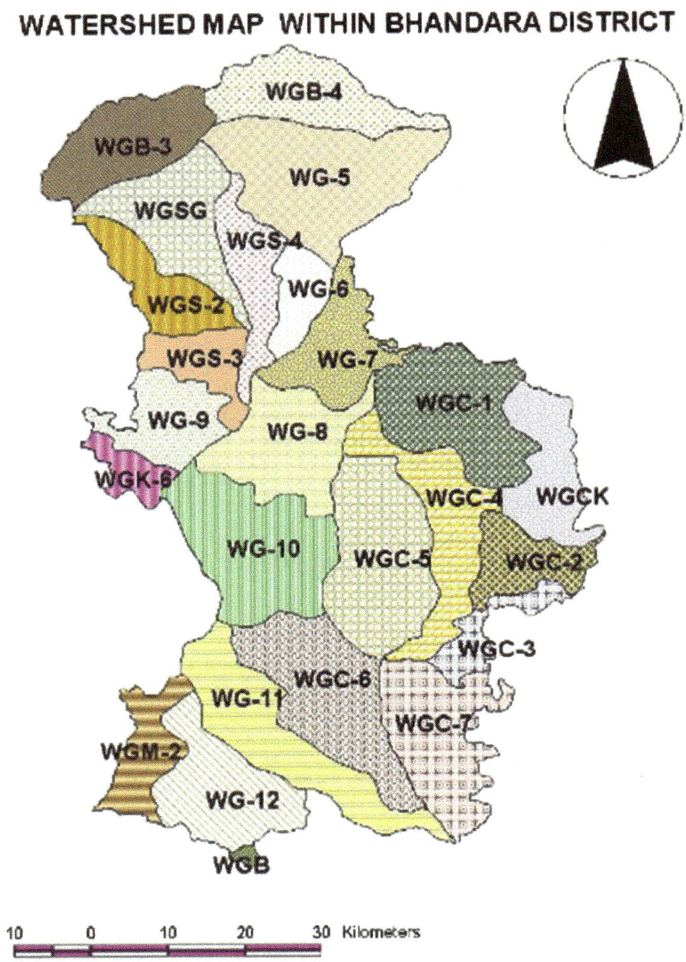

Fig. 3 Watershed map of study area

forest, outcrop and water bodies (only river). In this area most of the area is covered by land with scrub about 1155 km^2 and at the least by habitation about only 15 km^2. The land for agricultural purposes also gives contribution in this area. The dense forest does not cover in large area. So, from this result, it has been apparently understood that there is less human activities, the natural creation and its form is still not much destroyed; however, the development activities in this area in the forms of industries have been increasing day by day. Hence, in near future, the pattern and percentage of area covered by different landuse/landcover will be abruptly changed affecting the natural water resources and its phenomenon.

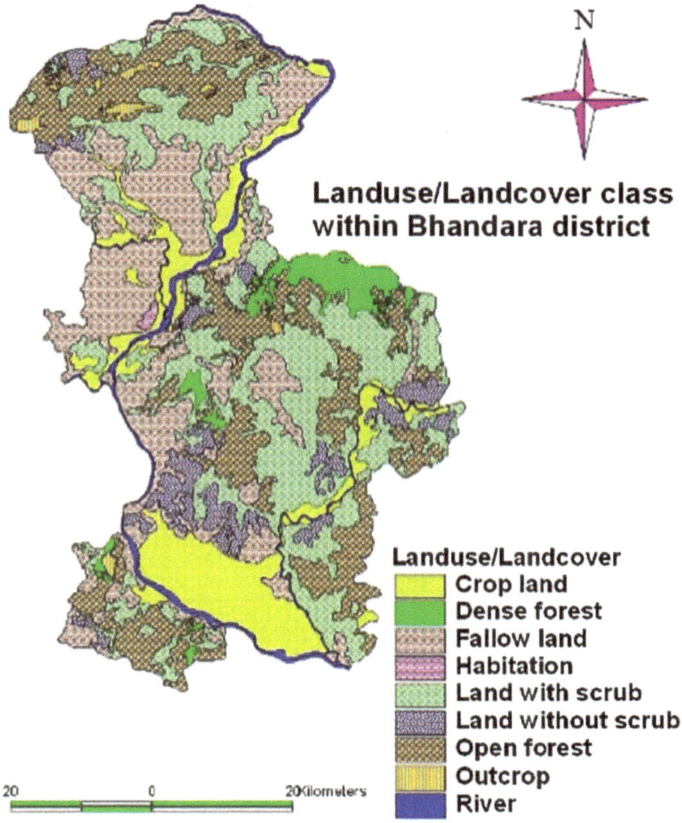

Fig. 4 Landuse/landcover map of study area

4.3 Soil Types and Hydrological Soil Group Map

Mostly, the soil type as investigated by the organization, All India Soil and Land Use and NBSS has been followed for the study. It is again noteworthy that clay soil is the basic characteristics of the soil types within the area, which is quite natural as the geological composition within the area is basaltic. Percentage soil cover of the study area is given in Fig. 5. Medium black soil is covered highest (53.78 %) followed by red sandy (33.46 %) and lowest by laterite only 2.67 %.

The study area is covered mostly by clay, clay loam and clay sandy having hydrological soil group C, B, A respectively and there is not found soil group D means in this study area moderately good infiltration should occur (Fig. 6).

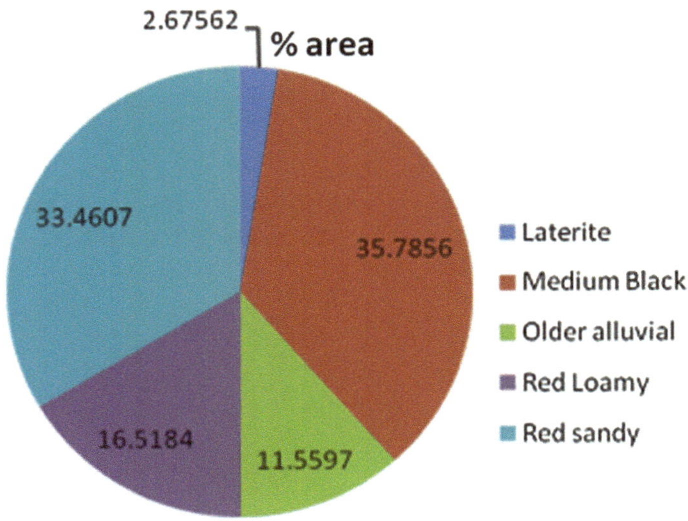

Fig. 5 Soil cover (%) of the study area

4.4 Curve Number Generated

Applicability of Curve Number method has been beyond discussion until the estimation of runoff depth within watersheds is done (Larry, 2001). This is because it takes into account the watershed characteristics of the landuse/landcover and vegetation etc. This has been corroborated through the present study where the results are found to be in accordance to the LU/LC and soil type in the study area. Weighted CN is calculated by adding all the CN of having different patches of landcover of that area and divided by total area. AMC II condition has been considered for generation of CN in this study. CN map for each watershed is presented in Table 1.

The CN is highest in the WGC-7 i.e. 78.26 covered by clay and clay loam having mix cover of land with scrub, open forest, small percentage of crop land, fallow and land without scrub which is naturally high in runoff. The smallest CN is in watershed WGC-2 i.e. 51.6 covered by equal proportion of crop land, land with scrub and land without scrub and 10 % by open forest, hence low runoff will occur in this watershed.

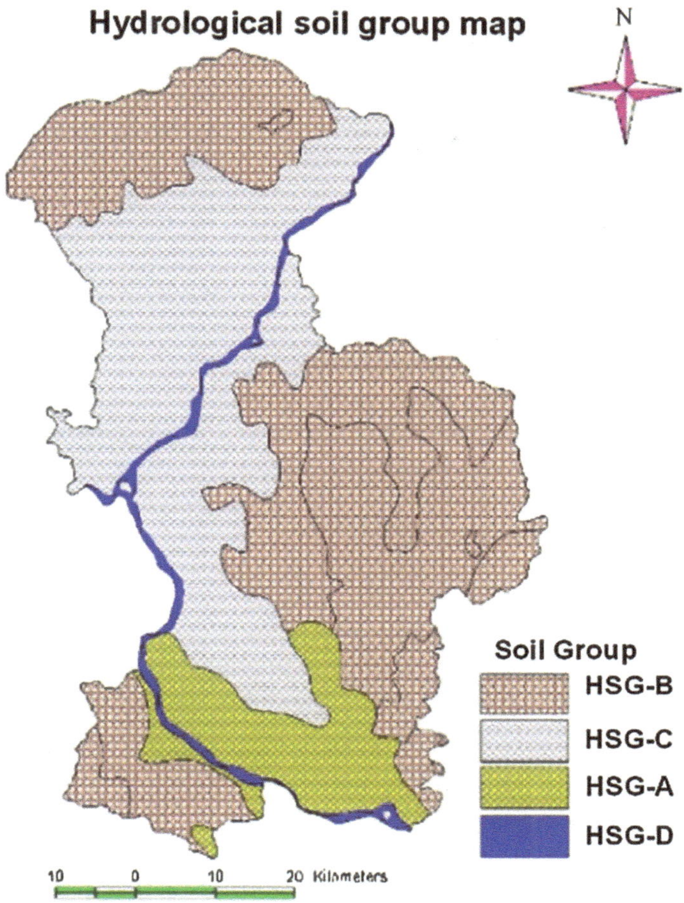

Fig. 6 Hydrological soil group map of study area

4.5 Estimated Surface Runoff Depth and Its Verification

Surface runoff depth estimated by using SCS-CN method in watershed wise is presented in Fig. 7. The values of surface runoff depth is ranges from 700 to 1000 mm and it has been classified into six classes as very low, low, medium low, medium high, high and very high. The entire study area mostly observed with very low to medium low and significantly also with medium high to high. A single watershed (WGC-7) contributes with very high surface runoff with the values of 950–1000 mm.

To verify this estimated results of surface runoff depth is not an easy task, however a significant verification is made with the AGLF (Fig. 8) of 70 number of wells exists in this study area. In this figure, the AGLF is ranges from 0 to 15 m and it is observed that most the wells in this region, AGLF is from 2 to 7 m, which can

Table 1 Watershed wise generated Weighted CN

S. No.	Watershed No.	S (cm)	$I_a = 0.3S$	CN
1	WG-5	16.116	48.350	61.18
2	WG-10	9.252	27.756	73.3
3	WGC-5	15.867	47.601	61.55
4	WGC-6	12.008	36.023	67.9
5	WG-11	13.378	40.135	65.5
6	WGC-1	17.469	52.407	59.25
7	WGC-7	7.055	21.167	78.26
8	WG-8	8.287	24.861	75.4
9	WG-12	15.442	46.327	62.19
10	WGC-4	12.968	38.905	66.2
11	WGSG	8.758	26.274	74.36
12	WGB-4	12.709	38.128	66.65
13	WGB-3	12.567	37.701	66.9
14	WGCK	12.146	36.438	67.65
15	WG-7	8.331	24.995	75.3
16	WG-9	9.110	27.332	73.6
17	WGS-4	8.739	26.219	74.4
18	WGS-2	9.394	28.183	73
19	WGC-2	23.824	71.474	51.6
20	WGM-2	14.349	43.048	63.9
21	WGS-3	10.374	31.123	71
22	WG-6	8.021	24.063	76
23	WGC-3	13.979	41.939	64.5
24	WGK-6	8.131	24.394	75.75
25	WGB	9.877	29.633	72

be considered as medium AGLF. Some few parts shown 8–12 m AGFL and which can be considered as high and the smallest area is shown AGLF value of 14–15 m. From the comparison of surface runoff with AGLF, is has been observed that the entire area is covered by low to medium in both the cases.

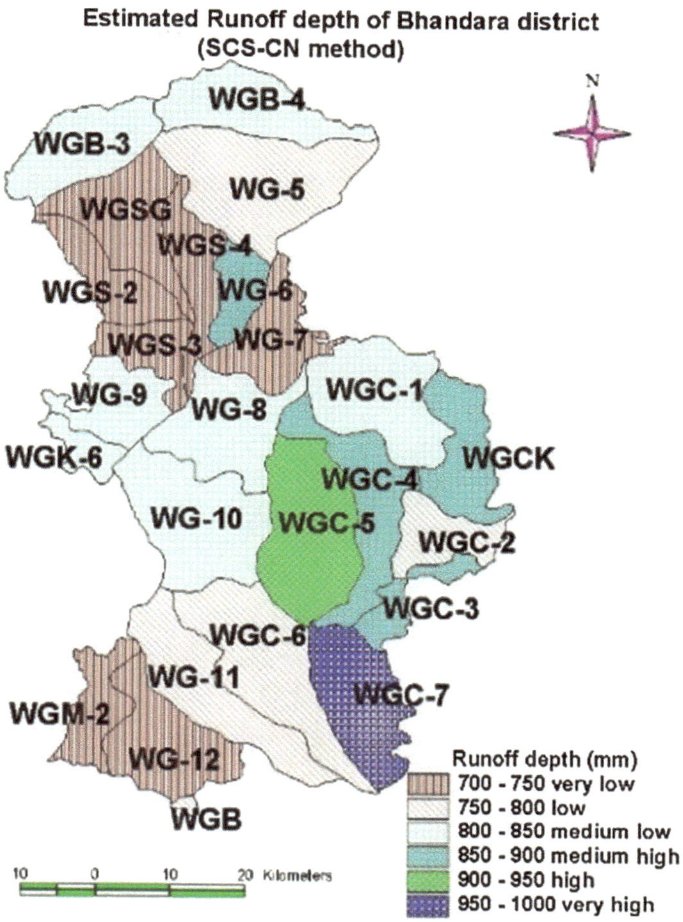

Fig. 7 Estimated surface runoff depth of study area by SCS-CN method

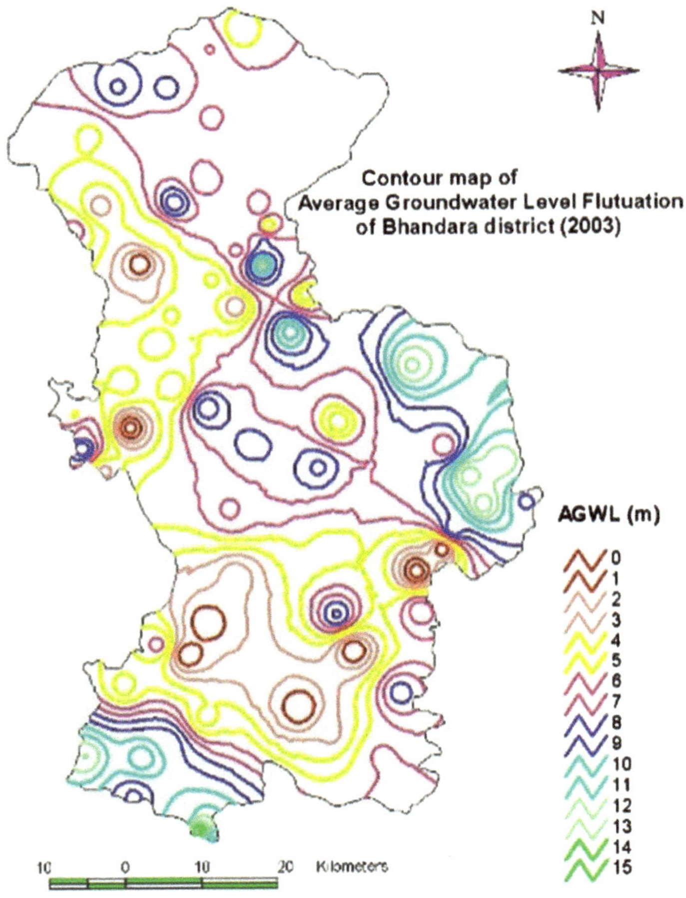

Fig. 8 Average groundwater fluctuation level (2003) of study area

5 Conclusion

SCS-CN method integrated with satellite image and GIS tools is a useful methodology for initial investigation of hydrological phenomenon without field observation and experimental work. This method required minimum input data and has the ability to clearly displays the required information about the inter-relationship of natural complex hydrological phenomenon giving the possible link for understanding one phenomenon in detail from others phenomenon. The surface runoff depth of study area has been estimated using SCS-CN method in watershed wise and verified with the field observations of AGLFs of wells exists in this area. From this comparison, it has been concluded that the entire study area indicates medium surface runoff apparently leading to moderate groundwater recharge potential.

References

Ahmad I, Verma V, Verma MK (2015) Application of curve number method for estimation of runoff potential in GIS environment. 2nd International Conference on Geological and Civil Engineering, 80, IACSIT Press, Singapore

Amutha R, Porchelvan P (2009) Estimation of surface runoff in Malattar sub-watershed using SCSCN method. J Indian Soc Remote Sens 37:291–304

ASCE (1975) Aspects of hydrological effects of urbanisation. J Hydraul Div ASCE 101:449–468

Bosznay M (1989) Generalization of SCS curve number method. Am Soc Civil Eng 1(139):139–144

Boughton WC (1989) A review of the USDA SCS curve number method. Aust J Soil Res 27 (3):511–523

Chattopadhyay GS, Choudhury S (2006) Application of GIS and remote sensing for watershed development project-a case study. Map India

Chow VT, David RM, Larry WM (1988) Applied hydrology. McGraw-Hill

Codner GP, Laurenson EM, Mein RG (1988) Hydrologic effects of urbanization: a case study. In: Proceedings of the hydrology and water resources symposium, Institution of Engineers, Australia pp 201–205

Jasrotia AS, Dhiman SD, Aggarwal SP (2002) Rainfall-runoff and soil erosion modeling using remote sensing and GIS technique—A case study on tons watershed. J Indian Soc Remote Sen 30(3):167–180

Larry WM (2001) Water resources engineering, 1st edn. Wiley, New York

Mahboubeh E, Ahmad Ainuddin B, Nuruddin, Mohd Amin, B, MohdSoom, Alias MohdSood, LiewJuNeng (2012) Runoff estimation in steep slope watershed with standard and slope-adjusted curve number methods. Pol J Environ Study, 21(5):1191–1202

Mein RG, Goyen AG (1988). Urban runoff. Civil Engineering Transactions, Institution of Engineers Australia CE30:225–238

Mishra SK, Singh VP, Sansalone JJ, Aravamuthan V (2003) A Modified SCS-CN Method: characterization and testing. Water Resour Manage 17(1):37–68

Ogrosky HO, Mockus V (1957) The hydrology guide. National Engineering Handbook, section 4, hydrology supplement, A.G.C.A., USDA

Pandey A, Dabral PP (2004) Estimation of runoff for hilly catchment using satellite data. J Indian Soc Remote Sen 32(2):236–240

Patil JP, Sarangi A, Singh OP, Singh AK, Ahmad T (2007) Development of a GIS interface for estimation of runoff from watersheds. Earth Env Sci, Water Resour Manage, Springer Sci 22 (9):1221–1239

Raina AK, Lubana PPS, Singh KG, Bhat VK (2009) Application of modified Soil Conservation Service Curve Number (SCS-CN) method for estimation-cum-comparison of moisture storage values of untreated and treated Shivalik micro-watersheds. J Soil Water Conserv 8(4):20–24

Ruslin AM (2011) The rainfall-runoff model using of the watershed physical characteristic approach. Int J Civil Environ Eng IJCEE–IJENS, 11(6)

Sarangi A, Bhattacharya AK, Singh AK, Sambaiha A (2005a) Performance of Geomorphologic Instantanous Unit Hydrograph (GIUH) model for estimation of surface runoff, International conference on recent advances in water resources development and management, 23–25 November 2005, IIT Roorkee, Uttaranchal, India, pp 569–581

Sarangi A, Madramootoo CA, Enright P, Prasher SO, Patel RM (2005b) Performance evaluation of ANN and geomorphology-based models for runoff and sediment yield prediction for a Canadian watershed. Curr Sci 89(12):2022–2033

Shwetha G, Maheshwara Babu B, Polisgowdar BS, Srinivas Reddy GV, Shanwad UK (2015) Estimation of surface runoff in timmapur watershed using SCS-CN method. Int J Agric Sci Res 5(6):1–8

SCS (1956, 1972, 1973) Hydrology, national engineering handbook, supplement A, Sect. 4, Chap. 10, Soil Conservation Service. USDA, Washington

Sindhu D, Shivakumar BL, Ravikumar AS (2013) Estimation of surface runoff in Nallur Amanikere watershed using SCS-CN method. Int J Res Eng Technol 404–409
Somashekar RK, Ravikumar P, Sowmya SV, Dar MA, Ravikumar AS (2011) Runoff estimation and morphometric analysis for Hesaraghatta Watershed using IRS-ID LISS III FCC satellite data. J Indian Soc Remote Sens 39:95–106
Viji R, Prasanna PR, Ilangovan R (2015) Modified SCS-CN and Green-Ampt methods in surface runoff modelling for the Kundahpallam Watershed, Nilgiris, Western Ghats, India. Aquatic Procedia 4:677–684
Yannian YU (1990) Hydrological effects of forest, The Hydrological Basis for Water Resources Management (Proceedings of the Beijing Symposium, October 1990), IAHS publ. no. 197
Yu B (2011) Validation of SCS Method for runoff estimation. J Hydrol Eng 17:1158–1163

Study on the Impact of Land Use Changes on Urban Hydrology of Cochin, Kerala, India

Archana M. Nair, Lekshmi Mohanlal, C.R. Ayishath Nabeela, T.D. Aneesh and Reji Srinivas

Abstract Urbanization and population pressure are the two main challenges for water resource management, especially in cities of developing countries. The process of urbanization results in land use alterations. Urban development can have a major impact on the local hydrology. The growth of urban areas results in significant changes in the physical properties of the land surface consequently increasing impervious surface area. Imperviousness is the most critical indicator to analyze the impact of urbanization on the hydrology. As a direct consequence, this results in increased surface runoff. Indirectly, this results in a reduction in infiltration, eventually altering the prevailing hydrologic system. The present study is an attempt to quantify the impact of land use changes due to urbanization on surface and subsurface hydrology. The study area is Cochin, one of the fast developing second tier metros in India. We estimated the runoff volume and ground water fluctuations for the study area. SCS-CN method was used to calculate runoff volume. The land use changes for the study area were prepared from Survey of India topographic map as well as satellite data. Runoff depth and volume was calculated for various land use categories using daily rainfall data for the past three decades. We also calculated the depth to ground water table for dug wells for last three decades in the study area. Depth to ground water table measured was used to generate water table contours using triangulation method. The water table contours were classified into four and the area under each class was measured since 1980. It was found that, the area for shallow water table gets decreased for the past three decades clearly depicting an increasing depth trend.

Keywords Surface runoff · Ground water · Cochin · Land use · Urbanization

A.M. Nair (✉)
Department of Civil Engineering, IIT Guwahati, Guwahati, India
e-mail: archana.nair@gmail.com

R. Srinivas
COP Group, National Centre for Earth Science Studies, Thiruvananthapuram, India

L. Mohanlal · C.R.A. Nabeela · T.D. Aneesh
NREM Group, National Centre for Earth Science Studies, Thiruvananthapuram, India

A.K. Sarma et al. (eds.), *Urban Hydrology, Watershed Management and Socio-Economic Aspects*, Water Science and Technology Library 73,
DOI 10.1007/978-3-319-40195-9_6

1 Introduction

Urbanization brings significant changes in physical properties of land surface leading to creation of more and more impervious surfaces. The consequence of this is enhanced surface runoff and reduced infiltration. The large rate of conversion of rural areas to cities or urban areas is affecting both surface and ground water systems. It alters topography and natural vegetation, stream flows and flooding characteristics, temperatures both above and below the land surface, and water quality of surface streams and ground water (Sharp 2010). The process of urbanization is unplanned and at a faster rate in developing countries like India. This makes the water resource management in India more challenging. Advent of satellite remote sensing and Geographical Information System (GIS) have opened new vistas for ground water studies. This is due to the fact that earth observing devices, both on space craft as well as on aircraft provide most up-to-date, accurate, unbiased, and detailed spectral, spatial, and temporal information on conditions of natural resources. The integration of satellite remote sensing and GIS is an effective approach for analyzing the direction, rate, and spatial pattern of land use change. Urbanization causes significant changes to the temporal characteristics of runoff from an area, such as shortening the runoff travel time resulting in flash floods, and more severe pollutant loadings. The amount of rainwater running off the surface into drains and sewers increases dramatically as new development covers previously permeable ground. Imperviousness is the most critical indicator for analyzing urbanization impacts on the water environment. According to Weng (2012), impervious surfaces are features of anthropogenic origin such as roads, driveways, sidewalks, parking lots, rooftops, etc., through which water cannot infiltrate into the soil. Hence, an increase in the impervious cover could lead to increase in the volume and intensity of urban runoff and an overall decrease in groundwater recharge (Weng 2012; Brun and Band 2000).

Increased surface runoff would eventually result in the alteration of the prevailing hydrologic system. Hence, urban imperviousness is a very important parameter in the management of urban watersheds and the related groundwater management. Climate models predict that heavier showers and hotter summers are expected to put more pressure on urban drainage. In this context, the present study is an attempt to quantify the impact of land use changes due to urbanization on surface runoff as well as on ground water. Berthier et al. (2006) explained that the hydrological behavior of urban areas can no longer be restricted to the runoff of rainwater on impervious surfaces, which constitutes the dominant flow component for design purposes. Ragab et al. (2003) exemplify that urban surfaces, such as road pavements and parking lots, are not completely impervious, and observed that 6–9 % of total annual rainfall on a paved street infiltrates and that 21–24 % of it evaporates.

In the present study, runoff estimation was carried out using a modified version of SCS-CN method for Indian scenario. Various researchers have used SCS-CN method for Indian conditions. Geetha et al. (2008) derived a conceptual model based on SCS-CN concept for long-term hydrologic simulation. The model was capable to derive a better match to the observed runoff as well capable to find out the dormant or dominant process involved in the runoff. Patil et al. (2008) estimated the surface Runoff using Curve Number techniques (ISRE-CN) and developed a model using the in-built macro programming language Visual Basic for Applications (VBA) of ArcGIS® tool to estimate the surface runoff by adopting one the most widely used NRCS-CN technique and its three derivatives. In the present study, runoff volume was calculated based on the land use pattern as derived for the year 1968 from topographic sheet, for 2006 and 2010 from Quick bird satellite images as well as IRS LISS IV and Landsat MSS images.

2 Materials and Methods

2.1 Study Area

The study area is Cochin which comprises of Mattanchery, Fort Cochin, part of the main land Ernakulum and a group of islands, which lies between 9° 55′N and 10° 04′N Latitude and 76° 14′E and 76° 20′E Longitude (Fig. 1). The western boundary of the study area is covered by Arabian Sea and Vembanad Lake. The study area is divided into two blocks by the Cochin backwater (Azhi) in the east–west direction with Vypeen on the northern side and Fort Cochin on the southern sides. Major rivers such as Periyar and Muvattupuzha drains to the Vembanad Lake at Cochin. The peculiarity of Cochin is that much of the area lies at sea level and is 'plain' land having natural facilities of drainage in the form of backwaters, canals, and rivers. Major part of the study area is covered by recent sediments or alluvium with sparse distribution of Charnockites and Gneiss. Since much of the area is low lying plain, flooding is a recurring natural calamity. Salt water intrusions destabilizing the existing fresh water table and shortage of portable drinking water during peak summer are added troubles encountered in the study area. Since, Arabian Sea and Vembanad Lake surrounds the western boundary of the city, the runoff travel time is very less. In spite of the fact that the study area receives a very heavy rainfall during the monsoon, the area faces shortage of portable drinking water during peak summer could be attributed to the short span of travel time needed for the surface runoff water to reach the Ocean. Consequently, with increase in percentage of impermeable surface, the drinking water shortage can be a grave problem in the near future. The ground water source in most parts of the study area is basically from unconfined aquifer with shallow water table. Hence, there is a direct impact on the recharge as more and more areas are converted to impervious surface.

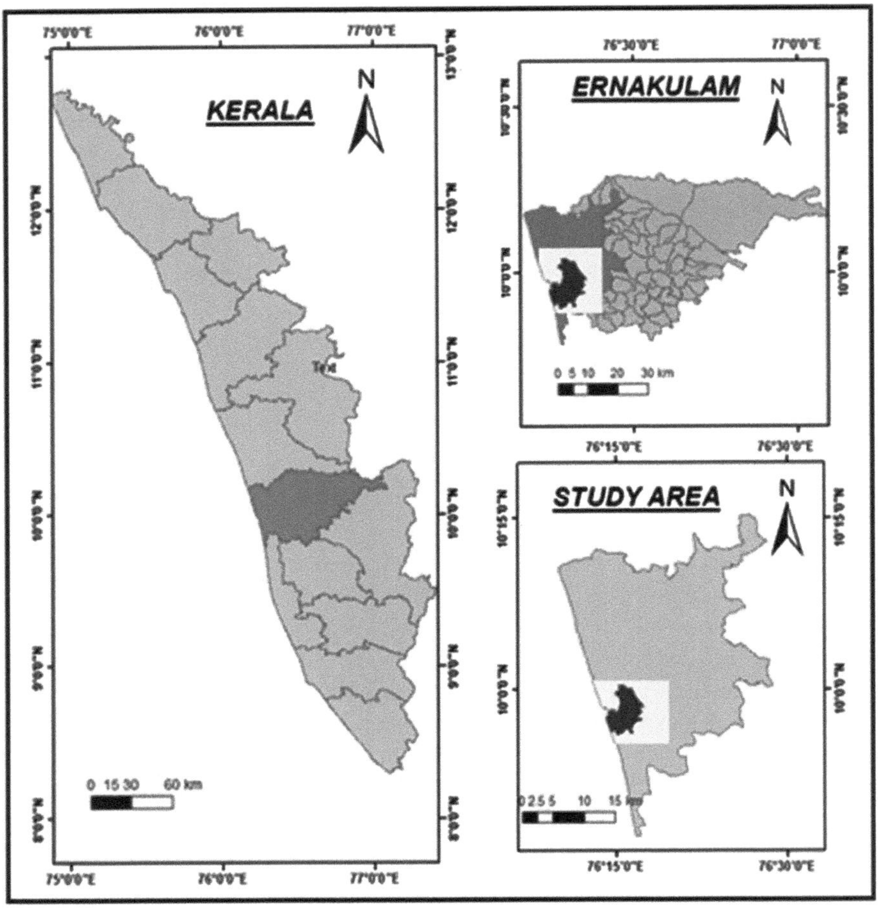

Fig. 1 Location map of the study area

2.2 *Methodology*

The process of transformation of rainfall to runoff is highly complex, dynamic, nonlinear, and exhibits temporal and spatial variability, further affected by many and often interrelated physical factors. In an urbanized area, the primary factor that influences the runoff is land use changes. In a growing or expanding city, many major changes occur to the land use and this result in an increased volume of surface runoff. In the present study, runoff was calculated not for a particular watershed; instead runoff was calculated based on the land use pattern. Surface runoff was calculated for each land use category using SCS-CN method. Survey of India topographic sheets of scale 1:50,000, was used to map land use classes and this map formed the base map for further analysis. Satellite images from IRS LISS IV, Quick Bird, and Landsat MSS were used to study the changes in land use

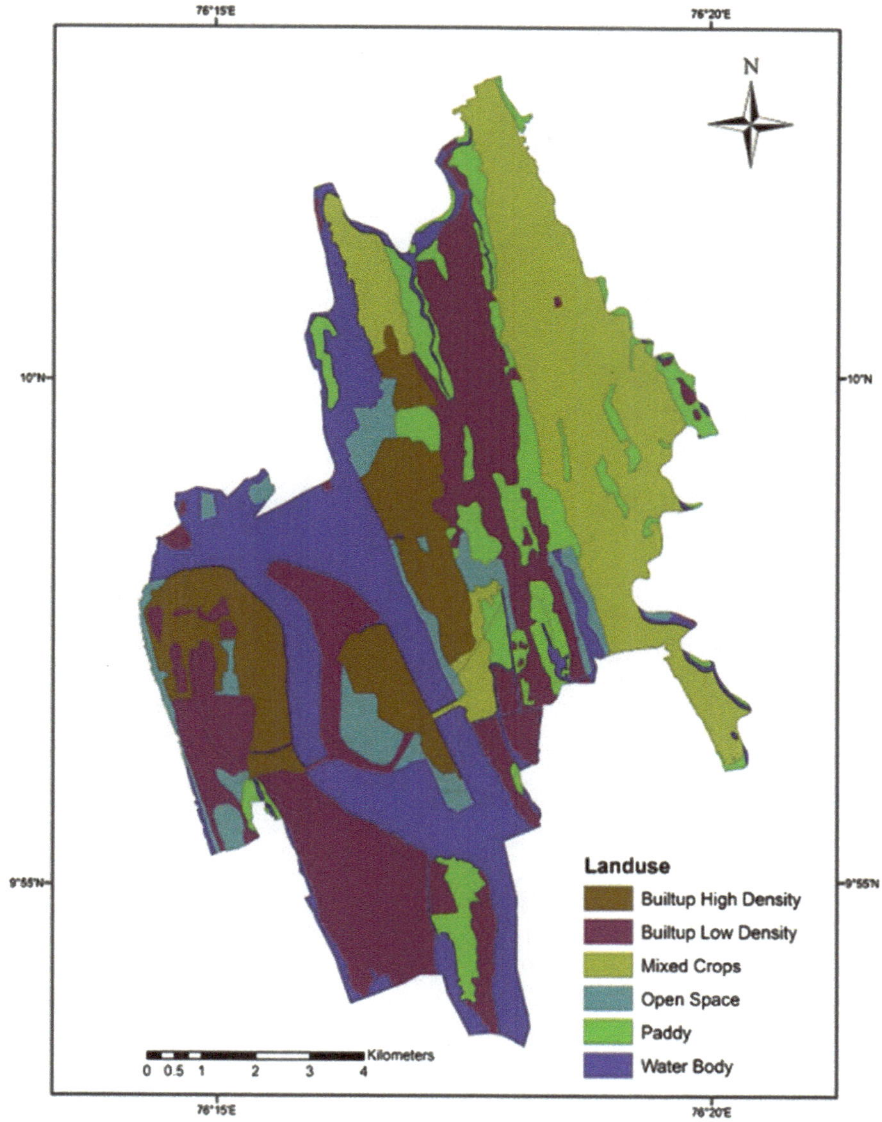

Fig. 2 Land use map of the study area from topographic sheet (1968)

variation for past ten years. The land use map for the years 1968 (Fig. 2), 2006 (Fig. 3), and 2010 (Fig. 4), for the study area was obtained from topographic sheet and satellite image LISS-1V (2006), Landsat (2010), respectively. The land use maps derived were used to find the total impervious area by reclassifying land use classes. The interconnected impervious area was then reclassified as the effective

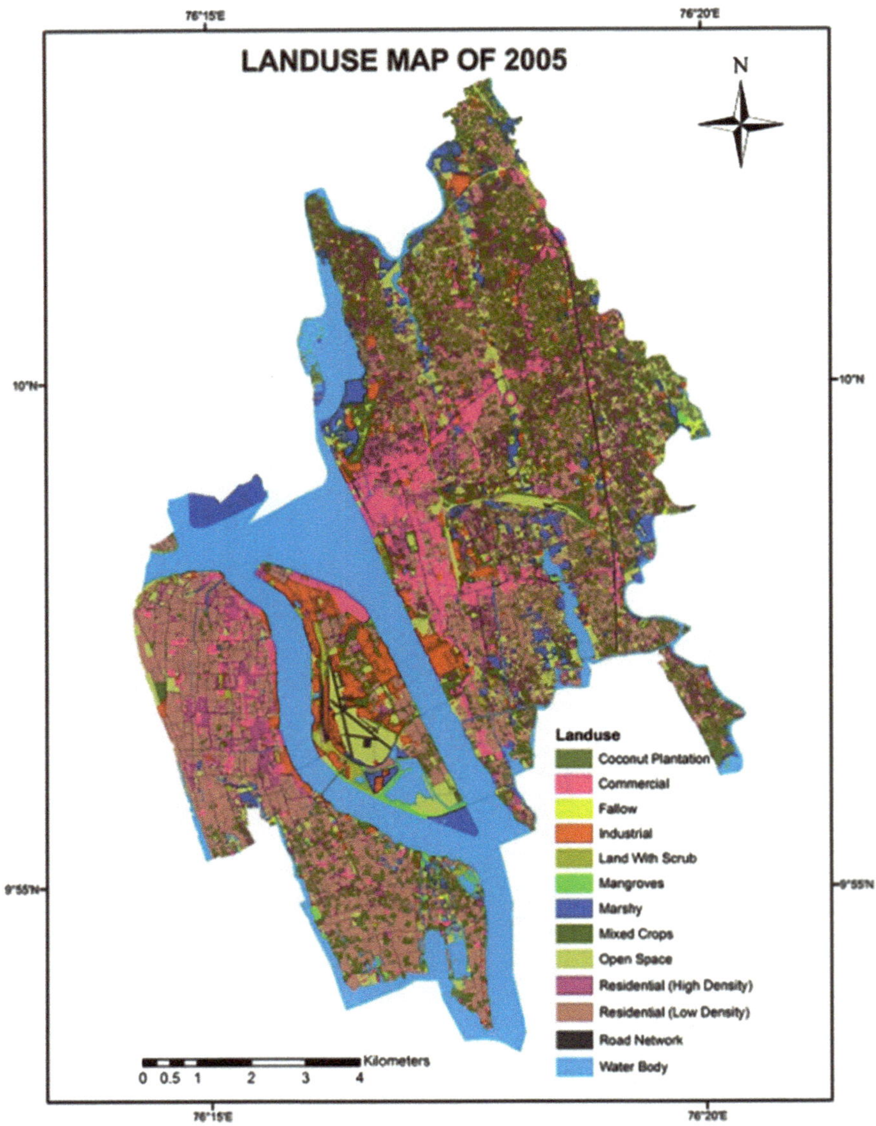

Fig. 3 Land use map of the study area for 2006

impervious area. Effective impervious area of Cochin for the year 1968 and 2010 is shown in Fig. 5.

Runoff is that portion of precipitation that flows over land surfaces toward larger bodies of water. There are two broad categories of factors that control runoff: rainfall (storm) characteristics and watershed physical conditions. Important rainfall

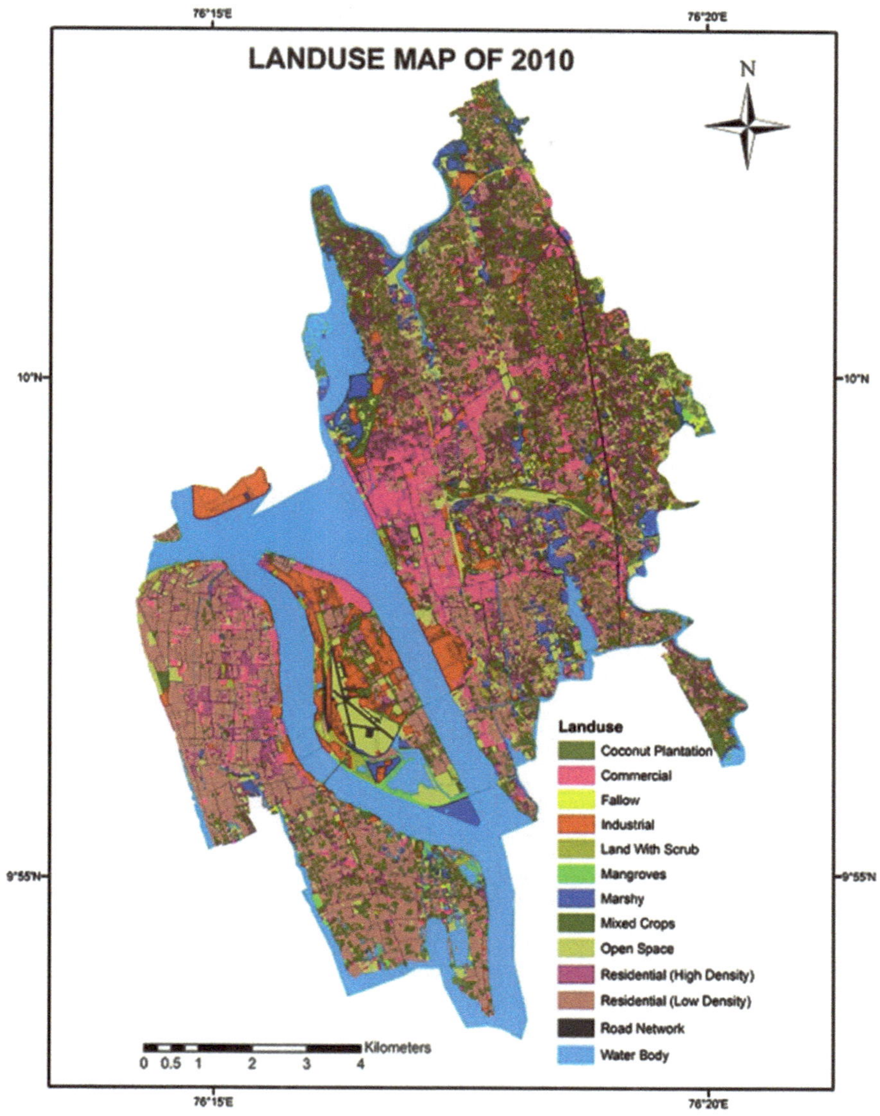

Fig. 4 Land use map of the study area for 2010

characteristics include duration, amount, intensity, and distribution. Rainfall must satisfy the immediate demands of infiltration, evaporation, interception, surface storage, and surface detention and/or channel detention before the start of runoff. In 1972, the U.S. Soil Conservation Service suggested an empirical model for rainfall abstractions which is based on the potential of the soil to absorb a certain amount of

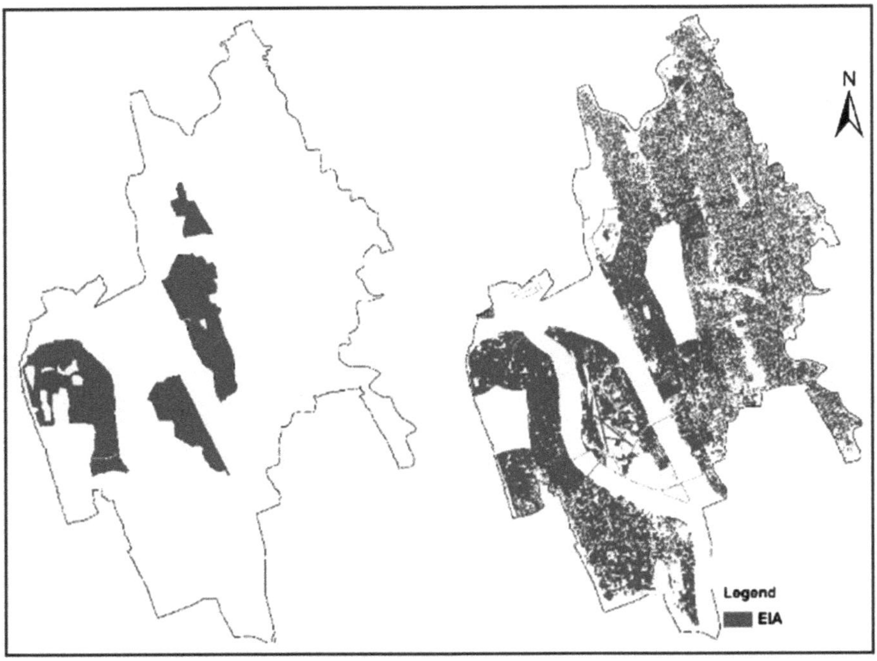

Fig. 5 Land use changes in the study area for last 42 (1968–2010) years in terms of effective impervious area (EIA)

moisture. On the basis of field observations, this potential storage S (millimeters or inches) was related to a 'curve number' CN which is a characteristic of the soil type, land use, and the initial degree of saturation known as the antecedent moisture condition. The curve number is based on the area's hydrologic soil group, land use treatment, and hydrologic condition. Although the method is designed for a single storm event, it can be scaled to find average annual runoff values. The major factors that determine CN are the hydrologic soil group (HSG), land use (Table 1), and AMC conditions. The CN value was calculated for antecedent moisture condition AMC II. The conversion of CNII to other two AMC conditions can be made through the following correlation equations. AMC refers to the moisture content present in the soil at the beginning of the rainfall–runoff event under consideration. It is well known that initial abstraction and infiltration are governed by AMC. The daily rainfall data for the study area from 1980 to 2010 was collected from Indian meteorological department. The daily rainfall data is necessary to calculate the antecedent moisture condition for the rainfall event under consideration. Antecedent moisture condition is determined by considering the 5 day rainfall prior to the day under consideration.

Table 1 Land use land cover description with corresponding curve number

Cover description	Curve numbers for hydrologic soil group			
Cover type and hydrologic condition	A	B	C	D
Fully developed urban areas (vegetation established)				
Open space (lawns, parks, golf courses, cemeteries, etc.) 3				
Poor condition (grass cover <50 %)	68	79	86	89
Fair condition (grass cover 50–75 %)	49	69	79	84
Good condition (grass cover >75 %)	39	61	74	80
Impervious areas				
Paved parking lots, roofs, driveways, etc. (excluding right-of-way)	98	98	98	98
Streets and roads				
Paved; curbs and storm sewers (excluding right-of-way)	98	98	98	98
Paved; open ditches (including right-of-way)	83	89	92	93
Gravel (including right-of-way)	76	85	89	91
Dirt (including right-of-way)	72	82	87	89
Western desert urban areas				
Natural desert landscaping (pervious areas only)	63	77	85	88
Artificial desert landscaping (impervious weed barrier, desert shrub with 1- to 2-inch sand or gravel mulch and basin borders)	96	96	96	96
Urban districts				
Commercial and business	89	92	94	95
Industrial	81	88	91	93
Residential districts by average lot size				
1/8 acre or less (town houses)	77	85	90	92
1/4 acre	61	75	83	87
1/3 acre	57	72	81	86
1/2 acre	54	70	80	85
1 acre	51	68	79	84
2 acres	46	65	77	82
Developing urban areas				
Newly graded areas (pervious areas only, no vegetation)	77	86	91	94

Most urban areas are only partially covered by impervious surfaces hence the soil remains an important factor in runoff estimates. Urbanization has a greater effect on runoff in watersheds with soils having high infiltration rates (sands and gravels) than in watersheds predominantly of silts and clays, which generally have low infiltration rates. Any disturbance of a soil profile can significantly change its infiltration characteristics. With urbanization, native soil profiles may be mixed or removed or fill material from other areas may be introduced. Hydrological soil groups were assigned to the study area based on comparison of the characteristics of unclassified soil profiles with profiles of soils already placed into hydrologic soil groups. Most of the groupings are based on the premise that soils found within a climatic region that are similar in depth to a restrictive layer or water table, transmission rate of water, texture, structure, and degree of swelling when saturated, will have similar runoff responses. The slope of the soil surface is not considered when assigning hydrologic soil groups.

The runoff equation is

$$Q = \frac{(P - I_a)^2}{P - I_a + S} \tag{1}$$

where Q is runoff; P is rainfall and S is the potential maximum soil moisture retention after runoff begins. I_a is the initial abstraction or the amount of water before runoff, such as infiltration, or rainfall interception by vegetation; and it is generally assumed that $I_a = 0.2S$.

The runoff curve number, CN, is then related

$$S = \frac{1000}{\text{CN}} - 10 \tag{2}$$

CN has a range from 30 to 100 with lower number indicating low runoff potential while larger numbers indicating higher runoff potential. In the present study, modified SCS-CN (CN-1, Patil et al. 2008) method was used considering the Indian scenario.

The study area has shallow aquifer and hence ground water is tapped mainly by dug well. Water level data for pre-monsoon and post-monsoon for past 35 years were collected from Central Ground Water Board that formed the basis for the analysis of historic trend in water table fluctuations. From the historic well data, the pre-monsoon and post-monsoon water depth level was plotted and the long-term trend analysis for each of the well falling within the study area was carried out. From the same data, the water depth contour map for pre-monsoon and post-monsoon period was prepared using triangulation method in ARC GIS. The water depth contour map was used to reclassify the map into four classes such as class I-below 2 m, class II—2–4 m, class III—4–6 m and class IV—above 6 m. The percentage change in water level for the years 1981, 1997, 2007, and 2010 for the study area was carried out from the reclassified contour map using area under curve method.

3 Results and Discussion

The rapid urbanization of the study area is particularly evident from the increase in residential, commercial, and industrial areas which ultimately resulted in an increase in impervious area. The land use pattern of Cochin showed a significant change from 1968 to 2010. Major changes in the land use pattern are visible by comparing the land use maps of 1968 (Fig. 2), 2006 (Fig. 3), and 2010 (Fig. 4).

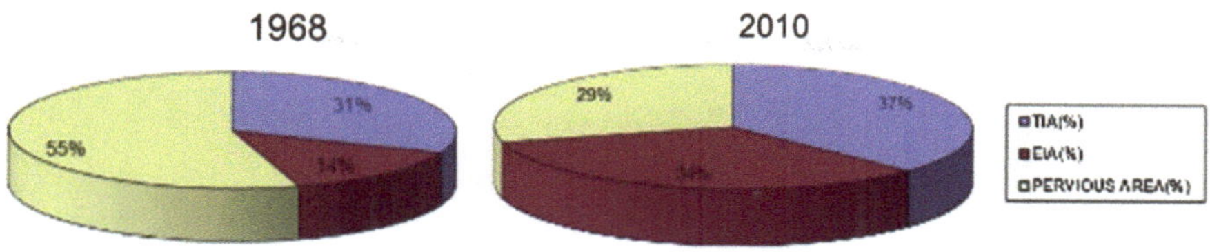

Fig. 6 Pie chart representing land use changes in terms of pervious nature (*TIA* Total Impervious Area; *EIA* Effective Impervious Area and Pervious Area)

The most significant changes were observed in residential, commercial, and industrial areas causing significant increase in the impervious surfaces in the study area. The open spaces and marshy areas are being converted into built-up areas, particularly into residential apartments. The built-up area has increased from 36 to 48 % over the period. In the period from 2005 to 2010, an increase in built-up area of 2 % is noticed. The areas occupied by different land use classes in 1968, 2006, and 2010 were reclassified into pervious and impervious areas. The percentage increase in the impervious area gives a direct indication of the changes over the last three decades. Compared to the area of impervious cover in the year 1968, a drastic change can be observed in 2010 due to urbanization (Fig. 5). A comparison between the impervious area for the years 1968 and 2010 showed that the effective impervious area increased from 14 % in 1968 to 34 % in 2010 (Fig. 6). The runoff volume is found to be highest in the months of June to September and it coincides with the southwest monsoon season which causes the highest rainfall in the region. The seasons "February to May" and "October to January" showed considerably lower runoff volumes. The runoff volume on a yearly and seasonal basis is calculated for each of the land use maps of 1968, 2006, and 2010, using daily rainfall data for a period of 30 years from 1980 to 2010. Using the daily rainfall data of a 30 year period negates the variations in rainfall over the years (Fig. 7). The calculation is done in such a way that the daily rainfall and its corresponding antecedent moisture condition are considered. The area within non-built up includes water body which forms almost 20 % of the study area. The built-up area has increased from 36.51 km^2 in 1968 to 48.07 km^2 in 2010. The most significant changes were observed in residential, commercial, and industrial areas causing significant increase in impervious surfaces in the study area. The vegetated regions such as mixed crop class and paddy class as well as open spaces were converted into built-up areas, particularly into residential apartments. This increase in built-up area has resulted in an increase in runoff volume of 23.5 Mm^3 in the last 45 years (Fig. 8).

With the increase in impervious area, the surface runoff also increased whereas recharge decreased. It was clear from water depth contour that depth to water table increased in pre-monsoon than that in the post-monsoon. Thus it is clear that the study area gets recharged mainly due to monsoon rainfall. Hence, the historic water depth trend for post-monsoon will give a picture of the recharge pattern. The post-monsoon and pre-monsoon water level data of the study area for last three

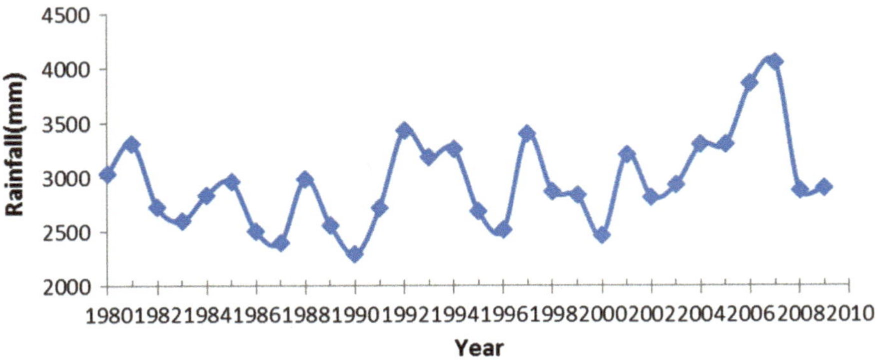

Fig. 7 Average yearly rainfall distribution for the study area

Fig. 8 Average runoff volume in terms of land use changes from 1968 to 2010

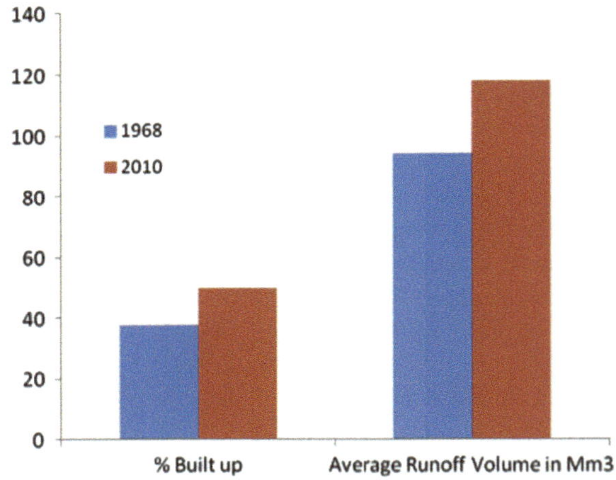

decades were analyzed separately for depth trends by calculating the percentage area under various classes of ground water table. Water table depth trend for post-monsoon is dealt here since a reduction in the recharge will be picked by post-monsoon water level than pre-monsoon water level. A yearly comparison of water depth for post-monsoon showed an increasing depth trend from 1980 to 2010 (Fig. 9). This can be attributed to increased urbanization because the yearly rainfall pattern for the same period shows an increasing trend. Thus, the long-term water depth trend analysis for unconfined aquifer in the study area showed a declining trend irrespective of an increasing trend in the average rainfall. This may be due to the reduced infiltration caused as a result of increase in impervious area. However, the present study has not carried out water budgeting or water balancing to account for variation in ground water consumption as well as minor variation in rainfall pattern for the last three decades. One of the assumptions made in the present study regarding groundwater consumption is that due to increased urbanization, the area

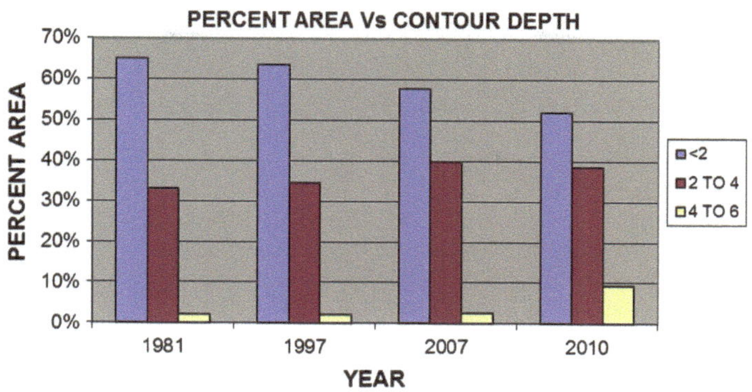

Fig. 9 Ground water depth trend based on area under curve (area under curve calculated for contour with water table depth below 2 m, water table depth between 2 and 4 m and water table depth between 4 and 6 m)

under study heavily dependent on centrally distributed water system and hence the groundwater consumption should be on negative trend.

4 Conclusions

The land use pattern in the study area has changed considerably since 1968. The most significant changes were observed in residential, commercial, and industrial areas causing significant increase in the generation of impervious surface in the study area. The effective impervious area of study area increased from 14 % in 1968 to 34 % in 2010. This is very much reflected in the comparison of percentage of built-up area for the last four decades. This increase in EIA has resulted in significant increase in run off volume. As a result, there is significant reduction in the amount of recharge. This reduction in recharge is well picked in the analysis of depth trend of ground water table. It was found that, the area for shallow water table gets decreased for the past three decades clearly depicting an increasing depth trend. Hence, the long-term water table depth analysis for unconfined aquifer in the study area showed a declining trend irrespective of an increasing trend in the average rainfall. This could be attributed to the increase in the impervious area which in turn reduces the amount of recharge. More analysis is required to establish the rate of reduction in recharge due to land use changes.

Acknowledgements The authors express their gratitude to Central Ground Water Board (CGWB), Thiruvananthapuram for giving ground water data. Authors acknowledge Dr. K.K. Ramachandran for providing digital data and Mr. Sreehari Sarangan, Scientist, CGWB for the support given during the execution of this work. The authors also express their sincere thanks to reviewers who helped to improve the manuscript.

References

Berthier E, Dupont S, Mestayer PG, Andrieu H (2006) Comparison of two evapotranspiration schemes on a sub-urban site. J Hydrol 328:635–646

Brun SE, Band LE (2000) Simulating runoff behavior in an urbanizing watershed. Comput Environ Urban Syst 24:5–22

Geetha K, Mishra SK, Eldho TI, Rastogi AK, Pande RP (2008) SCS-CN-based continuous simulation model for hydrologic forecasting. Water Resour Manag 22:165–190

Patil A, Sarangi OP, Singh AK, Singh TA (2008) Development of a GIS interface for estimation of runoff from watersheds. Water Resour Manage 22:1221–1239

Ragab R, Rosier P, Dixon A, Bromley J, Cooper JD (2003) Experimental study of water fluxes in a residential area: 2. Road infiltration, runoff and evaporation. Hydrol Process 17:2423–2437

Sharp JM Jr (2010) The Impacts of urbanization on ground water system and recharge. Aqua Mundi 1:51–56

Weng Q (2012) Remote sensing of impervious surfaces in the urban areas. Requirements, methods and trends. Remote Sens Environ 117:34–49

Flood Plain Characterization of a River in Lower Assam Using Digital Elevation Model Data

Diganta Barman and Arup K. Sarma

Abstract This paper conveys the strength of geospatial technology for characterization of flood plain in order to establish relation between river flood level and flood plain inundation. We have adopted geospatial approach to analyze and characterize flood plains coupled with profiles of river channels using CARTODEM. A case study on the downstream flood plain of Pagladia River in Lower Assam has been presented. The study shows that apart from conventional flooding reasons such as congestion due to high upstream contribution, embankment breach, etc., the flood plains of the aforesaid river is having a peculiar flooding genesis of backflow from receptor to feeder channel causing congestion in the confluence region. Here the reliability of CARTODEM for characterizing the flat terrain of the flood plain is exhibited.

Keywords CARTODEM · Congestion zone · Backflow · Hypsometry

1 Introduction

The Assam region of Brahmaputra basin mainly comprises of alluvial flood plains characterized by dominant land uses such as agriculture, rural and urban settlements, forest, etc. More than 30 tributaries in both North and South bank populate the 640 km stretch of the Brahmaputra river flowing through the relatively narrow Assam valley. Various parts of Assam valley get flooding almost every year due to high flood waves both in Brahmaputra main channel as well as the tributaries on North as well as South bank causing small, medium, and large scale inundation in the adjoining flood plains. The phenomenon of flooding in the flood plains of

D. Barman (✉) · A.K. Sarma
Department of Civil Engineering, Indian Institute of Technology Guwahati, Guwahati, India
e-mail: diganta_isro1@yahoo.co.in

A.K. Sarma
e-mail: aks@iitg.ernet.in

A.K. Sarma et al. (eds.), *Urban Hydrology, Watershed Management and Socio-Economic Aspects*, Water Science and Technology Library 73,
DOI 10.1007/978-3-319-40195-9_7

Assam is also characterized by frequent changes in river course, rapid bank line erosion, breaching of embankments, etc. As the flood plains of Assam is dominantly a flat alluvial terrain, a minor difference of elevation also plays important role in the extent and spread of flood inundation. Hence, a common but very difficult question often asked by decision-makers is the inundation spread for a particular magnitude of flood in terms of discharge and gauge or level. A precise characterization of the flood plain becomes necessary in order to address the aforesaid issue with a satisfactory level of confidence. With the advent of geospatial technology, it has become possible to characterize flood plains synoptically by studying its drainage, elevation, land use, etc. The horizontal and vertical resolution of satellite-based digital elevation model (DEM) plays important role in near realistic characterization of flood plains from inundation point of view. Hence, the appropriate selection of the digital elevation model data is of utmost importance for flood plain studies.

1.1 Study Area

The **Pagladia** watershed is one of the major watersheds of the north bank of Brahmaputra River. The Pagladia as the name implies, has been a chronic source of trouble due to annual flood and severe bank erosion leading to frequent changes in its flow course. The main river course originates at the foothills of Bhutan and finally terminates into the Brahmaputra near a village named Lowpara. In its entire course, the river consists of important tributaries like Mutunga, Nona, Baralia, and Chaulkhoa. This river system drains an area of 1674 km^2 and the drainage area lies between longitudes 91°20′E and 91°42′2E and between latitudes 26°14′N and 26°59′N. Out of the total catchment area of 1674 km^2, area within Indian territory is 1251 km^2 and the rest 423 km^2 lies in Bhutan. The hilly portion of the catchment area is about 465 km^2 of which 423 km^2 are in Bhutan and the rest 42 km^2 lies in Indian territory (Master Plan report of Brahmaputra Board, Ministry of Water Resources, Govt. of India 1996). The River flows for a length of 19 km in hilly tracts of the Bhutan territory and for the rest 178 km, it flows through the Nalbari district of Assam, India. In the hilly portion, slope of the river bed is very steep, being 1 in 75, in the middle reach it is 1 in 200 and in the lower reach, i.e., from Hajo-Nalbari road to outfall it is 1 in 2600. Figure 1 shows index map of the Pagladia Basin. Based on basin topography, river gradient and joining points of tributaries, the Pagladia river has been divided into three distinct reaches based on the hypsometry as shown in Fig. 2 such as (a) Bhutan hill range to Chowki in Baksa district of Assam, (b) from Chowki to NT road crossing in Nalbari district of Assam and (c) NT road crossing to the confluence of Brahmaputra. The second and third reaches mainly comprise the flood plain of Pagladia river whose characterization has been studied with the help of DEM data.

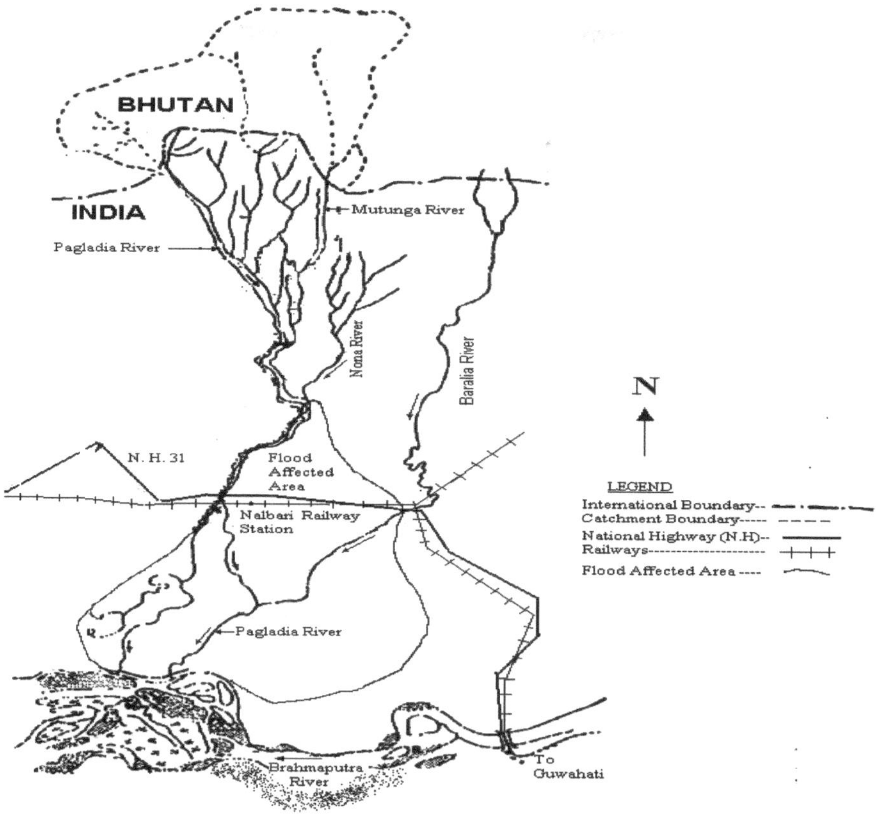

Fig. 1 Index plan of the river

2 DEM for Flood Plain Studies

Digital elevation models (DEM) are raster datasets that can be derived either from a pair of stereo photos or satellite images. DEM can also be created from a set of scattered points with elevation information, contour lines, and triangulated irregular networks (TIN). The main limitation of a regular gridded DEM appears to be the fixed grid cell size. Such DEM cannot accurately describe the flood plain topography especially on landscapes with varying complexity. Error are also introduced during interpolation through whatever method and most of the times these errors are spatially autocorrelated. Again for saving memory, pixel elevations are often rounded to nearest whole figure that creates flat areas with abrupt changes in altitude in regions of gentle slopes. While the slope gradient has the same degree of error as the original elevation data, slope aspect error is usually amplified during calculation (Dismet 1997). For most hydrological applications in flood plains the vertical resolution of a DEM is considered to be satisfactory if the ratio between the

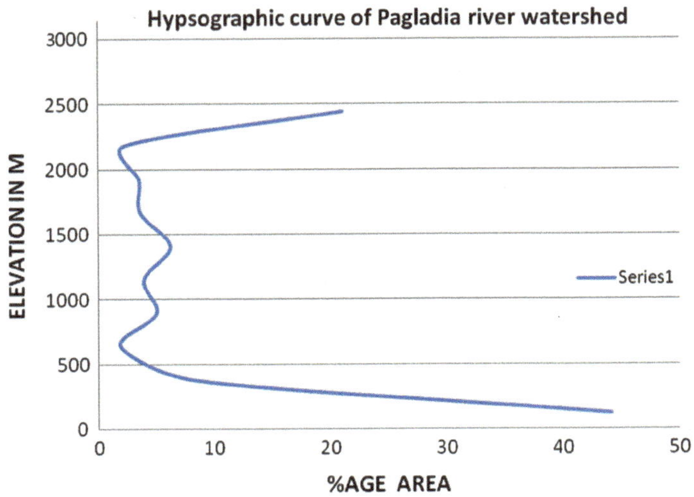

Fig. 2 CARTODEM derived basin hypsometry

average drop per pixel and the vertical resolution is greater than unity (Catchment delineation and characterization—A review, 2000).

$$\frac{d}{h} > 1 \tag{1}$$

where d is the average pixel drop and h is the vertical resolution of the DEM. The average drop per pixel is defined the difference between the elevation of a pixel and its next steepest descent (Walker and Willgoose 1999).

2.1 Problems of Flat Areas

The well known difficulty in a DEM is to discriminate between flat areas drained by incised channels and truly flat areas that carry water as sheet flow. This is a general problem that follows from both horizontal and vertical resolution of the DEM. It is clear that when the size of a drainage feature and the relative average undulation of elevation both are smaller than the pixel or grid size and elevation value, respectively, the channel and the undulations cannot be captured satisfactorily by the DEM. That is why a low to medium resolution DEM such as SRTM (90 m cell size), ASTER (30 m cell size), etc. though performs satisfactorily in hilly part of a watershed, fails measurably when it comes to capturing of drainages and local terrain undulations in flat alluvial flood plains (Fig. 3).

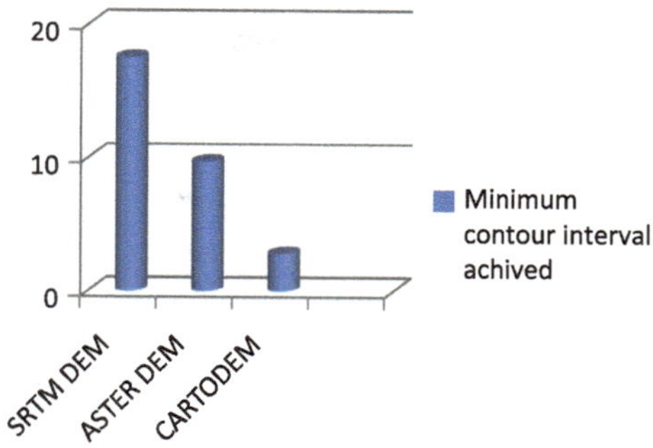

Fig. 3 Average vertical resolution achieved (m)

2.2 *Parameterization of Alluvial Flood Plains*

A typical analysis of alluvial flood plains requires important characteristics such as drainage (pattern and cross-section), surface topography, land use (for ascertaining roughness) apart from other parameters such as soil characteristics, vegetation, etc. where the DEM has no role to play. Comparison of applicability of different DEMs such as SRTM (Gorkovich and Voustonioke 2006), ASTER & CARTODEM for characterizing the above-mentioned vital parameters and their limitations in case of alluvial flood plains has been the focus of this paper where the **Pagladia** river flood plain has been considered as a case study as this particular flood plain has got good degree of similitude with the flood plains of number of other important parallel flowing north bank tributaries of the Brahmaputra river so that the findings can be representative for a number of flood plains of similar nature.

2.3 *Applicability and Limitations of CARTODEM*

CARTOSAT-I DEM is generated by combination of two stereo images of 2.5 m cell size in 26° and 5° inclination with the true vertical. Its horizontal and vertical accuracy determines its capability to effectively capture the land features for correctly characterizing flood plain parameters as mentioned under heading 2.2. Alluvial floods plains of Assam especially on the north bank of river Brahmaputra comprises of vast area within a very narrow elevation range. In the present study area 460 km^2 has been estimated as the total flood plain of Pagladia river system. Total river stretch along the flood plain is about 64 km (source: Departmental field maps). From a survey done with total station in few points just at the bottom of the embankment all along the 64 km stretch of the flood plain a total drop of 26.9 m (from 71.2 to 44.3 m) has been found, i.e., an average drop of 0.42 m/km. If we further upscale up to the cell size (2.5 m) of a CARTODEM pixel, the drop per 2.5 m becomes $\frac{0.42}{1000} * 2.5 = 0.00105$ m. Had this been actual rate of drop of elevation in the ground, the normal conclusion would have been that a precision of less than or equal to 0.00105 m (1.05 mm) is needed for a CARTODEM in order to accurately generate the surface profile of the flood plain from the highest point to the lowest point. Since 1.05 mm vertical precision is an unrealistic precision to expect from any high resolution DEM especially from space-based and arial platform, it is simply an issue not worth discussing. But as 1.05 mm is only a computed average and does not represent the actual undulation of the surface, the usefulness and limitations of CARTODEM for the purpose may well be discussed and worth attempting.

3 Data and Methodology

3.1 Data Sources

Keeping in view the research focus, three data sets as mentioned above, namely, SRTM, ASTER, and CARTOSAT-I generated DEM have been used to characterize the alluvial flood plain surface of the river concerned. Apart from that, spot elevations have been surveyed through total station in critical locations wherever history of occurrences of flood inundation is authentically documented (Reports of Water Resources Department, Assam). Moreover, field maps of the drainage system from various sources have also been referred to for ground validation of land features derived from the three DEM datasets. Survey of India (SOI) reference maps of 1:25,000 scale has also been used for identification of various subchannels of the main river and correlating them with the rivers names mentioned in other references such as field maps, departmental reports, etc.

3.2 Methodology and Workflow

The present study involves assimilation of multisource information and datasets to characterize the flood plain of the river concerned. The entire workflow comprises of the following prime steps:

Step 1: Delineation of the floodplain boundary by integrating satellite data with SOI ref maps, departmental reports and other field maps, etc.

Step 2: DGPS survey in critical point of the flood plain for collection Ground Control Points (GCP) for generation of CARTODEM.

Step 3: Processing of CARTOSAT—I stereo pair in LPS plate form incorporating the GCPs collected through DGPS survey for generation of CARTODEM of the floodplain area.

Step 4: Comparison of two online DEMs of SRTM and ASTER with CARTOSAT-I in terms extraction of various land features relevant to characterization of the flood plain.

Step 5: Morphometric analysis of the entire watershed with special focus on the drainage of the floodplain area.
Step 6: Comparison of ground elevations at critical locations such as embankment base with CARTODEM derived elevation values.

4 Results and Discussion

4.1 Morphometry of the Watershed

The analysis of the hypsometry reveals that around 50 % of the total watershed belongs to the flat alluvial flood plains (Fig. 2). While 20 % is in the upper elevation ranges, rest 30 % belongs to the mid elevation ranges. This steep drop in elevation covering the mid elevation ranges results in bed scour in the interface region between hills and plains producing good amount of silt load further downstream which affects the channel configuration across the flood plain. Following table reveals the other important morphometric parameters of the watershed. The values of both circularity ratio and elongation ratio suggest a fern shaped watershed which also has been found to be the shape when the basin is delineated automatically from CARTODEM by defining only the outlet. This clearly exhibits the suitability of the above DEM to delineate the basin boundary not only of the hilly upstream areas (where other coarse resolution DEM such as SRTM, ASTER, etc., also performs satisfactorily) but also of the flat alluvial flood plain areas. The dendritic drainage pattern as revealed in Fig. 4 suggests stable tectonic behavior of the watershed. This indicates that all frequent geomorphologic changes are purely fluvial in nature (Table 1).

4.2 Analysis of the Digital Surface Model

The flood plain of Pagladia river starts at a place known as Thalkuchi where the main channel of the river is joined by two more rivers namely Mutanga and Nona. The average elevation of the flood plain at embankment base drops from 71.2 to about 44.3 m at the outfall at Brahmaputra. The main channel traverses a distance of around 64 km for this drop. Elevation values of CARTODEM were compared with various spot heights from departmental field maps and the result is presented in Fig. 5. A correlation of 0.9759 has been observed between CARTODEM elevation values and the corresponding spot heights on the field. Moreover, further analysis of the DSM also reveals values of threshold elevations below which are low lying drainage congestion zones. Thus, few major flooding and congestions zones have been identified.

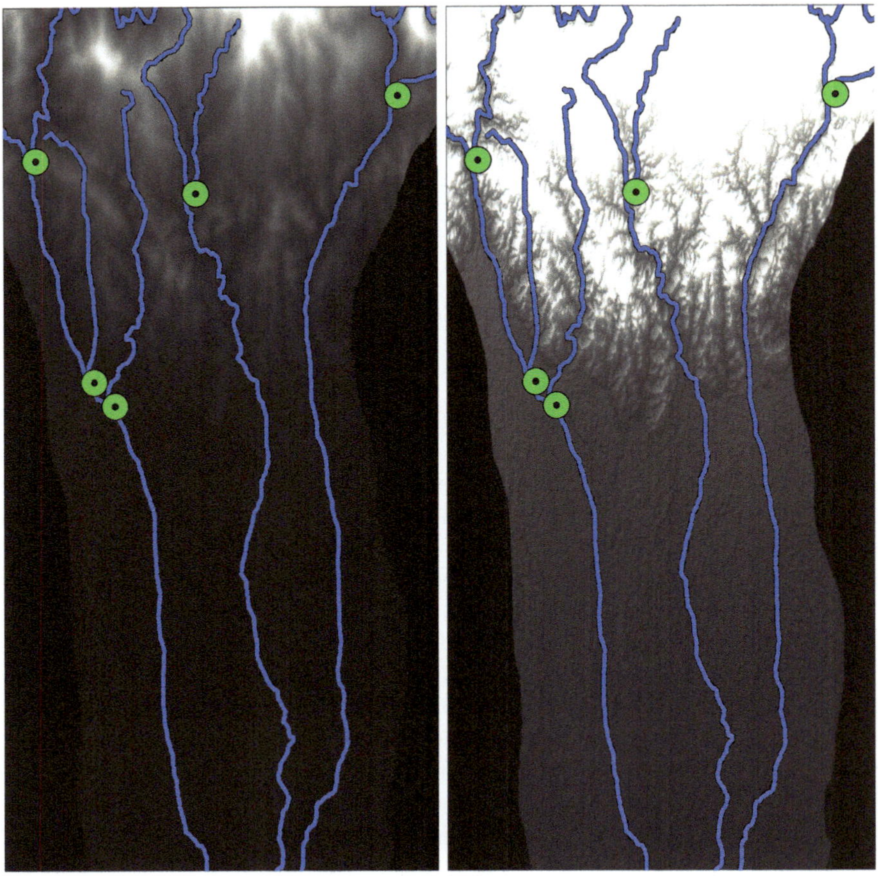

Fig. 4 Surface visualization of the flood plain in ASTER and CARTOSAT DEM

Table 1 Morphometric parameters

Sl No.	Morphometric parameter	Values
1	Form factor	7.26
2	Circularity ratio	0.26
3	Elongation ratio	0.42
4	Drainage pattern	Dendritic

Among them the upstream portion of the flood plain, i.e., the upper middle reach and the lower middle reach of the river the genesis of flood has been found to mainly due to excess run-off and inadequate channel capacity downstream of the confluences. But the lower most part of the flood plain toward the river's outfall with river Brahmaputra, Figs. 5 and 6 reveals that some portion of the downstream areas are at higher elevation that its preceding upstream areas which results in congestion due to backflow from the receptor to the feeder channel. Such a peculiar

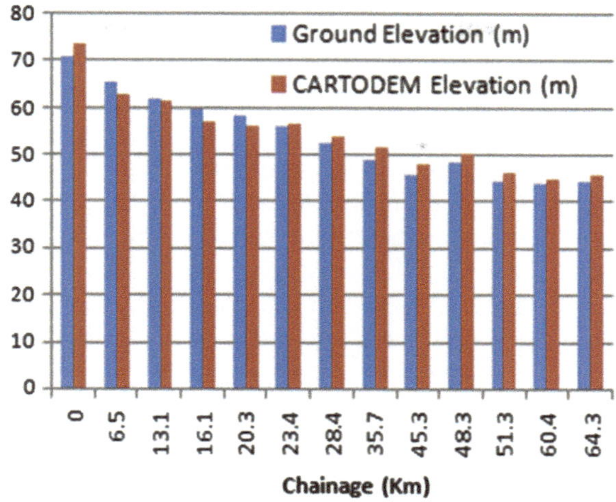

Fig. 5 Comparison of CartoDEM elevation values with field heights

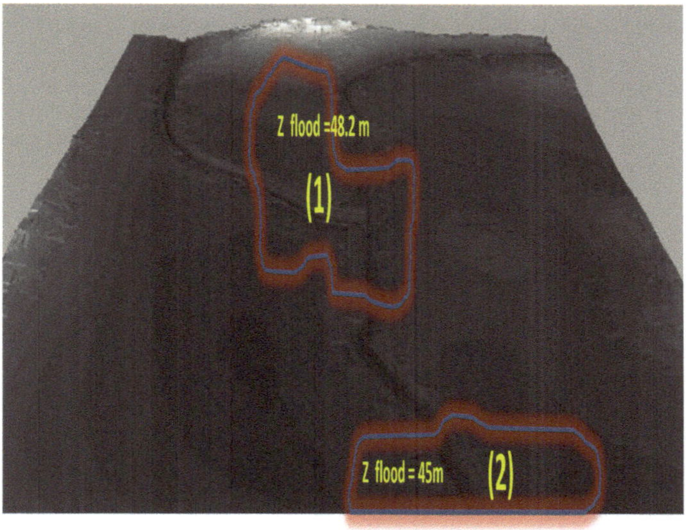

Fig. 6 CARTODEM derived lower flood plain showing congestion zones

low lying area has been found in the lower most portion of the Pagladia river flood plain locally known as Barbhag. This area comes under frequent flooding zone not only due to heavy rainfall and subsequent flood discharge in its own upstream catchment but also due to back flow from river Brahmaputra whenever its water level remains higher than that of river Pagladia. This genesis of flood is not very common in other north bank tributaries of river Brahmaputra especially in upper Assam.

5 Conclusion

This study has revealed the suitability and limitations of DEM generated from CARTOSAT I stereo data for geomorphologic and surface characterization of fluvial floodplains. An average vertical precision of 2.5 m as achieved has been found to be adequate to capture major elevation drops and lifts along the 64 km long downstream flood plain, whereas the computed hypothetical local relief of 1.05 mm as discussed in Sect. 2.3 above has been found to be far beyond its capability. Hence whenever an application demands capturing of very minor local undulations within a very small area, alternatives such as field leveling with total station, Airborne or terrestrial laser scanning (Stratmaa and Baptist 2009), etc. should be the obvious choice for terrain mapping.

Moreover processing of CARTOSAT stereo data is another research area where there is enough scope for exploring new methods and techniques in order to improve its accuracy as required for different hydrological applications.

Apart from the present river under study, there are other similar rivers on the lower Assam part of the Brahmaputra valley, which also have their origin in the kingdom of Bhutan. Similar approach for flood plain characterization of those rivers will also result in interesting revelations about their flooding characteristics.

Acknowledgments The authors humbly acknowledge the support received from various state and central government agencies regarding various information and data sharing during an ongoing study that has led the way for preparation of this manuscript. The authors also sincerely thank various authors of other relevant publications from where important references have been drawn.

References

Brahmaputra Board, Ministry of Water Resources (1996) Master plan report

Gorkovich Y, Voustonioke A (2006) Accuracy assessment of processed SRTM based elevation data by CGIAR using field data from USA and Thailand and its relation to the terrain characteristics. Remote Sens Environ 104:409–415

Smith L (1997) Satellite remote sensing of river inundation area, stage and discharge: a review. Hydrol Process 11:1427–1439

Stratmaa MW, Baptist MJ (2009) Flood plain roughness parametrization using airborne laser scanner and spectral remote sensing. Remote Sens Environ 112(3):1062–1080

Walker JP, Willgooze GR (1999) On the effect of digital elevation model accuracy on hydrology and geomorphology. Water Resour Res 35:7

Part III
Ground and Subsurface Water Hydrology of Urban Areas

Coastal Aquifer Management Models: A Comprehensive Review on Model Development

Rajib K. Bhattacharjya and Triptimoni Borah

Abstract The demand for fresh water is accelerating as the world population is increasing in alarming rate. In order to cope with the increased demands, the overexploitation of groundwater resources has become unavoidable in many parts of the world. It has been reported that at least 70 % of the world population is living in coastal areas. The main sources of fresh water for these people are the freshwater aquifer near the coastal region. The unplanned exploitation of freshwater from coastal aquifers, hydraulically connected with sea or ocean may cause saltwater intrusion into coastal aquifers. The saltwater intrusion in coastal aquifers contaminates the aquifers and makes the coastal aquifers unusable for further human utilization. The contamination of coastal aquifers may also cause serious consequences on environment, ecology, and the economy of the region. The remediation of contaminated aquifers is generally very expensive and time-consuming. In order to protect the vital resource, it is necessary to protect coastal aquifers from further contamination by saltwater intrusion. Saltwater intrusion can be controlled by suitable management policies. The main objective of a coastal aquifer management model is to evolve planned operational strategies to meet required demand of fresh water while maintaining the salinity of water within permissible limit. In order to evolve a physically meaningful strategy, the flow and transport processes need to be simulated within the optimization-based management model. Different methodologies like embedded optimization method, response matrix approach, linked simulation optimization method, etc., have been developed to incorporate the aquifer simulation models with the management model. These methods have their own advantages and disadvantages and none of the methods can be declared as the best method for solving coastal aquifer management models. As such, a suitable method should be selected based on the data availability and other related local advantages. The study presents a review on development of coastal aquifer management models.

R.K. Bhattacharjya · T. Borah (✉)
Department of Civil Engineering, Indian Institute of Technology Guwahati, Guwahati 781 039, Assam, India
e-mail: btriptimoni@gmail.com

R.K. Bhattacharjya
e-mail: rkbc@iitg.ernet.in

A.K. Sarma et al. (eds.), *Urban Hydrology, Watershed Management and Socio-Economic Aspects*, Water Science and Technology Library 73,
DOI 10.1007/978-3-319-40195-9_8

Keywords Aquifer · Saltwater intrusion · Management model · Simulation · Salinity

1 Introduction

The demand for fresh water is accelerating as the world population is increasing in alarming rate. The overexploitation of groundwater resources has become unavoidable in many parts of the world. It is reported that at least 70 % of the world population live in coastal areas (Bear et al. 1999). The main sources of fresh water for those people are the freshwater aquifers near the coastal region. The unplanned exploitation of freshwater from coastal aquifers, hydraulically connected with sea or ocean may cause saltwater intrusion into coastal aquifers. The exploitation of coastal aquifers is restricted many times because of saltwater intrusion. The contamination of coastal aquifers in different parts of the world due to saltwater intrusion has been reported by many researchers. Rouve and Stoessinger (1980) reported degradation of groundwater reservoirs due to saltwater intrusion in the costal aquifer in Madras, India. Sherif et al. (1988) reported contamination of Nile delta aquifer in Egypt. Degradation of the costal aquifer in Yun Lin basin in southwestern Taiwan was reported by Willis and Finney (1988). Cheng and Chen (2001) also reported the degradation of the coastal aquifer in Jahi river basin, Shandong province in China due to saltwater intrusion. The main causes of saltwater intrusion into the coastal aquifers are mainly excessive exploitation of groundwater and improper arrangement of pumping wells. The problem of saltwater intrusion in the Middle East has been reported by Yakirevich et al. (1998) in Gaza Strip, Kacimov et al. (2009) in the coast of Oman, Shammas and Jacks (2007) in Salalah plain aquifer, Oman, etc.

Saltwater intrusion generally occurs, when excessive pumping from coastal aquifers lowers hydraulic potential. This allows seawater to move into the freshwater aquifers near coastal region. The saltwater intrusion in coastal aquifers contaminates the aquifers and makes the coastal aquifers unusable for further human utilization. The contamination of coastal aquifers may cause serious consequences on environment, ecology, and the economy of the region. The remediation of contaminated aquifers is generally very expensive and time-consuming. In order to protect a vital resource, it is necessary to protect coastal aquifers from further contamination by saltwater intrusion. Saltwater intrusion can be controlled by suitable management policies. The philosophy of coastal aquifer management policy is to maintain a balance between exploitation of coastal aquifers for withdrawing freshwater and restricting the movement of saltwater in coastal aquifers. Therefore, the main objective of a coastal aquifer management model is to evolve planned operational strategies to meet required demand of fresh water, while maintaining the salinity of water within permissible limits. In order to evolve a physically meaningful strategy, the flow and transport processes need to be simulated within the optimization-based management model. This would ensure that the prescribed optimal strategies obtained as solutions of the management model would accurately represent the response of the aquifer. In this paper, an attempt has been made to discuss the various issues related to the development of the saltwater intrusion management model. The issues discussed here are: saltwater intrusion process in a coastal aquifer, simulation of saltwater intrusion process in a coastal aquifer, incorporation of simulation model with optimization-based management model, optimization algorithms, and coastal aquifer management model.

1.1 Saltwater Intrusion Process in Coastal Aquifer

Saltwater intrusion is generally caused by unplanned exploitation of freshwater from coastal aquifers. The heavier salt water has the tendency to underlie freshwater because of the hydrodynamic mechanism. At the same time, a mixing zone of varying density also exists between freshwater and saltwater due to hydrodynamic dispersion. This zone is known as the transition zone or, the zone of dispersion or diffusion. In this zone, the density of the mixed fluid increases gradually from freshwater to saltwater. Figure 1 shows the vertical section of an unconfined coastal aquifer with the transition zone. The thickness of the transition zone may vary from few meters to more than hundred meters (Todd 1980). The transition zone may be ignored during the development of simulation model when the thickness of the zone is relatively small. In this case, salt water and fresh water may be considered as immiscible fluid and the interface would be a sharp interface. Figure 2 shows the vertical section of an unconfined coastal aquifer with a sharp interface between saltwater and freshwater. The sharp interface does not provide any information about the nature of the zone of dispersion. For a large thickness of the transition zone, sharp interface model may give an erroneous result. Also, the sharp interface modeling approach may not be adequate when the mixing zone is large, and saline concentrations in between freshwater and saline water cannot be incorporated into the management model. Therefore, for real world saltwater intrusion problem, the

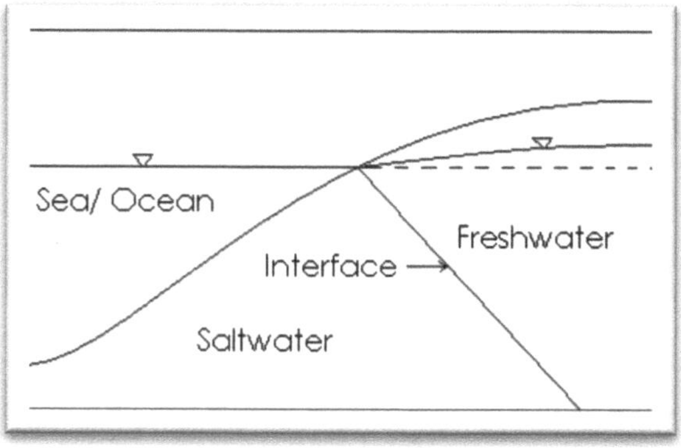

Fig. 1 Vertical cross section of an unconfined coastal aquifer (without transition zone)

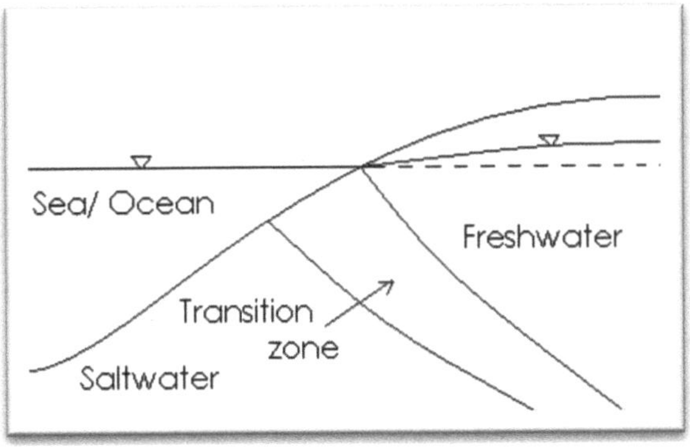

Fig. 2 Vertical cross section of an unconfined coastal aquifer (with transition zone)

transition zone needs to be considered for simulating saltwater intrusion process in coastal aquifers. Incorporation of the transition zone makes the simulation of saltwater intrusion process highly complex. In this case, both flow and transport processes become density dependent.

1.2 Simulation of Saltwater Intrusion Process

Saltwater intrusion is caused by overexploitation of groundwater from coastal aquifers. The higher density saltwater generally stays separate from low-density fresh water. However, as discussed above a mixing zone exists because of hydrodynamic dispersion. The mixing zone is also known as transition zone or zone of diffusion or dispersion. For narrow transition zone, the interface may be assumed as a sharp interface (Fig. 2). In this case, the effect of hydrodynamic dispersion is considered as negligible. Analytical solutions of sharp interface model for single homogeneous aquifers have been reported by Henry (1959), Bear and Dagan (1964), Dagan and Bear (1968), Hantush (1968), Schmorak and Mercado (1969), and Strack (1976). Analytical solutions for layered aquifers were also presented by Mualem and Bear (1974), and Collins and Gelhar (1971). All of these researchers applied Dupuit approximation to obtain analytical solutions. Numerical simulations of sharp interface model were presented by Mercer et al. (1980), Polo and Remis (1983), Taigbenu et al. (1984), Essaid (1990), Huyakorn et al. (1996), Guvanasen et al. (2000), etc.

Moreover, sharp interface approximation does not give any idea about the transition zone. This sharp interface approach is not applicable when permissible limits on saline concentration are imposed on management strategies so that some portion of the transition zone satisfies these restrictions. As such for simulating the real-world scenarios, the transition zone has to be considered during the development of the simulation model. In this case, the flow and transport equations are coupled together by density coupling term. As a result, of this coupling, the simulation of the process becomes complex and time-consuming. Henry (1959) presented the first analytical solution considering transport of salt in density-dependent fluid flow condition for steady state condition in a confined coastal aquifer. The numerical solution of the Henry's problem was presented by Pinder and Cooper (1970) for transient condition using the method of characteristics. Lee and Cheng (1974) presented the first finite element-based simulation model to solve saltwater intrusion problem in coastal aquifers. They adopted elliptical type governing equations. They used stream function and considered salt concentration as dependent variables. Galerkin finite element technique was used by Segol et al. (1975) to solve saltwater intrusion problem dominated by advective transport. Huyakorn and Taylore (1976) presented a model using reference hydraulic head and concentration as dependent variables. The simulation of saltwater intrusion process for an aquifer–aquitard system was presented by Kawatani (1980) and Frind (1982) using Galerkin finite element technique. Volker and Rushton (1982) used finite difference technique, and Rubin (1983) used boundary layer approximation technique to simulate saltwater intrusion process in coastal aquifers. Voss (1984) developed finite element-based SUTRA code for simulation of density depended on saltwater intrusion process in two-dimensional coastal aquifers. Till date, SUTRA has been considered as one of the best simulation models and has been used by many researchers to solve the density-dependent saltwater intrusion process in a coastal aquifer. Some other significant contributions in simulation of saltwater intrusion process are Sherif et al. (1988), Galeati et al. (1992), Cheng et al. (1998), Tejeda et al. (2003), Huyakorn et al. (1987), Putti and Paniconi (1995), Das and Datta (2000), Cheng and Chen (2001), etc.

1.3 Incorporation of Simulation Model with Optimization Model

In order to obtain physically meaningful management strategies, the coastal aquifer simulation model has to be incorporated into the management model. Embedding technique and response matrix approach (Gorelick 1983) are the two methods generally used to incorporate governing equations within the management model. The embedded optimization technique incorporates finite difference or finite element approximation of the governing equations as equality constraints within the management model, along with the other physical and managerial constraints. Incorporation of the highly nonlinear flow and transport equations within the optimization model as equality constraints convert the optimization problem into a non-convex nonlinear problem. The solution of such non-convex nonlinear optimization problem is difficult as there are several local optimal solutions. The applicability of embedding technique for saltwater intrusion problem (Das 1995; Das and Datta 1999a, b) is limited, especially for large-scale aquifer systems. This approach is also numerically inefficient, especially when applied to large aquifer systems with considerable heterogeneity. The response matrix approach is based on the principle of superposition and linearity. In this case, an external simulation model is used to generate the response matrix. This method is reported to be unsatisfactory for highly nonlinear systems (Rosenwald and Green 1974) such as density-dependent saltwater intrusion in coastal aquifers.

As an alternative to the embedding technique and the response matrix approach, the numerical simulation model may also be incorporated into the management model as an external module. In this approach, an external simulation model is linked to the optimization model (Finney et al. 1992; Emch and Yeh 1998). The optimization model calls the simulation model as and when it requires any information from the simulation model. The methodology has been applied effectively for solving large-scale groundwater management models (Bhattacharjya and Dutta 2009). The main disadvantage of this approach is that numerous repetitive iterations between the simulation model and the optimizer are required to arrive at an optimal solution. As the simulation of flow and transport processes are highly nonlinear and time-consuming in the case of coastal aquifers, this linked simulation optimization approach would take a considerable computational time to obtain the optimal solution. The computational time can be substantially reduced by utilizing parallel processing capabilities of advanced computers. This would enable the use of rigorous numerical models for simulations and its linkage to an optimization model. Also, the time requirement for iterative solutions of the optimization model and the simulation model can be drastically reduced. However, this requires appropriate computer hardware and numerical simulation models specially tailored to explicit parallel processing capabilities. Another alternative for reduction of computational time is the use of an approximate simulation model in place of the actual numerical simulation model. The approximate simulator is generally less costly in terms of computational time and computer hardware requirements. Regression analysis and artificial neural networks (ANN) model are generally used for approximating the flow and transport processes in groundwater. The ANN model is considered as a better approximator than regression analysis since it can handle partial information. Also, the performance of the approximate simulation model does not degrade much with noisy data (Bhattacharjya et al. 2007). Applications of regression analysis for approximating an aquifer processes within a management model are reported by Alley (1986), and Lofkoff and Gorelick (1990). Applications of ANN model for approximate simulation of aquifer processes are reported in Rogers and Dowla (1994), Morshed and Kaluarachchi (2000), Aly and Parelta (1999b), Johnson and Rogers (2000), Bhattacharjya et al. (2007), etc. It may also be mentioned here that the Genetic Programming can also be applied to develop an approximate coastal aquifer simulation model. Incorporation of these approximate simulation models would drastically reduce the computational time of the management model. However, care has to be taken in evaluating the performance of the approximate simulation model.

2 Optimization Algorithms

An optimization algorithm is required for solving the aquifer management models. Classical optimization techniques, e.g., linear programming, nonlinear programming, mixed integer programming were used extensively for solving saltwater intrusion management problems (Shamir et al. 1984; Willis and Finney 1988; Finney et al. 1992; Hallaji and Yazicigil 1996, Emch and Yeh 1998; Loaiciga and Leipnik 2000; Das and Datta 1999a, b, etc.). The main disadvantage with most of the classical methods is their dependence on gradient search technique. Most of the time, these gradients are calculated numerically. The numerical estimation of gradients is often the most time-consuming part of an optimization problem. Moreover, numerical calculation of gradients sometimes may lead to severe errors. Another disadvantage of classical optimization techniques is that many times it is inefficient in avoiding local optimal solutions, especially when the optimization problem is a non-convex one and response surface is highly irregular. One possible remedy is the use of multiple solution points as an initial solution. The other limitations of classical methods are point-to-point search, the necessity of initial guesses, deterministic transition rule, the assumption of unimodality, etc. (Deb 2001). It may be noted that classical optimization techniques are not efficient to solve multiple objectives coastal aquifer problems. The classical optimization techniques require several runs to obtain the Pareto-optimal solutions of a multiobjective optimization problem. As such, some researchers have used non-classical optimization techniques, such as Genetic Algorithm for solving coastal aquifer management models. Genetic Algorithms is a search technique based on the concept of natural selection inherent from natural genetics. It is relatively more efficient in obtaining the global optimal solution even when the response surface is highly irregular. GA is also efficient in handling multiple objectives optimization problems, as it can generate the entire nondominating front (Pareto-optimal solutions) in a single run. Another advantage of GA is the relative ease in solving an externally linked simulation optimization model. Due to the various advantages over classical optimization methods, genetic algorithm has been used efficiently for solving various groundwater management models (Ritzel et al. 1994; Mckinney and Lin 1994; Rogers and Dowla 1994; Cieniawski et al. 1995; Aly and Peralta 1999a, b; Morshed and Kaluarachchi 2000; Bhattacharjya and Dutta 2009, etc.).

3 Coastal Aquifer Management Models

The coastal aquifers management models are generally formulated to achieve specific objectives, such as maximum exploitation to cope with the increasing demands, in a best possible way without causing salt intrusion into the aquifers. The management models may have various managerial and physical limitations. These managerial and physical restrictions are incorporated within the management models as constraints. Development of management models has been reported by many researchers. Most of the models reported earlier used sharp interface approximation of the interface between saltwater and freshwater. Shamir et al. (1984) presented a linear programming model to determine the optimal annual operation of a coastal aquifer. Willis and Finney (1988) also adopted sharp interface approximation of transition zone for optimal control of saltwater intrusion in Yun Lin aquifer in Taiwan. They used finite difference approximation of the governing equations to simulate the response of the aquifer to management strategies. The developed models were solved by influence coefficient method allied with quadratic programming technique, and reduce-gradient methods in conjunction with a quasi-Newton algorithm. Finney et al. (1992) presented a quasi three-dimensional optimization model for control of saltwater intrusion in Jakarta multiple aquifer system. They used finite difference approximation to generate hydraulic-response equations of the groundwater basin by relating freshwater and saltwater heads, the location of the interface, and the pumping and recharge schedules. They considered saltwater and freshwater interface as sharp as the zone of diffusion was relatively narrow. They solved their optimization model using MINOS. Hallaji and Yazicigil (1996) presented groundwater management model to determine optimal planning and operational policies of a coastal aquifer in southern Turkey. They used response matrix approach and the finite-element-based flow and transport simulation model SUTRA to generate the response matrix. They considered three management objectives. Those were the maximization of agricultural water withdrawals, minimization of drawdown, and minimization of pumping cost. They presented their results in the form of tradeoff curves relating optimal pumping rates and pumping costs. Emch and Yeh (1998) presented nonlinear multiple objectives management model for management of water used within coastal aquifers. They used linked simulation optimization technique similar to Finney et al. (1992). They assumed sharp interface approximation of the transition zone. They linked the quasi three-dimensional finite-difference-based groundwater flow model SHARP (Essaid 1990) with the optimization model. SHARP simulates both saltwater and freshwater dynamics for layered aquifer for steady and transient conditions. The optimization model was solved using MINOS. The two conflicting objectives considered in their study were cost-effective allocation of surface and groundwater supplies and minimization of saltwater intrusion. They adopted constraint method to convert the multiobjectives optimization model to a single objective optimal management model. Loaiciga and Leipnik (2000) derived closed form solutions of a groundwater management model for coastal aquifers. The objective of the model was the maximization of long-term revenue subjected to climatic, hydrologic, and environmental constraints. The proposed model is applicable to relatively homogeneous aquifer systems and limited size. The climatic and environmental conditions were assumed as uniform. Cheng et al. (2000) presented an optimization model to optimize pumping from saltwater-intruded coastal aquifers for steady state condition. They have considered sharp interface aquifer simulation model by considering saltwater and freshwater as immiscible fluid. They employed Dupuit assumptions to convert three-dimensional problem to a two-dimensional problem. They presented an analytical solution for one well, two wells, and one well with recharge canal problems. For the problem with multiple wells, a structured messy Genetic Algorithm was used to search for an optimal solution.

The sharp interface approximation of saltwater and freshwater interface is not acceptable especially when the thickness of the transition zone is large. Moreover, sharp interface approximation does not give any idea about the transition zone. The sharp interface approach is not applicable when permissible limits on saline concentration are imposed on management strategies so that some portion of the transition zone satisfies these restrictions. Das and Datta 1999a, b) presented a number of nonlinear optimization-based multiple objectives management models for sustainable utilization of coastal aquifers. The objectives considered by them were the maximization of total pumping from the coastal aquifer, minimization of maximum salt concentration in the monitoring wells and minimization of barrier pumping. It may be noted that barriers pumping is a very effective measure to control seawater intrusion. In this method, a hydraulic barrier is constructed artificially to restrict the intrusion of saltwater into the inland side of the aquifer. Das and Datta 1999a, b) used the embedded optimization technique. The nonlinear finite difference approximation of density-dependent miscible flow and transport equations were added as embedded constraints to the management models. They have reported the difficulties of the embedded approach especially for large-scale real-world coastal aquifer problem and also reported the necessity of large computational time. In order to reduce the computational time of the simulation optimization model, Bhattacharjya and Datta (2009) presented linked simulation optimization model using ANN and GA for deriving multiple objectives management strategies for coastal aquifers. They have approximated the three-dimensional simulation model using ANN and the ANN model is then linked with the GA-based coastal aquifer management model. In this process, they have shown that the computational time of the simulation optimization model can be reduced drastically. They have solved different single and multiple objectives optimization problems for demonstrating the efficiency of the developed methodology. Sreekant and Datta (2011) applied Genetic programming to approximate the saltwater intrusion process.

4 Conclusions

A review on development of saltwater intrusion management models in a coastal aquifer is presented. The development of saltwater intrusion simulation models, coastal aquifer management models, and multi-objective coastal aquifer management models are discussed. The use of approximate simulation model in developing coastal aquifer management model is also reviewed. It is hoped that the discussed made here will be useful for the new research to develop coastal aquifer management model.

References

Alley WM (1986) Regression approximations for transport model constraint sets in combined aquifer simulation-optimization studies. Water Resour Res 22(4):581–586

Aly AH, Peralta RC (1999a) Comparison of a genetic algorithm and mathematical programming to the groundwater cleanup systems. Water Resour Res 35(8):2415–2425

Aly AH, Peralta RC (1999b) Optimal design of aquifer cleanup systems under uncertainty using a neural network and genetic algorithms. Water Resour Res 35(8):2523–2532

Bear J, Dagan G (1964) Some exact solutions of interface problems by means of the hodograph method. J Geophys Res 69:1563–1572

Bear J, Cheng AH-D, Sorek S, Ouazar D, Herrera I (1999) Seawater Intrusion in coastal aquifers—concepts, methods and practices. Kluwer Academic Publishers, Dordrecht

Bhattacharjya RK, Datta B (2009) ANN-GA based multiple objective management of coastal aquifers. J Water Resour Plann Manage 135(5):314–322

Bhattacharjya RK, Datta B, Satish MG (2007) Artificial neural networks approximation of density dependent saltwater intrusion process in coastal aquifers. J Hydrol Eng 12(3):273–282

Cheng JM, Chen CX (2001) Three-dimensional modeling of density-dependent salt water intrusion in multilayered coastal aquifers in Jahe River Basin, Shandong Province, China. Ground Water 39(1):137–143

Cheng JR, Strobl RO, Yeh G-T, Lin H-C, Choi WH (1998) Modeling of 2D density-dependent flow and transport in the subsurface. J Hydrol Eng ASCE 3(4):248–257

Cheng AH-D, Halhal D, Naji A, Ouazar D (2000) 'Pumping optimization in saltwater-intruded coastal aquifers'. Water Resour Res 36(8):2155–2165

Cieniawski SE, Eheart W, Ranjithan S (1995) Using genetic algorithms to solve multiobjective groundwater monitoring problem. Water Resour Res 31(2):399–409

Collins MA, Gelhar LW (1971) Seawater intrusion in layered aquifers. Water Resour Res 7 (4):971–979

Dagan G, Bear J (1968) Solving the problem of interface upconing in a coastal aquifer by the method of small perturbations. J Hydraul Res 6(1):15–44

Das A (1995) Development of nonlinear optimization based single and multiple objective management models for transient 3-D density dependent seawater intrusion in coastal aquifers. A Ph.D. Thesis to department of Civil Engineering, IIT, Kanpur

Das A, Datta D (1999a) Development of multiobjective management models for coastal aquifers. J Water Resour Plann Manage ASCE 125(2):76–87

Das A, Datta B (1999b) Development of management models for sustainable use of coastal aquifers. J Irrg Drainage Eng ASCE 125(3):112–121

Das A, Datta B (2000) Optimization based solution of density dependent seawater intrusion in coastal aquifers. J Hydrol Eng ASCE 5(1):82–89

Deb K (2001) Multi-objective optimization using evolutionary algorithms. Wiley, UK

Emch PG, Yeh WW-G (1998) Management model for conjunctive use of coastal surface water and ground water. J Water Resour Plann Manage ASCE 124(3):129–139

Essaid HI (1990) A multilayered sharp interface model of coupled freshwater and saltwater in coastal systems: model development and application. Water Resour Res 27(7):1431–1454

Finney BA, Samsuhadi S, Willis R (1992) Quasi-three dimensional model of Jakarta basin. J Water Resour Plann Manage ASCE 118(1):18–31

Frind EO (1982) Seawater intrusion in continuous coastal aquifer-aquitard systems. Adv Water Resour 5:89–97

Galeati G, Gambolati G, Neuman SP (1992) Coupled and partially coupled Eulerian-Lagrangian model for freshwater-seawater mixing. Water Resour Res 28(1):149–165

Gorelick SM (1983) A review of distributed parameter groundwater management modeling methods. Water Resour Res 19(2):305–319

Guvanasen V, Wade SC, Barcelo MD (2000) Simulation of regional groundwater flow and saltwater intrusion in Hernando County, Florida. Ground Water 38(5):722–783

Hallaji K, Yazicigil H (1996) Optimal management of a coastal aquifer in southern Turkey. J Water Resour Plann Manage ASCE 122(4):233–244
Hantush MS (1968) Unsteady movement of freshwater in thick unconfined saline aquifers. Bull Int Assoc Sci Hydrol XIII 2:40–60
Henry HR (1959) Saltwater intrusion into freshwater aquifers. J Geophys Res 64:1911–1919
Huyakorn PS, Taylor C (1976) Finite element models for coupled groundwater and convective dispersion. In: Proceeding of the 1st international conference on finite element in water resources. Pentech Press, London, pp 1.131–1.151
Huyakorn PS, Andersen PF, Mercer JW, White HO (1987) Saltwater intrusion in aquifers: development and testing of a three-dimensional finite element model. Water Resour Res 23 (2):293–312
Huyakorn PS, Wu YS, Park NS (1996) Multiphase approach to the numerical solution of a sharp interface saltwater intrusion problem. Water Resour Res 32(1):93–102
Johnson VM, Rogers LL (2000) Accuracy of neural network approximations in simulation-optimization. J Water Resour Plann Manage ASCE 126(2):48–56
Kacimov A, Sherif M, Perret J, Al-Mushikhi A (2009) Control of sea-water intrusion by salt-water pumping: coast of Oman. Hydrogeol J 17(3):541–558
Kawatani T (1980) Behavior of seawater intrusion in layered coastal aquifers. Paper presented at 3rd international conference on finite element in water resources. University of Mississippi, pp 19–23
Lee CH, Cheng RT (1974) On seawater encroachment in coastal aquifers. Water Resour Res 10 (5):1039–1043
Lefkoff LJ, Gorelick SM (1990) Simulating physical processes and economic behavior in saline, irrigated agriculture: model development. Water Resour Res 26(7):1359–1369
Loaiciga HA, Leipnik RB (2000) Closed form solution for coastal aquifer management. J Water Resour Plann Manage ASCE 126(1):30–35
McKinney DC, Lin M-D (1994) Genetic algorithms solution of groundwater management models. Water Resour Res 30(4):1897–1906
Mercer JW, Larson SP, Faust CR (1980) Simulation of salt-water interface motion. Ground Water 18(4):374–385
Morshed J, Kaluarachchi J (2000) Enhancements to genetic algorithm for optimal ground-water management. J Hydrol Eng ASCE 5(1):67–73
Mualem Y, Bear J (1974) The shape of the interface in steady flow in a stratified aquifer. Water Resour Res 10(6):1207–1215
Pinder GF, Cooper HH (1970) A numerical technique for calculating the transient position of the saltwater front. Water Resour Res 6(3):875–880
Polo JF, Ramis FJR (1983) Simulation of saltwater freshwater interface motion. Water Resour Res 19(1):61–68
Putti M, Paniconi C (1995) Picard and Newton linearization scheme for the coupled model of saltwater intrusion in aquifers. Adv Water Resour 18(3):159–170
Ritzel BJ, Eheart JW, Ranjithan S (1994) Using genetic algorithms to solve a multiple objective groundwater pollution containment problem. Water Resour Res 30(5):1589–1603
Rogers LL, Dowla FU (1994) Optimization of groundwater remediation using artificial neural networks with parallel solute transport modeling. Water Resour Res 30(2):457–481
Rosenwald GW, Green DW (1974) A method for determining the optimal location of wells in a reservoir using mixed-integer programming. Soc Res Eng J 14:44–54
Rouve G, Stoessinger W (1980) Simulation of the transient position of the seawater intrusion in a coastal aquifer near Madras coast. In: Proceedings of the 3rd international conference finite element water resources. University of Mississippi, Oxford
Rubin H (1983) On the application of the boundary layer approximation for the simulation of density stratified flows in aquifers. Adv Water Resour 6:96–105
Schmorak S, Mercado A (1969) Upconing of freshwater-seawater interface below pumping wells, Field study. Water Resour Res 5(6):1290–1311

Segol G, Pinder GF, Gray WG (1975) A Galerkin finite element technique for calculating the transport position of the saltwater front. Water Resour Res 11(2):343–347

Shamir U, Bear J, Gamliel A (1984) Optimal annual operation of a coastal aquifer. Water Resour Res 20(4):435–444

Shammas M, Jacks G (2007) Seawater intrusion in the Salalah plain aquifer, Oman. Environ Geol 53(3):575–587

Sherif MM, Singh VP, Amer AM (1988) A two-dimensional finite element model for dispersion (2D-FED) in coastal aquifer. J Hydrol 103:11–36

Sreekanth J, Datta B (2011) Coupled simulation-optimization model for coastal aquifer management using genetic programming-based ensemble surrogate models and multiple-realization optimization. Water Resour Res 47(4)

Strack ODL (1976) A single-potential solution for regional interface problems in coastal aquifers. Water Resour Res 12(6):1165–1174

Taigbenu AE, Liggett JA, Cheng AH-D (1984) Boundary integral solution to seawater intrusion into coastal aquifers. Water Resour Res 20(8):1150–1158

Tejeda I, Cienfuegos R, Munoz JF, Duran M (2003) Numerical modeling of saline intrusion in Salar de Atacama. J Hydrol Eng ASCE 8(1):25–34

Todd DK (1980) Groundwater hydrology. Wiley, New York

Volker R, Rushton K (1982) An assessment of the importance of some parameters for seawater intrusion in aquifers and comparison of dispersive and sharp interface modeling approaches. J Hydrol 56:239–250

Voss CI (1984) SUTRA: a finite element simulation model for saturated-unsaturated, fluid density dependent groundwater flow with energy transport of chemically reactive sample species solute transport. US Geol Surv Water Resour Invest 84(436):409

Willis R, Finney BA (1988) Planning model for optimal control of saltwater intrusion. J Water Resour Plann Manage ASCE 114(2):163–178

Yakirevich A, Melloul A, Sorek S, Shaath S, Borisov V (1998) Simulation of seawater intrusion into the Khan Yunis area of the Gaza Strip coastal aquifer. Hydrogeol J 6(4):549–559

Linked Optimization Model for Groundwater Monitoring Network Design

Deepesh Singh and Bithin Datta

Abstract Groundwater is a major source of water supply for irrigation, industrial use, and other public consumptions due to its quality, local accessibility, and relative cost. Groundwater resource is also equally vulnerable to contamination and depletion as the surface water. The detection of these contaminants is difficult, as they are not visible as the surface water systems. The detection and monitoring of the contaminants is very much important for the prediction of the contaminant transport process, and for designing efficient remedial measures. This study involves the development of methodologies for optimal design of groundwater contamination monitoring networks for detection the contaminant movement in groundwater systems, based on the application of simulated annealing as an optimization tool, and geostatistical kriging. This methodology is developed for single and time-varying optimal network design for different management periods. A budgetary constraint is used to limit the number of monitoring wells to be installed in a particular management period. The kriging linked simulated annealing (SA)-based optimization model essentially utilizes a numerical flow and transport simulation model (MODFLOW and MT3DMS) to simulate the physical and geochemical processes. It searches for an optimal set of permissible number of monitoring wells. The specified objective function of minimizing the contaminant mass estimation error is found to be quite suitable. The methodologies developed are applied to an illustrative study area comprising of homogeneous unconfined aquifer. The performance of the methodology is evaluated for the illustrative study area and the limited evaluation results demonstrate the potential applicability of the methodologies developed.

D. Singh (✉)
Department of Civil Engineering, Harcourt Butler Technological Institute,
Kanpur, U.P., India
e-mail: dr.deepeshsingh@gmail.com

B. Datta
Discipline of Civil Engineering, College of Science Technology and Engineering,
James Cook University, Townsville 4811, Australia
e-mail: bithin.datta@jcu.edu.au

A.K. Sarma et al. (eds.), *Urban Hydrology, Watershed Management and Socio-Economic Aspects*, Water Science and Technology Library 73,
DOI 10.1007/978-3-319-40195-9_9

Keywords Groundwater monitoring network design · Simulated annealing · Kriging · Groundwater contamination · Contaminant transport simulation

1 Introduction

Groundwater, under most conditions, is safer and more reliable for use than surface water. Groundwater my also be contaminated by some of the common activities which cause groundwater contamination like: illegal and unmonitored injection of pollutants in the aquifer, leakage from underground tanks, and pipelines carrying sewage and other toxic contaminants; insufficient knowledge about the application of pesticides and fertilizers in agricultural fields, land disposal of wastes, etc. The detection of groundwater contamination is difficult as it is not openly visible unlike surface water systems. Prediction of the transport of contaminants in groundwater systems also becomes difficult due to various associated uncertainties. The detection and remediation of groundwater contamination involves monitoring of the contaminant plume in groundwater systems. This study is aimed at developing methodologies for optimal monitoring of groundwater contamination. A groundwater model can have two distinct components: (i) groundwater flow and (ii) groundwater contaminant transport components. The uncertainties involved in the prediction of plume movement and the economic constraint to limit the number of monitoring well installations necessitates the design of an optimal monitoring network design. The economic efficiency is incorporated by limiting the maximum permissible number of monitoring locations as an upper limit.

Especially during last two decades, the problem of designing an optimal monitoring network under condition of uncertainty has been addressed by number of researchers. Meyer and Brill (1988) developed a method that involves the use of two independent but linked models, a groundwater contaminant transport simulation model and an optimization model. Loaiciga (1989), Loaiciga et al. (1994, 1995) formulated groundwater monitoring (GWM) optimal sampling plan as mixed integer programming (MIP) problem. Meyer and Brill (1988) also utilized MIP to determine optimal location of a network of groundwater monitoring wells under conditions of uncertainty. Datta and Dhiman (1996) developed a groundwater quality monitoring network using a groundwater contamination transport simulation and optimization linked model. Lin et al. (2001) estimated the spatial maps of transmissivity and simulated using ordinary kriging (OK) and simulated annealing. Reed and Minsker (2004) demonstrated the use of high-order Pareto optimization on a long-term monitoring optimization (LTM) application. Chadalavada and Datta (2008) developed optimal and efficient sampling locations for plume detection. Dhar and Datta (2007) developed a methodology based on the solution of optimization models for optimal design of groundwater quality monitoring networks. Reed et al. (2000) developed a methodology that combines a fate and transport model, plume interpolation, and a genetic algorithm to identify cost-effective sampling plans that accurately quantify the total mass of dissolved contaminant for plume interpolation. Nunes et al. (2004) proposed three optimization models to select the best subset of stations from a large GMN. Zheng et al. (2005) did a similar work in concept to the work of Reed et al. (2000) using OK and inverse distance weighting (IDW).

Monitoring networks are integral to effective, efficient, and economical groundwater management. The objectives of designing optimal monitoring network may be different and vary as per site-specific conditions (Prakash and Datta 2015). To account for the transient and dynamic nature of groundwater flow and pollution transport, monitoring networks need to be designed and implemented sequentially with time. Sequential monitoring network design for transient transport process was considered by Grabow et al. (2000). Other previously reported works include sampling strategy in space and time using Kalman filter (Kollat et al. 2011) and integer programming for sequential monitoring network design (Dhar and Datta 2007; Mahar and Datta 1997). Kollat et al. (2008) developed a new multiobjective evolutionary algorithm (MOEA) to solve large, long-term groundwater monitoring (LTM) design problems. Time-varying dynamic monitoring network design methodology and its evaluation is reported in Chadalsavada and Datta (2008). Most of the optimization techniques for monitoring network design were aimed at detection of plume movement, with an implicit objective of minimizing the monitoring cost.

There exists a large body of literature dealing with design of optimal monitoring networks using heuristic optimization tools like genetic algorithm (GA) (Cieniawski et al. 1995; Wu et al. 2005; Yeh et al. 2006; Chadalavada et al. 2011) and Simulated Annealing (SA) (Prakash and Datta 2012, 2015) and genetic programming (GP) (Prakash and Datta 2013; Datta et al. 2013, 2014). MIP as the optimization algorithm together with chance constraints to define reliability of the monitoring network design was proposed by Datta and Dhiman (1996). Data interpolation techniques like geostatistical kriging have been used in groundwater monitoring network design (Yeh et al. 2006; Feng-guang et al. 2008; Chadalavada et al. 2011; Prakash and Datta 2012). Chadalavada and Datta (2007) proposed models for determining optimal and efficient sampling locations for contamination plume detection. Dhar and Datta (2010) developed an optimization-based solution for reducing the redundancy in a groundwater quality monitoring network. Bashi-Azghadi and Kerachian (2010) developed a new methodology for optimally locating monitoring wells using Monte Carlo technique. Other proposed methodologies include optimizing the groundwater monitoring network using probabilistic support vector machines (PSVMs) (Bashi-Azghadi et al. 2010), many-objective long-term groundwater monitoring (LTGM) network design and its trade-offs (Reed and Kollat 2012), and groundwater quality monitoring network using vulnerability mapping and geostatistics (Husam 2010). The main aim of all the above-mentioned methods was to reduce the cost due to redundancy in the monitoring, even then improved the detection of pollutants under conditions of uncertainty in a dynamic scenario. However, these methods did not address the need for an optimal sampling network that can improve the accuracy of pollutant source identification. A number of methodologies have been proposed using different optimization algorithms for improving the source identification results as reported in Chadalavada et al. (2011).

Initial attempt to design an optimal sampling network that can improve the accuracy of pollutant source identification was addressed by Mahar and Datta (1997), Datta et al. (2009a, b), Prakash and Datta (2013, 2014). Datta et al. (2013) used GP-based monitoring factors for design of optimal monitoring network to improve the accuracy of pollutant source identification. Potential applicability of GP in groundwater problems was first discussed by Sreekanth and Datta (2012). These methodologies used trained GP models to calculate the impact factor of the sources on the candidate monitoring locations. However, these monitoring networks were not dynamic in design or implementation. The issue of sequentially designing monitoring networks to gather feedback information regarding the compliance with implemented management strategies for saltwater intrusion in coastal aquifers is discussed in Sreekanth and Datta (2014, 2015). Datta and Singh (2014) presented a methodology using kriging linked optimization model for contaminant monitoring network design by incorporating uncertainty.

The optimal monitoring network design methodology developed in this study utilizes a kriging-linked SA-based optimization model. It essentially utilizes a numerical flow and transport simulation model (MODFLOW and MT3DMS) to simulate the physical and geochemical processes. The next section describes the methodology adopted in developing the linked model.

2 Methodology

Kriging is a geostatistical estimation technique especially used in geologic and geophysical systems. The kriging linked SA model has three main components: (a) a groundwater flow and transport simulation, (b) a global mass estimation using Geostatistics, and (c) optimization using SA.

2.1 Groundwater Flow and Transport Simulation

The equation describing the transient, two-dimensional areal flow of groundwater through a nonhomogeneous, anisotropic, and saturated aquifer can be written in Cartesian tensor notation (Pinder and Bredehoeft 1968) as

$$\frac{\partial}{\partial x_i}\left(T_{ij}\frac{\partial h}{\partial x_j}\right) = S\frac{\partial h}{\partial t} + W; \quad i,j = 1,2 \tag{1}$$

where T_{ij} = transmissivity tensor (L^2T^{-1}) = $K_{ij}b$; K_{ij} = hydraulic conductivity tensor (LT^{-1}); b = saturated thickness of aquifer (L); h = hydraulic head (L); W = volume flux per unit area (LT^{-1}); and x_i, x_j = Cartesian coordinates (L). In this study, the flow model is simulated using MODFLOW (1996).

The partial differential equation describing the fate and transport of contaminants of species k in 3-D, transient groundwater flow systems can be written as follows (MT3DMS 1999):

$$\frac{\partial(\theta C^k)}{\partial t} = \frac{\partial}{\partial x_j}\left(\theta D_{ij}\frac{\partial C^k}{\partial x_j}\right) - \frac{\partial}{\partial x_i}(\theta v_i C^k) + q_s C_s^k + \Sigma R_n \quad (2)$$

where θ = porosity of the subsurface medium, dimensionless; C^k = dissolved concentration of species k, ML^{-3}; t = time, T; $x_{i,j}$ = distance along the respective Cartesian coordinate axis, L; D_{ij} = hydrodynamic dispersion coefficient tensor, L^2T^{-1}; v_i = seepage or linear pore water velocity, LT^{-1}; it is related to the specific discharge or Darcy flux through the relationship, $v_i = q_i/\theta$; q_s = volumetric flow rate per unit volume of aquifer representing fluid sources (positive) and sinks (negative), T^{-1}; C_s^k = concentration of the source or sink flux for species k, ML^{-3}; ΣR_n = chemical reaction term, $ML^{-3}T^{-1}$. MT3DMS (1999) has been used as contaminant transport simulation.

The contaminant plume simulated by flow and transport model represents the actual field measurements in the future time periods.

2.2 *Geostatistics*

ASCE task committee, (1990a, b) has defined the Geostatistics as a collection of techniques for making inferences about properties that vary in space. Geostatistics offers a collection of deterministic and statistical tools that provide understanding and modeling spatial variability. Ordinary kriging (OK) is the most commonly used variant of the simple Kriging (SK) algorithm. Kriging (SK or OK) has been performed to provide a "best" linear unbiased estimate (BLUE) for unsampled values (Deutsch and Journel 1998). Different semivariogram models are used in the kriging that are selected by performing a sensitivity analysis using different sets of parameters. For this study, spherical variogram model was selected and the package provided by GSLIB (1998) was modified to estimate the plume concentration at all the unsampled locations within the study area.

2.3 *Optimization Algorithm: Simulated Annealing*

The optimization algorithm used for the design of the monitoring network is based on SA. Annealing is the cooling process of molten metals. At the high temperature atoms with high energy move freely, and when the temperature is reduced get ordered and finally form crystals having minimum possible energy. This crystalline stage will not be achieved if the temperature is reduced at fast rate (Deb 2002).

Corana et al. (1987) presented a modified algorithm of Simulated Annealing. The SA parameters- temperature reduction factor, initial temperature, number of function evaluations for termination criteria are based on the sensitivity analysis as well as guidelines available in the literature (Deb 2002; Nunes et al. 2004).

3 Optimization Model Formulation

The objective is to determine the optimal set of the monitoring locations for which the normalized mass estimation error is minimum, while constraining the total number of monitoring wells (Singh 2008). The objective function is defined as

$$\text{Minimize}: \left|\left(\frac{M_{\text{cal}} - M_{\text{est}}^{k}}{M_{\text{cal}}}\right)\right| \tag{3}$$

Subject to

$$M_{\text{est}}^{k} = F\left\{\underset{k}{KR}\left(C_{ij}^{s}\right)\right\} \quad \forall i, j \in k \tag{4}$$

$$C_{ij}^{s} = f(I, BC, S) \quad \forall i, j \in k \tag{5}$$

$$\sum_{m=1}^{M} W_m \leq W_M \quad \forall m \in M \tag{6}$$

where M_{cal} = total contaminant mass present in the study area; M_{est}^{k} = total estimated contaminant mass based on *k*th subset of candidate (optimal) monitoring locations; C_{ij}^{s} = simulated contaminant concentration at spatial location *i, j* belonging to the *k*th subset of the potential monitoring locations *N* at the end of the management period; $\underset{k}{KR}\left(C_{ij}^{s}\right)$ = spatially kriged concentrations at all nodes of the study area based on simulated concentrations at the *k*th subset potential monitoring locations; $F\left\{\underset{k}{KR}\left(C_{ij}^{s}\right)\right\}$ = function of the spatially kriged concentrations C_{ij}^{s}; $f(I, BC, S)$ = predicted concentration at end of a management period obtained as solution of the simulation model based on defined initial conditions (*I*, observed at the implemented monitoring network at the end of last management period) at the beginning of the management period, boundary condition (*BC*) as specified, and source characteristics (*S*) if any known and specified. W_m = a binary decision variable, *1* indicates a monitoring well is selected at potential monitoring location, and *0* indicating otherwise; W_M = maximum permissible number of installed monitoring wells in a given management period; M = total number of potential monitoring well locations; and m = maximum permissible number of potential locations in a design set.

In the first step of the proposed methodology, before applying the monitoring network design model to the field, it is essential to have a calibrated flow and transport model for the study area. The simulation model is used to simulate the contaminant scenario of the site from initial time t_0 to some expected time t_n. It is assumed to represent the future conditions to be monitored using the given initial conditions, boundary conditions, flow and transport parameters, contaminant source characteristics, and potential monitoring locations. The simulated concentration value at the potential monitoring locations are assumed to be known, and these known values of the contaminant concentrations are used to construct the contaminant plume using spatial extrapolation.

In the second step, the concentration values at the potential monitoring wells and their coordinates are used to randomly generate the specified number of the groundwater monitoring well-designed plans given as a maximum permissible number of wells in that particular analysis. In the third step, the optimization process starts with the given sets of the initialized parameters. At a given iteration, the SA algorithm chooses a subset of monitoring locations from the set of potential locations. This subset is then used as an input to the kriging model which estimates the mass of the contaminant based on the set provided by the SA. All these different sets are evaluated for the objective function and the algorithm terminates when the final function value at the current temperature differs from the current optimal function value by less than error tolerance of termination, and the optimal design set is evolved satisfying the imposed constraints.

4 Model Application

The size of the study area is 2800 m × 2700 m. The study area for illustrative application of the proposed monitoring network models are discretized into identical grids of 100 m × 100 m, as shown in Fig. 1.

The study area is assumed to have three continuous sources each with an injection mass flux of value 27.4 g/s. The injection rate of the contaminant sources is assumed to be 45 m^3/day which is approximately half liters per second. One hundred and eight observation wells are distributed in the area, for possibly collecting the concentration data in all management periods. The sources are running for all time periods. All other advection and dispersion parameters are described in Table 1.

The sensitivity analysis has been performed on SA model and Kriging model to find out the best sets of parameters utilized for the linked optimal model (Singh 2008).

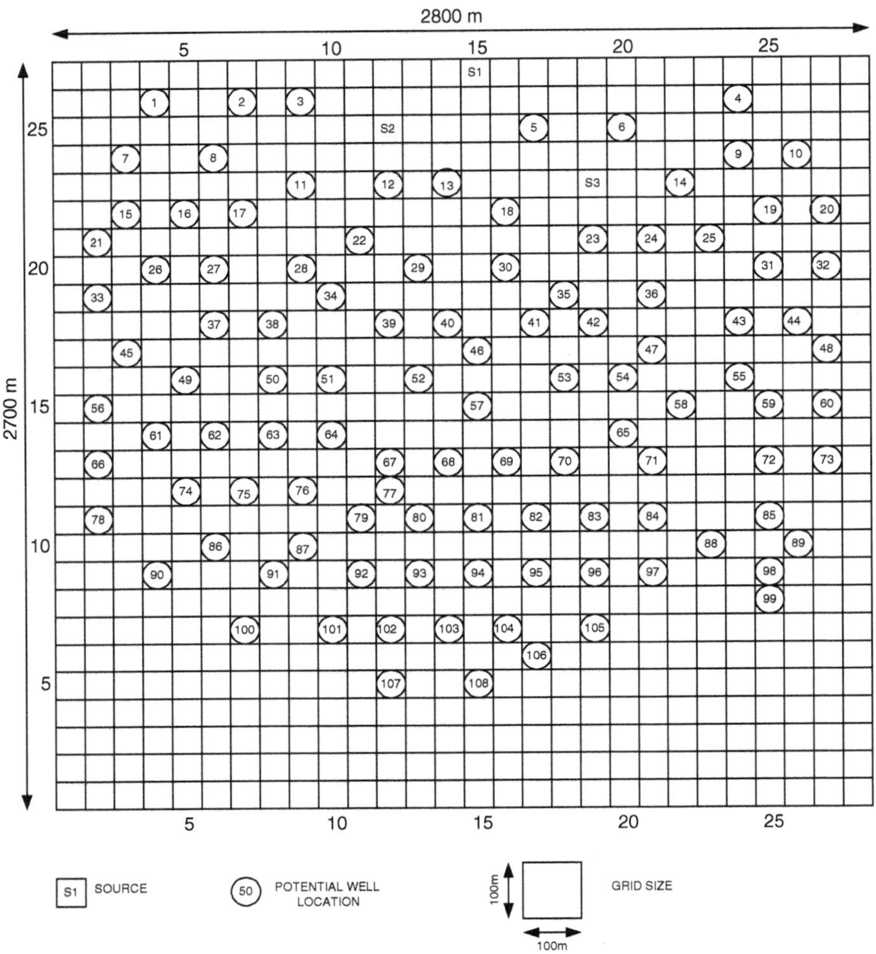

Fig. 1 Illustration of study area showing its physical parameters and potential monitoring well locations

4.1 Performance Evaluation

The performance of the proposed methodology is evaluated for an illustrated study for different contamination and management Scenarios.

4.1.1 Scenario 1

The study area as shown in Fig. 1 is considered as the aquifer which is unconfined, homogeneous, and isotropic in nature. All the flow and transport parameters are

Table 1 Flow and transport simulation model parameters

Parameters	Values
Porosity of the aquifer, θ	0.29
Grid size in *x*-direction, Δx	100 m
Number of nodes in *x*-direction, *NX*	28
Grid size in *y*-direction Δy	100 m
Number of nodes in *y*-direction, *NY*	27
Management period, *t*	365 days
Management time interval 1, Δt_1	90 days
Management time interval 2, Δt_2	275 days
Number of pumping wells	756
Number of injection wells	3
Mass flux of the injection wells	27.4 g/s
Injection rate, q_s	45 m^3/day
Storage coefficient	0.2
Longitudinal dispersivity, α_L	40 m
Ratio of the horizontal to the longitudinal dispersivity, α_{TH}/α_L	0.1

assumed to remain constant over time. The recharge rate, pumping rate, and boundary conditions change with time, contaminant sources are continuous, and mass flux rate is constant over time for all the sources. Three contaminant sources *S1*, *S2*, *S3,* and 108 potential monitoring well locations are considered. The monitoring network is designed for a management period of 1-year duration. Each 1-year management period is divided into two time intervals. Total mass estimate based on concentration values simulated at the end of the 1-year management period at the specified 108 potential monitoring locations constitute an input to the kriging linked SA model. The objective function which minimizes the mass estimation error is first utilized for solution of model. The numbers of monitoring wells are restricted to be 35, 45, and 55.

Table 2 gives the mass estimation error in percentage with the optimal number of wells as 35, 45, and 55. The error decreases with the increase in maximum permissible number of wells. This is expected, and it can be seen that the contaminant mass estimation error is not varying much from 45 to 55 wells.

Figures 2, 3, and 4 show the optimal locations of the 35, 45, and 55 wells respectively. The well locations chosen as solution of the design model are spatially spread over the study area the performance in terms of the mass estimation errors as shown in Table 1 can be judged as satisfactory. However, the order of the mass estimation errors decreases with 35 to 45 and 55 optimal monitoring locations.

Table 2 Number of wells and their corresponding mass estimation error in percentage in Scenario 1

No. of wells	Mass estimation error in %
35	0.2322
45	0.0021
55	0.0018

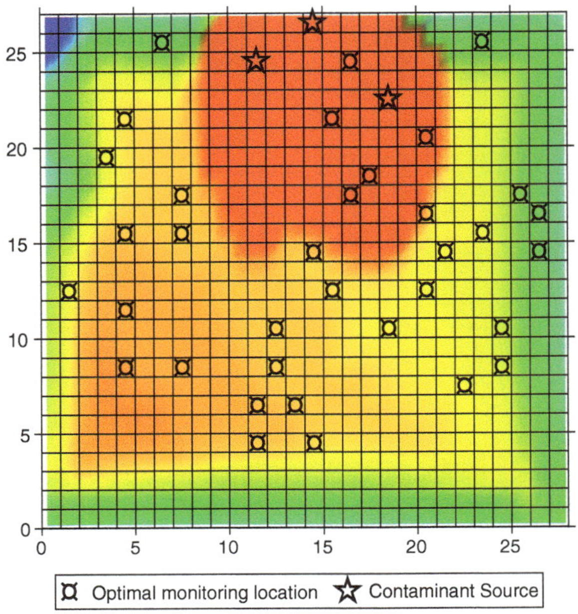

Fig. 2 Optimal location of monitoring wells for design set of 35 wells in Scenario 1

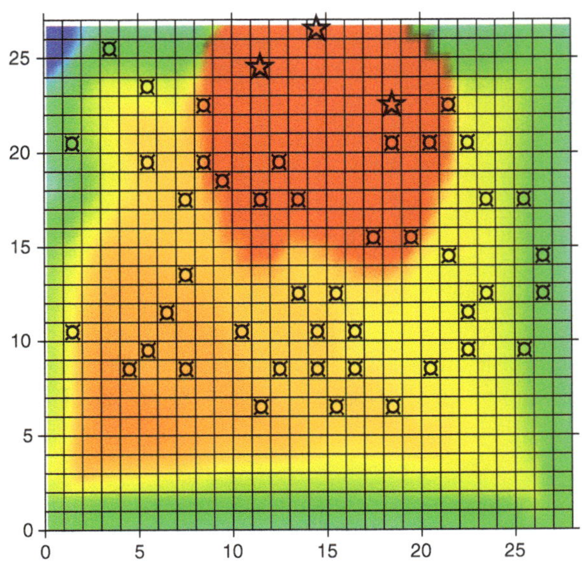

Fig. 3 Optimal location of monitoring wells for design set of 45 wells in Scenario 1

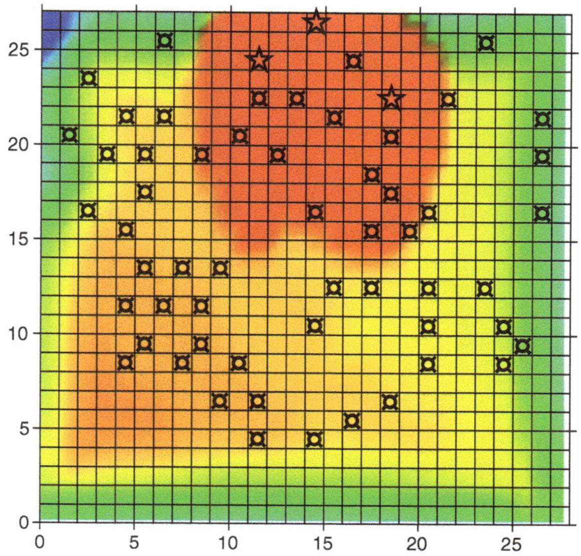

Fig. 4 Optimal location of monitoring wells for design set of 55 wells in Scenario 1

4.1.2 Scenario 2

All the conditions are same as in Scenario 1 except with the modifications in the contaminant source strength is considered uncertain. The mean contaminant source strengths are specified as 52608.0 mg/l. Uncertainties in terms of percentage are incorporated on the source concentration. Uncertainties in the source strength are incorporated by specifying a range of values for the source strengths. The range of upper bound and lower bound are shown in Table 3. These lower bound and upper bounds are utilized to generate different realizations of the source concentration using a uniform distribution. Ten realizations are generated from the uniform distribution for each of the sources. The simulation of the concentration plumes is performed for all 10 sets realizations. The mean concentration values at all the 108 potential monitoring locations obtained for these 10 realizations are utilized for spatial estimations of the concentration over the entire study area by kriging. This total contaminant mass estimate based on a number of realizations of the contaminant plume is used as input to the kriging linked SA model. Monitoring design model is solved to obtain the optimal monitoring network design.

Table 3 Range of contaminant source strengths for scenario 2

Uncertainty level (%)	Lower bound in mg/l	Upper bound in mg/l
10	47347.2	57868.8
20	42086.4	63129.6
30	36825.6	68390.4

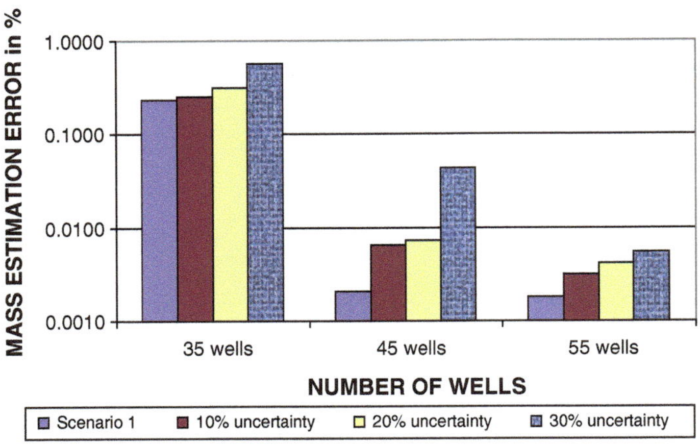

Fig. 5 Comparison of contaminant mass estimation error for scenario 1 and 2

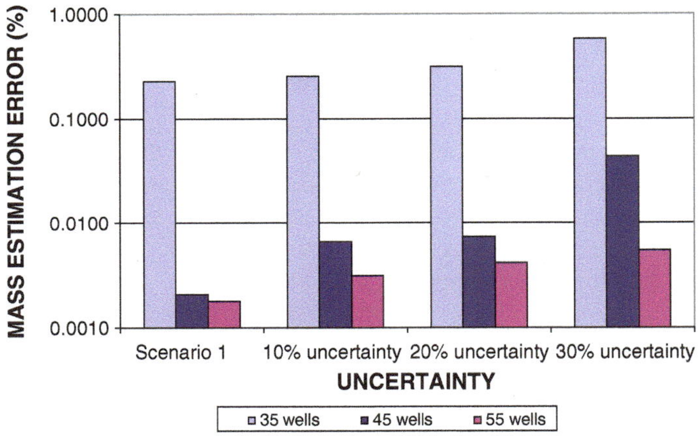

Fig. 6 Comparison of mass estimation errors with different levels of uncertainty

For each of the uncertainty levels, the optimal design set is obtained for 35 wells, 45 wells, and 55 maximum numbers of permissible wells. Figure 5 shows that for each set of wells, the mass estimation error increases with increase in the uncertainty level. The error values are compared with the values obtained for Scenario 1 without uncertainty. Figure 6 also shows that the mass estimation error decreases as the number of monitoring wells increases. These values are also given in Table 4.

Table 4 Mass estimation error in percentage for scenario 2

No. of wells	Uncertainty			
	Scenario 1	10 %	20 %	30 %
35	0.2322	0.2541	0.3139	0.5772
45	0.0021	0.0065	0.0073	0.0431
55	0.0018	0.0031	0.0042	0.0054

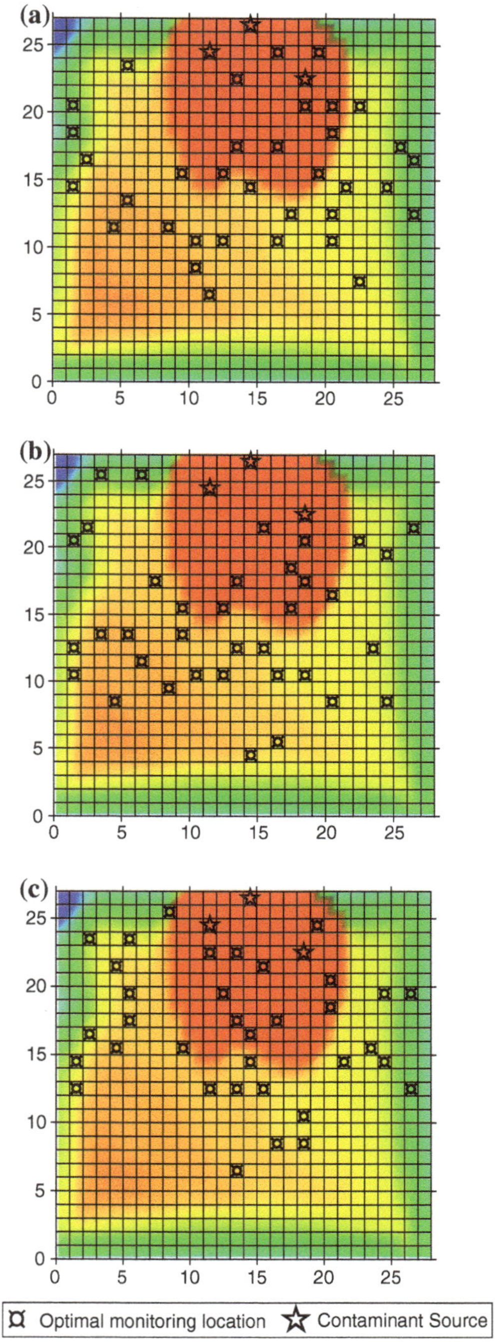

Fig. 7 Optimal monitoring locations for 35 monitoring wells with different uncertainties levels in scenario 2

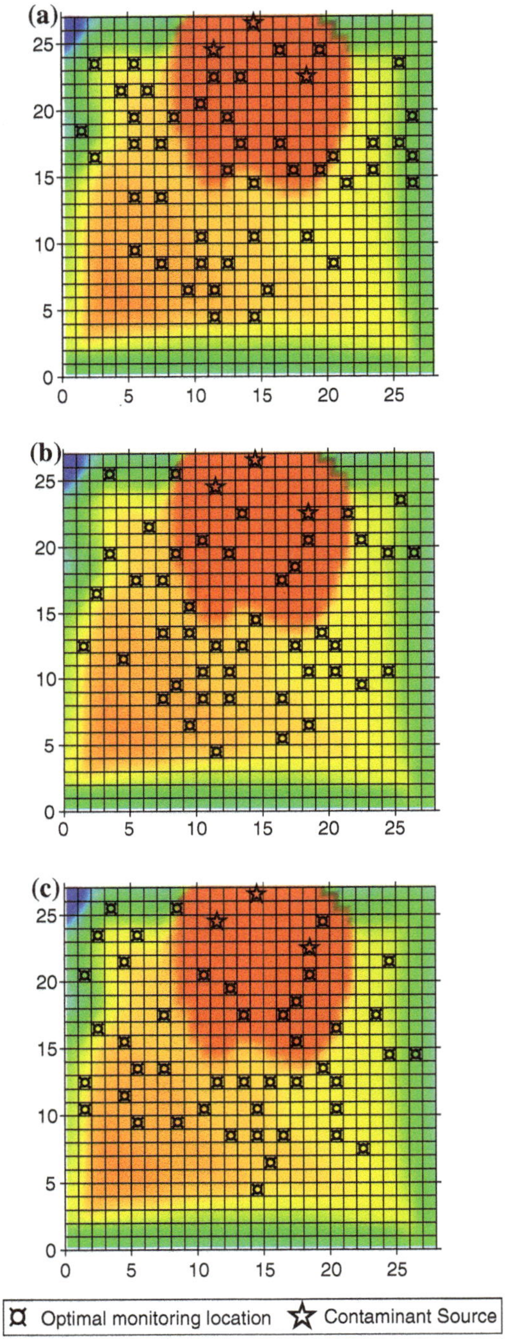

Fig. 8 Optimal design locations for 45 monitoring wells with different uncertainties in scenario 2

Fig. 9 Optimal design locations of 55 monitoring wells with different uncertainties in scenario 2

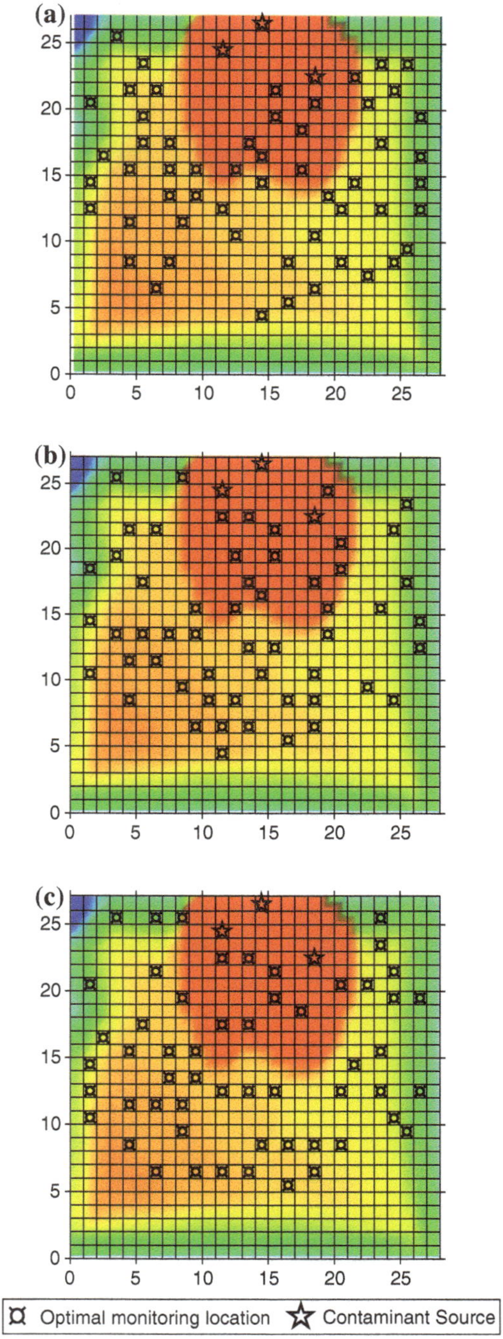

For the optimal design set of 35 wells the level of error is very high as shown in the Table 3. Some of the locations are common to all these designs. It can be noted from Table 3, that the mass estimation errors decrease with the increase in total number of monitoring wells, and it increases with the increase in the level of uncertainty in estimating the contaminant sources and therefore, the total mass of the contaminant. The optimal monitoring design locations for the different maximum permissible number of monitoring wells are shown in Figs. 7a–c, 8a–c, 9a–c. These figures show the variations in the designed networks for different permissible number of monitoring wells.

5 Conclusions

The developed optimization-based methodology for design of an optimal groundwater contamination monitoring network is solved for different flow and transport scenarios in an illustrative contaminated aquifer study area. The solution results are primarily useful for evaluation of the performance of the proposed models. These results appear to be consistent with intuitive solutions and therefore acceptable. However, much more rigorous performance evaluations may be necessary to establish the applicability of the proposed methodologies. The methodologies developed are applied to homogeneous unconfined aquifer for uniform parameters with transient contaminant transport process. Contamination sources are assumed to be conservative and continuous in nature. However, some of the following issues can be addressed in the future: (a) different types of contaminant sources varying in time and space may be considered for further evaluations; (b) the proposed methodology can be extended to non conservative and multi species contaminants in a 3-D system; (c) actual field data may be necessary before establishing the applicability of the developed methodologies. Also, the impact of measurement data reliability can be an important factor to consider.

References

ASCE Task Committee on Geostatistical Techniques (1990a) Review of geostatistics in geohydrology I: Basic concepts. J Hydraul Eng ASCE 116(5):612–632

ASCE Task Committee on Geostatistical Techniques (1990b) Review of geostatistics in geohydrology. II: Applications. J Hydraul Eng ASCE 116(5):633–658

Bashi-Azghadi BNS, Kerachian R, Lari MRB, Solouki K (2010) Characterizing an unknown pollution source in groundwater resources systems using PSVM and PNN. Expert Syst Appl 37 (2010):7154–7161

Bashi-Azghadi NS, Kerachian R (2010) Locating monitoring wells in groundwater systems using embedded optimization and simulation models. Sci Total Environ 408(10):2189–2198

Corana A, Marchesi M, Martini C, Ridella S (1987) Minimizing multimodal functions of continuous variables with the 'Simulated Annealing' algorithm. ACM Trans Mathematical Software 13(3):262–280

Chadalavada S, Datta B (2008) Dynamic optimal monitoring network design for transient transport of pollutants in groundwater aquifers. Water Resour Manag 22:651–670

Chadalavada S, Datta B, Naidu R (2011) Optimizations approach for pollution source identification in groundwater: an overview. Int J Environ Waste Manage 8(1–2):40–61

Cieniawski SE, Eheart WJ, Ranjithan S (1995) Using genetic algorithm to solve a multiple objective groundwater monitoring problem. Water Resour Res 31(2):399–409

Datta B, Dhiman SD (1996) Chance-constrained optimal monitoring network design for pollutants in ground water. J Water Resour Plan Manage ASCE 122(3);180–188

Datta B, Chakrabarty D, Dhar A (2009a) Optimal dynamic monitoring network design and identification of unknown groundwater pollution sources, Water Resour Manage 23 (10):2031–2049

Datta B, Chakrabarty D, Dhar A (2009b) Simultaneous identification of unknown groundwater pollution sources and estimation of aquifer parameters. J Hydrol 376(1, 2);48–57 (Elsevier)

Datta B, Prakash O, Campbell S, Escalada G (2013) Efficient identification of unknown groundwater pollution sources using linked simulation-optimization incorporating monitoring location impact factor and frequency factor. Water Resour Manage

Datta B, Prakash O, Sreekanth J (2014) Application of genetic programming models incorporated in optimization models for contaminated groundwater systems management. Adv Intell Syst Comput 288:183–199

Datta, B, Singh D (2014) Optimal Groundwater Monitoring Network Design for Pollution Plume Estimation with Active Sources. J GEOMATE 6(2) (Sl. No. 12);864–869

Dhar A, Datta B (2007) Multiobjective design of dynamic monitoring networks for detection of groundwater pollution. J Water Resour Plan Manage ASCE 133(4):329–338

Dhar A, Datta B (2010) Logic-based design of groundwater monitoring network for redundancy reduction. J Water Resour Plan Manage 136:88

Deb K (2002) Optimization for engineering design. Prentice Hall of India, New Delhi

Deutsch CV, Journel AG (1998) GSLIB: geostatistical software library and user's guide, 2nd edn. Oxford University Press, New York

Feng-guang Y, Shu-you C, Xing-nian L, Ke-jun Y (2008) Design of groundwater level monitoring network with ordinary kriging. J Hydrodyn 20:339–346

Grabow G, Mote RC, Yoder CD (2000) An empirically-based sequential ground water monitoring network design procedure. J Am Water Resour Assoc 36(3):549–566

GSLIB (1998) Geostatisical Software Library and User's Guide 1998 developed by C. V. Deutsch and A. G. Journel, Oxford University Press, New York

Husam B (2010) Assessment of a groundwater quality monitoring network using vulnerability mapping and geostatistics: a case study from Heretaunga Plains. NZ Agricul Water Manage 97 (2):240–246

Kollat JB, Reed PM, Kasprzyk JR (2008) A new epsilon-dominance hierarchical bayesian optimization algorithm for large multi-objective monitoring network design problems. Adv Water Resour 31(5):828–845

Kollat JB, Reed PM, Maxwell R (2011) Many-objective groundwater monitoring network design using bias-aware ensemble kalman filtering, evolutionary optimization, and visual analytics. Water Resour Res 47:W02529

Lin YP, Tan YC, Rouhani S (2001) Identifying spatial characteristics of transmissivity using simulated annealing and kriging methods. Environ Geol 41:200–208
Loaiciga HA (1989) An optimization approach for groundwater quality monitoring network design. Water Resour Res 25(8):1771–1782
Loaiciga HA, Fethi BJ, Marino MA (1994) Multivariate geostatistical design of ground-water monitoring networks. J Water Resour Plan Manage ASCE 120(4):505–522
Loaiciga HA, Hudak PF, Marino MA (1995) Regional-scale ground water quality monitoring via integer programming. J Hydrol Elsevier 164:153–170
Mahar PS, Datta B (1997) Optimal monitoring network and ground-water pollution source identification. J Water Resour Plan Manage 123(4):199–207
Meyer PD, Brill ED (1988) A method for locating wells in a groundwater monitoring network under conditions of uncertainty. Water Resour Res 24(8):1277–1282
MODFLOW (1996) U.S. Geological survey modular finite-difference ground-water flow model developed by Arlen W. Harbaugh and Michael G. McDonald
MT3DMS (1999) A modular Three-dimensional multispecies transport model for simulation of advection, dispersion, and chemical reactions of contaminants in groundwater systems developed by Chunmiao Zheng, P. Patrick Wang
Nunes LM, Cunha MC, Ribeiro L (2004) Groundwater monitoring network optimization with redundancy reduction. J Water Res Plan Manage ASCE 130(1):33–42
Pinder GF, Bredehoeft JD (1968) Application of the digital computer for aquifer evaluations. Water Resour Res 4(5):1069–1093
Prakash O, Datta B (2012) Sequential optimal monitoring network design and iterative spatial estimation of pollutant concentration for identification of unknown groundwater pollution source locations. Environ Monit Assess. doi:10.1007/s10661-012-2971-8
Prakash O, Datta B (2013) Multi-objective monitoring network design for efficient identification of unknown groundwater pollution sources incorporating genetic programming based monitoring. J Hydrol Eng ASCE HE. 1943, 5584, 0000952, 04014025
Prakash O, Datta B (2014) Optimal monitoring network design for efficient identification of unknown groundwater pollution sources. Int J Geomate 6(1):785–790
Prakash O, Datta B (2015) Optimal characterization of pollutant sources in contaminated aquifers by integrating sequential-monitoring network design and source identification: methodology and an application in Australia. Hydrogeol J 23(6):1089–1107
Reed P, Minsker B, Valocchi AJ (2000) Cost-effective long term groundwater monitoring design using a genetic algorithm and global mass interpolation. Water Resour Res 36(12):3731–3741
Reed PM, Minsker BS (2004) Striking the balance: long-term groundwater monitoring design for conflicting objectives. J Water Resour Plan Manage ASCE 130(2);140–149
Reed PM, Kollat JB (2012) Save now, pay later? Multi-period many-objective groundwater monitoring design given systematic model errors and uncertainty. Adv Water Resour 35:55–68
Singh D (2008) Optimal monitoring network design for contamination detection and sequential characterization of contaminant plumes with feedback information using simulated annealing and linked kriging. Ph.D. thesis submitted at Indian Institute of Technology, Kanpur, India
Sreekanth J, Datta B (2012) Genetic programming: efficient modeling tool in hydrology and groundwater management. New Genetic Programming-new approaches and successful applications. InTech, Rijeka, Croatia, pp 227–240
Sreekanth J, Datta B (2014) Design of an optimal compliance monitoring network and feedback information for adaptive management of saltwater intrusion in coastal aquifers. J Water Resour Plan Manage 140(10):04014026
Sreekanth J, Datta B (2015) Review: simulation optimization models for the management and monitoring of coastal aquifers. Hydrol J 23(6):1155–1166

Wu J, Zheng C, Chien CC (2005) Cost-effective sampling network design for contaminant plume monitoring under general hydrogeological conditions. J Contam Hydrol 77:41–65

Yeh MS, Lin YP, Chang LC (2006) Designing an optimal multivariate Geostatistical groundwater quality monitoring network using factorial Kriging and genetic algorithm. J Environ Geol 50:101–121

Zheng C, Wu J, Chien CC (2005) Cost-effective sampling network design for contaminant plume monitoring under general hydrogeological conditions. J Contam Hydrol 77:41–65 Elsevier

Suction–Water Content Relationship for Hill Soil of North-East India

C. Malaya and S. Sreedeep

Abstract Hill-slopes of north-east (NE) India mainly constitute of unsaturated red soil. While analyzing the stability of hill-slope, hydrology, i.e., infiltration and surface drainage becomes indispensable. Defining this suction–water content relationship (SWR) of the soil is important. It is generally expressed in terms of a graphical relationship between soil suction and water content (or saturation). SWR of red soil has been obtained using equitensiometer, relative humidity sensor, and volumetric water content sensor. From the measured SWR, its parameters were defined. The parameters defining this SWR are key input for analyzing hill-slope hydrology and stability. The details of the methodology adopted in this study are presented in this paper.

Keywords Hill-slope · Unsaturated soil · Suction–water content relationship · Parameters · Hydrology · Stability

1 Introduction

The hill-slopes of north-east (NE) India consist of unsaturated red soil (RS). In NE India, hill-slope developmental activities have increased the occurrence of landslides during the rainy season due to the heavy rainwater infiltration and surface drainage. Accurate description of infiltration and surface drainage for unsaturated soil remains a fundamental problem in hydrology. Modeling of rainfall infiltration and surface drainage is needed for the analysis of slope failure induced by heavy rainfall (Chen and Lee 2003). The parameters of soil suction–water content relationship (SWR) are crucial inputs for this. SWR is defined as the relationship

C. Malaya (✉)
Department of Civil Engineering, Assam Engineering College, Guwahati, India
e-mail: mchetia.ce@aec.ac.in

S. Sreedeep
Department of Civil Engineering, Indian Institute of Technology Guwahati, Guwahati,, India
e-mail: srees@iitg.ernet.in

A.K. Sarma et al. (eds.), *Urban Hydrology, Watershed Management and Socio-Economic Aspects*, Water Science and Technology Library 73,
DOI 10.1007/978-3-319-40195-9_10

between soil suction and water content (either, gravimetric or volumetric) or soil saturation (Fredlund and Rahardjo 1993).

This study deals with the measurement of the SWR of RS with the help an equitensiometer (EQT) and a relative humidity sensor (RHS). SWR equations reported in the literature Fredlund and Rahardjo (1993) are fitted to the measured SWR for RS for determining its SWR parameters.

1.1 Theoretical Background

Soil specific parameters for a typical drying (desorption) SWR are: (1) the volumetric water content at saturation, θ_s, which is the water content at which the soil is completely saturated and typically depicts the initial condition for the drying path, (2) air entry value (AEV), which is defined as the suction at which air starts entering the largest pores present in the soil sample, and (3) residual water content (θ_r), defined as the water content below which water content changes are minimal as suction changes.

In this study, the SWR is represented by van Genuchten (1980) and Fredlund and Xing (1994) equation represented by Eqs. 1 and 2, respectively

$$\theta(\psi) = \theta_r + (\theta_r - \theta_s)\left[\left[1 + \left(\frac{\psi}{a_{vg}}\right)^{n_{vg}}\right]^{m_{vg}}\right]^{-1} \tag{1}$$

$$\theta(\psi) = \theta_s\left[1 - \frac{\ln\left[1 + \frac{\psi}{h_r}\right]}{\ln\left[1 + \frac{10^6}{h_r}\right]}\right]\left[\left[\ln\left[\exp(1) + \left(\frac{\psi}{a_f}\right)^{n_f}\right]\right]^{m_f}\right]^{-1} \tag{2}$$

where $\theta(\psi)$ is the volumetric water content at any suction, ψ; θ_r is the residual volumetric water content; θ_s is the volumetric water content at saturation; a_{vg} and a_f are fitting parameters primarily dependent on the air entry value (AEV); n_{vg} and n_f are fitting parameters that are dependent on the rate of extraction of water from the soil; m_{vg} and m_f are fitting parameters which depend on θ_r; h_r is the suction (in kPa) corresponding to residual state.

The computer code RETC, version 6.02 (Van Genuchten et al. 1991) has been used for estimation of SWR parameters. RETC uses nonlinear least-squares optimization technique for curve fitting.

2 Materials and Methods

A locally available fine-grained red soil from hill-slope designated as RS was used in this study. The soil is characterized for its specific gravity, grain size distribution, liquid limit, and plastic limit by following the guidelines presented in the literature

Table 1 Physical properties and classification of the soil used in the study

Property	Soil RS
Specific gravity	2.62
Particle size characteristics (%)	
Sand (4.75–0.075 mm)	26
Coarse sand (4.75–2 mm)	0
Medium sand (2–0.425 mm)	5
Fine sand (0.425–0.075 mm)	21
Silt (0.075–0.002 mm)	67
Clay (<0.002 mm)	7
Atterberg limits (%)	
w_L	46
w_P	27
I_P	19
Classification*	CL

*ASTM 2008

(ASTM 2005a, 2006, 2007). The details of the characterization are listed in Table 1. It can be noted that the soil is clayey soil of low compressibility.

The equitensiometer (EQT) consists of a precision soil moisture sensor, the ThetaProbe, whose measuring rods are embedded in a porous material, the equilibrium body. The porous material has a known, stable relationship between water content and matric suction (ψ_m). When the (EQT) is inserted into the soil for matric suction (ψ_m) measurement, it equilibrates with the surrounding soil. The water content of the equilibrium body is measured directly by the ThetaProbe and gives the output in millivolt (mV) (Soil Matric Potential Sensor, User Manual 1999), and this is converted to ψ_m of the surrounding soil using the calibration curve supplied by the manufacturer (Delta-T Devices, UK).

The EC-TE volumetric water content sensor (VWCS) is low cost sensor manufactured by Decagon Inc. USA. This probe determines 'θ' based on the dielectric constant or permittivity of the material in which it is inserted. EC-TE is a three-pronged probe. The details of this probe are reported in the literature (Kizito et al. 2008; Malaya and Sreedeep 2010). The measurements using equitensiometer and EC-TE probes are performed employing the test set up as shown in Fig. 1. The test set up essentially consists of a perspex container with a base plate into which the soil is compacted. The ψ_m and θ measuring probes are connected to a computer through respective data loggers.

The air-dried soil is compacted at required compaction state in the perspex mold of 200 mm diameter and 160 mm height. The EC-TE probe is inserted directly into the soil, and the sharpened tip of the probe facilitates the insertion process. For inserting equitensiometer, dummy hole is made (whose size is smaller than the probe size) and then the probe is inserted. It must be noted that the soil next to the probe has the strongest influence on the readings, and hence sufficient care has been

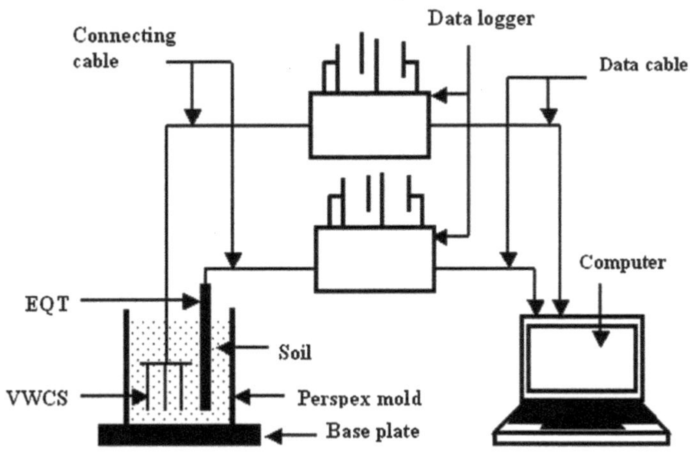

Fig. 1 Details of the EQT test set up used in the study

taken to avoid air gap around the probe. The soil has been initially saturated so that $\psi_m \approx 0$. The probes are inserted into the soil sample and the readings recorded over a period of time. The measured ψ_m and θ values are used for developing SWR.

The RHS essentially consists of two parts: (a) polymer capacitance sensor, which measures relative humidity (0–100 %) and temperature (−40 to 85 °C); and (b) wire filter, which protects the sensor from air-borne particles. The body of the RHS immediately above the polymer sensor is filled with a special resin to quickly remove condensation after measuring high RH samples (Hygroclip Manual 2011). Such a provision reduces the time interval between two successive measurements considerably, as the RHS regains ambient RH rapidly. The connecting cable of the RHS is attached to the read out unit for measurement of RH and temperature. After placing the soil sample in the relative humidity box (RHB), the RH and temperature readings are noted down at regular interval (generally every 5 min) until constant equilibrium value is obtained. In most of the cases, equilibrium was achieved within 40 min of the start of the measurement. The RH and temperature measured at equilibrium are used for determining total suction (ψ) of the soil sample using Eq. 3.

$$\psi = -\frac{R.T}{\upsilon_{w0}.\omega_v} \ln\left(\frac{\overline{u}_v}{\overline{u}_{v0}}\right) \tag{3}$$

where

ψ soil suction or total suction (kPa)
R universal (molar) gas constant [i.e., 8.31432 J/(mol K)]
T absolute temperature in Kelvin [i.e., T = (273 + t_0) (K)]
t_0 temperature (°C)
v_{w0} specific volume of water or the inverse of the density of water [i.e., $1/\rho_w$ (m^3/kg)]
ρ_w density of water (i.e., 998 kg/m^3 at t_0 = 20 °C)

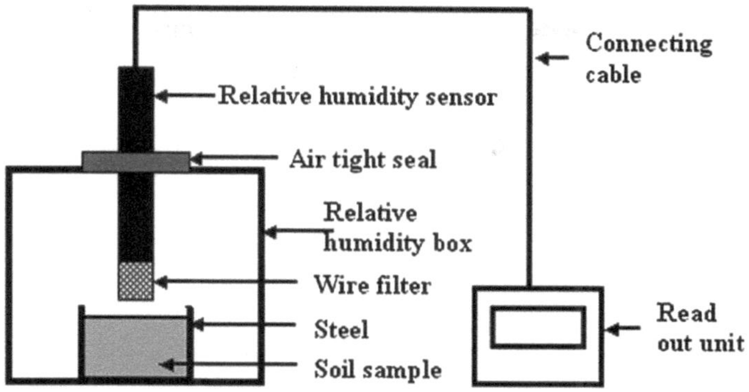

Fig. 2 Details of the RHS test set up used in the study

ω_v molecular mass of water vapor (i.e., 18.016 kg/kmol)
$\bar{u}_v$ partial pressure of pore-water vapor (kPa)
$\bar{u}_{v0}$ saturation pressure of water vapor over a flat surface of pure water at the same temperature (kPa)

The term $\bar{u}_v/\bar{u}_{v0}$ is called relative humidity, RH (in %). Depending upon the purity of pore water, $\bar{u}_v$ will change. Since there are no salts present in the soil used in this study, Eq. 3 will give $\psi = \psi_m$.

The relative humidity test set up employed for the present study is depicted in Fig. 2. A sharp edged stainless steel ring of internal diameter less than the unconfined compressive strength sample maker and height 15 mm has been used to extract compacted soil sample. The steel ring with compacted soil sample is placed in an in-house fabricated Teflon relative humidity box (RHB), which is perfectly airtight. The RHB is 65 mm in internal diameter and 60 mm in height. The top cap of the RHB is fitted with HygroClip S (ROTRONIC AG, Grindelstrasse 6) relative humidity sensor (RHS) (as shown in Fig. 2) in such a way that its tip (sensing portion) remains just above the top surface of the soil sample. The ψ measurements were conducted on "as compacted" soil samples. The water content, w of compacted soil sample has been determined by following the guidelines presented in ASTM 2005b. The measured ψ and θ values are used for developing SWR.

3 Results and Discussion

The ψ_m and θ values obtained using EQT and VWCS are plotted as depicted in Fig. 3.

The ψ (=ψ_m) values measured using RHS are plotted as shown in Fig. 4. θ was computed corresponding to ψ measurements from the measured w and dry unit weight of the soil sample. The entire data obtained from EQT and RHS measurements were used to plot SWR of RS as depicted in Fig. 5. The combined data

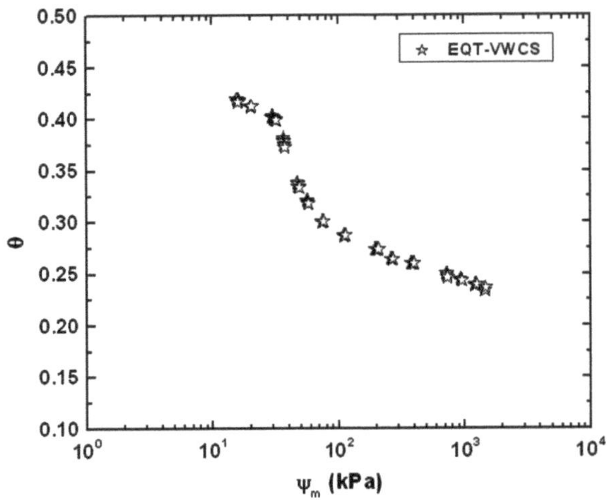

Fig. 3 SWR obtained from EQT measurement for RS

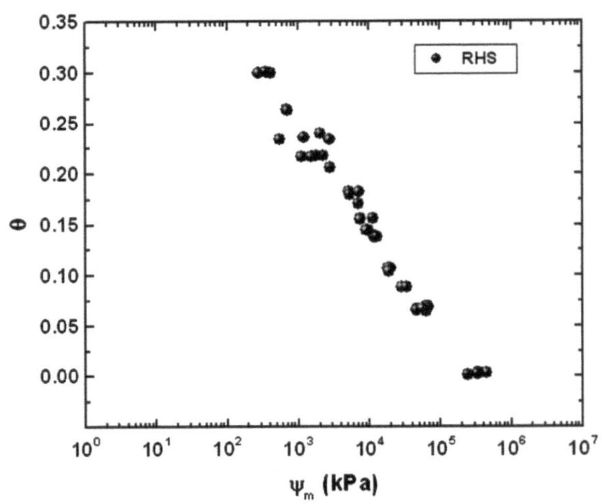

Fig. 4 SWR obtained from EQT measurement for RS

points bridge together well for soil RS to give a broad range of measured ψ_m. The combined SWRs of RS were used for obtaining SWR parameters. SWR Eqs. 1 and 2, designated as VG and FX, respectively, have been fitted to the experimental data using the computer code RETC. The SWR equation parameters obtained from this analysis are listed in Table 2. It can be observed that the fitting functions yield a regression coefficient (R^2) close to unity, which indicates an excellent fit to the experimentally obtained data.

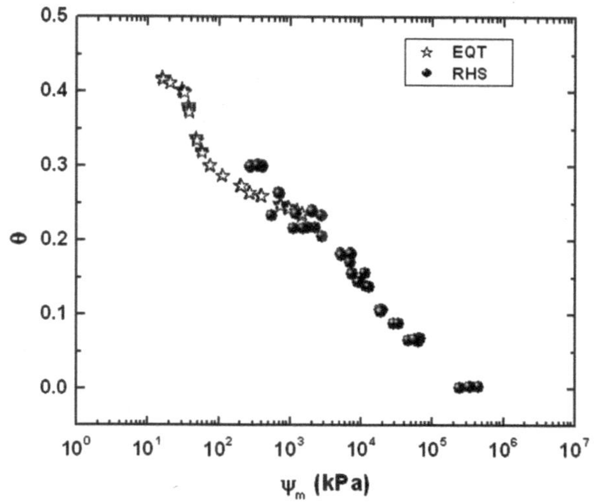

Fig. 5 Combined SWR obtained from EQT and RHS measurement for RS

Table 2 Details of the SWR equation fitting parameters for RS

Fitting function	Parameter	Value
VG	a_{vg} (kPa)	16.44
	n_{vg}	13.38
	m_{vg}	0.0114
	θ_r	0
	θ_s	0.4167
	R^2	0.9310
FX	a_f (kPa)	31.10
	n_f	14.64
	m_f	0.1135
	h_f (kPa)	819.92
	θ_s	0.4167
	R^2	0.9856

4 Conclusion

The study deals with the measurement and parameterization of SWR of a locally available red soil from a hill-slope. The SWR was measured using EQT and RHS. The SWR parameters of van Genuchten and, Fredlund and Xing models were obtained by fitting the SWR equations to the experimental data using computer code RETC, version 6.02. The SWR parameters are important inputs for modeling rainwater infiltration and surface drainage in hill-slopes for the analysis of slope failure induced by heavy rainfall.

References

ASTM D 2216-05 (2005b) Standard test methods for laboratory determination of water (moisture) content of soil and rock by mass. Annual Book of ASTM Standards, vol 04.08. ASTM International, West Conshohocken

ASTM D 2487-06$^{\varepsilon 1}$ (2008) Standard practice for classification of soils for engineering purposes (Unified Soil Classification System). Annual Book of ASTM Standards, vol 04.08. ASTM International, West Conshohocken

ASTM D 4318-05 (2005a) Standard test methods for liquid limit, plastic limit, and plasticity index of soils. Annual Book of ASTM Standards. ASTM International, West Conshohocken

ASTM D 854-06 (2006) Standard test method for specific gravity of soil solids by water pycnometer. Annual Book of ASTM Standards. ASTM International, West Conshohocken

ASTM D 422-63 (2007) Standard test method for particle-size analysis of soils. Annual Book of ASTM Standards. ASTM International, West Conshohocken

Chen JF, Lee CH (2003) An analytical solution on water budget model in unsaturated zone. J Chin Inst Eng 26(3):321–332

Fredlund DG, Rahardjo H (1993) Soil mechanics for unsaturated soils. Wiley, New York

Fredlund DG, Xing A (1994) Equations for the soil-water characteristic curve. Can Geotech J 31 (3):521–532

Hygroclip Manual (2011) http://www.rotronic-humidity.com/_upload/gen_downloads/in-hyclip-DI-v1.0.pdf, website visited on 3-3-2011

Kizito F, Campbell CS, Campbell GS, Cobos DR, Teare BL, Carter B, Hopmans JW (2008) Frequency, electrical conductivity and temperature analysis of a low-cost capacitance soil moisture sensor. J Hydrol 352:367–378

Malaya C, Sreedeep S (2010) A study on the influence of measuring procedures on suction-water content relationship of a sandy soil. J Test Eval 38(6):1–9

Soil Matric Potential Sensor (1999) User Manual EQ2-UM-1.3, Delta-T Devices, UK

Van Genuchten MT (1980) A closed form equation for predicting the hydraulic conductivity of unsaturated soils. Soil Sci Soc Am J 44(5):892–898

Van Genuchten MT, Leij FJ, Yates SR (1991) The RETC code for quantifying the hydraulic functions of unsaturated soils, Report No. EPA/600/2-91/065. R.S. Kerr Environmental Research Laboratory, U.S. Environmental Protection Agency, Ada, OK

Ground Water Management—Sustainability and Methodology

Vishwajit Anand and Sumit Kumar

Abstract One of the major environmental issues of concern in an urban ecosystem is the sustainability of ground water for future generations. To ensure this, one needs to understand the impacts of present and future utilization practices, and to adopt the most appropriate one. This paper proposes to explore the needs and methodology of ground water management, pertaining to qualitative, quantitative, environmental, nutritional, and economic aspects. While ground water storage is vast, its rate of replenishment is finite and mainly limited to shallower aquifers, whose quality can be seriously and even irreversibly degraded. Rapid urbanization, leading to high rates of deforestation and industrialization has resulted into degradation of these aquifers. The ground water recharge system has got disrupted. Further, lowering of the water table has led to ecological and economical imbalances. Food security and health of individuals have been subjected to potential threats. Rain water harvesting and afforestation are among the traditional water management strategies. Individual pumpings may be replaced by social pumping to avoid over-exploitation of ground water. Common drainage system treading the sewage water to recycling plants may be brought up. Use of chemicals and synthetic additives may be minimized. Use of concrete in pavements should be replaced with interlocking fly-ash pavers to support water recharge system. Building bye-laws should be followed so that enough area is open for water recharge. Remote sensing and Geographic Information System (GIS) may be used to monitor the quality and quantity of ground water. There is an urgent need for community monitoring and bringing up water management solutions from efforts toward increasing water availability, and not from demand management. Sustainable ground water development can bring major benefits such as food security, safe drinking water, high value agricultural produce, and inland fisheries, and must be pursued through technical and management approaches.

V. Anand (✉)
Department of Civil Engineering, Indian Institute of Technology Madras, Chennai 600036, India
e-mail: a.vishwajit@iitg.ernet.in

S. Kumar
Bhabha Atomic Research Centre, Trombay, Mumbai 400085, India
e-mail: kr.sumit@iitg.ernet.in

A.K. Sarma et al. (eds.), *Urban Hydrology, Watershed Management and Socio-Economic Aspects*, Water Science and Technology Library 73,
DOI 10.1007/978-3-319-40195-9_11

Keywords Degradation of aquifers · Ecological balance · Social pumping Water recycling · Community monitoring · Awareness among masses

1 Introduction

Evidences from history reveal that every human settlement and civilization has prospered near the water sources. This emphasizes the fact that life cannot be imagined without water. Not only do we humans need it, but every living being needs it to live. However, the utilization practices show a gradual decline in approach to sustainability of water resource for future generations. Ancient texts and epics give a good insight into the water storage and conservation systems prevalent in those days. The Indus Valley Civilization, that flourished along the banks of river Indus about 5000 years ago, had one of the most sophisticated urban water supply and sewage systems in the world. Similar systems may be observed in the ruins of Nalanda University and ancient Rome Civilization. One of the oldest water harvesting systems has been found about 130 km from Pune in India along Naneghat in Western Ghats. Mesopotamians used to have stone rainwater channels for harvesting rainwater. Moreover many civilizations give evidences of existing rooftop water harvesting systems. Such systems directed rainwater from rooftops into underground storages.

The changing trend exhibits a major shift of water usage from surface water to ground water. Today almost 50 % of all urban water use worldwide is attributed to ground water sources due to its relatively low cost and generally high quality. However over the years, rising populations, growing industrialization, and expanding agriculture have pushed up the demand for water. Rapid urbanization has promoted deforestation which has affected the health of ground water aquifers both qualitatively and quantitatively. Earlier dense vegetation used to break the fall and surface runoff promoting the percolation of ground water into the ground and reviving these aquifers. Impermeabilisation caused due to soil compaction, paving, and roofing significantly disrupts the ground water recharge system. Rapid industrialization and use of chemicals in agriculture has deteriorated the quality of ground water affecting the health of all the organisms. Further heavy ground water abstraction has resulted into falling water tables leading to many ecological and economical imbalances, sometimes land subsidence also. Natural disasters like famines have crept in leading to food insecurity. The health of individuals has also got impaired. The effects of urbanization get transmitted to surrounding peri-urban areas too. Overall urbanization affects the underlying ground water system in two ways:

- by radically changing patterns and rates of aquifer recharge, and
- by adversely impacting the quality of this recharge.

Today the situation has turned quite alarming in various countries of the world like Sudan and Venezuela. The water tables are falling in scores of countries including Northern China, the US, and India due to widespread over-pumping using powerful diesel and electric pumps. The advent of plastics has led to over-exploitation and qualitative degradation of ground water. About 80 % of the world's population live in areas with threats to water security.

This future water security dilemma has posed a need for efficient urban ground water management strategies. The goal of urban ground water management should be to strike a reasonable balance between maintaining water supply availability and quality, preserving the urban infrastructure and ensuring the safe disposal of wastes and practical steps. Foster (2001) emphasized on this very interdependence of ground water and urbanization and provided a framework for a systematic consideration in urban management, based on detailed investigations in six different cities in Latin America and Asia. With well-planned and sustainable ground water managerial techniques, we can achieve major goals like food security, safe drinking water, and high agriculture produce. Stone (2001) discusses how to achieve this middle ground of ground water sustainability. On the other hand, if the present alarming situation of ground water is ignored, future generation may get deprived of ground water and be forced to use polluted water for consumption. This will lead to outcrop of many existing and even newer diseases which will be a threat to life on this planet.

2 Potential Aspects of Ground Water Sustainability

Rapid growth of population has significantly pushed up the demand for water of which ground water forms a major fraction. Rapid and uncontrolled urbanization have triggered the process of degradation of ground water, pertaining to qualitative, quantitative, environmental, health, and economic aspects. Exploitation of ground water aquifers in excess of natural recharge or their contamination poses a real risk of hydrologic, economic, and social crisis. The challenge is to meet the growing needs from finite resources while maintaining the planet's fragile life support system. This is achievable only with a holistic understanding of the science behind local and regional water policy, which has been explained by Stone (2005).

2.1 Quantitative Aspects

Ground water is present in the aquifers below the Earth surface. Water in these aquifers gets recharged by the percolation through the soil layers. When as a part of hydrological cycle, precipitation occurs in the form of rains or snow, water percolates through the soil pores. Presence of dense vegetation slows down the surface runoff and hence promotes the water recharge system. Our rapidly increasing population and ever-increasing wants have created a situation of crisis of natural resources. Over-exploitation of these aquifers leads to quantitative degradation of these aquifers. Large scale deforestation has significantly affected both water recharge system and hydrological cycle. Thus, deforestation has affected the sustainability of ground water in two-way process. The increased rate of urbanization has made the situation even graver. Almost all spaces in urban areas have been paved using concrete or some other impermeable materials leaving very little space open for water recharge system. In addition, uncontrolled industrialisation has caused global warming which has disrupted the hydrological cycle. Over-abstraction of ground water has led to falling water levels in many regions such as Sana'a Basin in Yemen.

The most efficient solution to this menace is afforestation. Afforestation on a large scale can promote recharge of ground water along with reviving the hydrological cycle. Water harvesting structures may be of our help in areas where ground water has suffered high levels of quantitative degradation. The notion of social pumping in lieu of individual ones may reduce the over-exploitation of ground water. Additionally surface water may be used for non-potable uses in order to reduce burden on ground water. Self-interlocking fly ash pavers may be used in pavements so that water is able to percolate through the pores between them. Building bye-laws should be followed and enough spaces should be kept open for water recharge. These days at waste disposal sites, a layer of clay on account of its permeability is being laid before waste disposal in order to avoid land pollution. But this has not met the real objective. At the same time, it has blocked the water recharge system at that site. The situation thus demands a better waste management strategy.

Since many regions of the world are facing scarcity of ground water, there is a need for community awareness and bringing up proper ground water management plan.

2.2 *Qualitative Aspects*

Rapid urbanization and industrialisation have led to high levels of water and land pollution. These pollutants get dissolved in water during its recharge through the soil pores and reach the aquifers. Since ground water aquifers are interconnected, contamination at one point leads to contamination of nearby aquifers. Increased use of chemicals has triggered the qualitative degradation of ground water. In some places, the unplanned drainage systems and open bore holes are sources of ground water pollution. The qualitative degradation of ground water has lead to spread of many water-borne diseases and significant decline in agricultural produce. Saline ingression in coastal areas and high levels of arsenic and fluoride in ground water are common manifestations of qualitative degradation of ground water.

Awareness among the masses is the most efficient solution. People should be made aware of the adverse effects of using contaminated water for potable purposes. Provision of safe drinking water and continuous quality monitoring of ground water should be done by the governments of various nations. The use of chemicals should be minimized. Synthetic additives may be replaced by natural extracts. Some countries are practising economically nonviable methods of cleaning highly contaminated aquifers. In the case of arsenic, methods for in situ treatment have already been used in developed countries. In the United States, zerovalent iron, permeable reactive barriers (PRBs) are used in situ to remove chromium and several chlorinated solvents in ground water and are tested successful for removing arsenic.

An urgent need for qualitative and quantitative community monitoring of ground water has emerged. Remote sensing and geographic information system (GIS) may be used to monitor the quality and quantity of ground water. The availability and present quality of ground water may be determined using these techniques. They can even be used for cleaning the aquifers highly contaminated with arsenic, fluoride, and other toxic metals. A careful study and management of ground water using GIS can thus alleviate the problems of degradation of ground water reserves.

2.3 Environmental Aspects

In addition to degradation in ground water levels, deforestation inhibits the water recharge system which further deteriorates the situation. As a result of falling water table, trees in a region get dried triggering an ecological imbalance. When the imbalance so created goes beyond the adaptability of a species, it may get endangered and face threats of being extinct in near future. When there is an acute water shortage in a region for a considerably long period, the soil starts losing its fertility. Salinity ingression and other contaminations in ground water also results into soil infertility. The situation may turn into famines and droughts. Sometimes even in the river basins, as a result of a prolonged famine, large cracks appear in the ground. Extreme situations may turn into land subsidence as in the case of Mexico.

Rainwater harvesting is one of the traditional and effective methods to prevent falling water levels. Afforestation can tread our steps towards ecological balance. Awareness among people about the possibility of an ecological imbalance and methods to tackle it can turn to be quite helpful.

2.4 Nutritional and Health Aspects

Ground water is a prominent source of fresh water. Almost 70 % of the fresh water is used for agricultural purposes. With rapid population growth, the demand of food as well as fresh water has rapidly increased. Since last few decades, over-abstraction of ground water has led to falling water tables and acute water scarcity in many parts of the world. When there is an acute water shortage in a region for a long time, the soil starts losing its fertility. The situation may take form of famines. Food security is threatened and people become prey of malnutrition and many diseases. Under cases of prolonged famines and droughts, farmers suffering from severe economic tragedies commit suicides.

Qualitative degradation of ground water reserves has made safe drinking water inaccessible in many regions of the world. Residents are forced to drink contaminated water which leads to many water-borne diseases of which some are incurable. In fact 80 % of all diseases in the world can be traced to drinking and washing with unsafe water supplies. People in some coastal areas face water contaminated by saline ingression. Arsenic and fluoride contaminated ground water are, respectively, major threats to health of residents of Nellore and parts of Jharkhand in India. High death rates as a result of consumption of arsenic containing water have been observed in Bangladesh. Under extreme circumstances when some species are unable to adapt, they become endangered and even extinct in that region.

The most efficient way to curb these nutritional and health threats is to bring awareness among the masses. However, at the same time, the governments of various nations should guarantee their every citizen accessibility to safe drinking water. Integration of water supply provision and community health under same agency can ensure provision of scientifically tested drinking water to every individual. Other solutions may comprise of watersheds and rain water harvesting structures. Afforestation can ensure ground water availability and food security. The notion of social pumping may help us attain quality control over potable water. This aspect of ground water sustainability attracts proper attention in order to sustain the life support system on this planet.

2.5 *Economic Aspects*

Over-abstraction of ground water in order to sustain the rapidly growing population has led to sharp decline in the ground water levels in many regions round the globe. The falling water level has led to increased cost of ground water abstraction which in real sense adds to the economic burden of a nation. Moreover, the non-accessibility of safe drinking water forces people purchase water from various water agencies which in turn adds to the problems of commoners. Qualitative degradation of ground water leads to many water-borne diseases. This fall in the quality of human resource leads to a sharp fall in GDP of a nation. In regions where a major fraction of people are involved in agriculture, fall in agricultural produce due to degradation of water resources poses serious impacts on the GDP. Moreover, water plays an important role in the world economy as it functions as a solvent in most of the industries and facilitates industrial cooling and transportation.

A common and effective approach to water management issues is to regard water as a commodity and view its value in light of its overall economic contribution to the local, regional and national economy. Social pumping is highly economical as compared to individual ones. As estimated by us in a peri-urban area in Bihar in

India, we can have a cost saving of over 50 % in case of social pumping. Provision of safe drinking water may be made in return of waste water. This waste water should be treated in recycling plants in order to make it reusable for non-potable activities. This may reduce the burdens on the current economy of various nations. The economic benefits of social pumping, usually referred to as water markets, has been well justified by a data envelopment analysis approach by Manjunatha et al. (2011).

The rapid and uncontrolled rates of population growth, urbanization and industrialisation have posed serious threats to the sustainability of ground water resources. It has been estimated that by 2030, in some developing regions of the world, water demand will exceed its supply by 50 %. By then more than half of the world population will be facing water-based vulnerability. There is a great educational challenge to help communities understand the situation and to achieve balance of long term ecological needs against immediate economic and social requirements. Stone (2000) has immensely focussed upon this challenge.

3 Results and Discussion

The efficiency and sustainability of any ground water management technique depends solely on its suitability to the existing situation and awareness among the masses. This is quite evident in context to Cherrapunji and Rajasthan in India. Cherrapunji, located in East Khasi Hills receives an annual rainfall of approximately 12,000 mm which is next only to nearby Mawsynram. However, the residents there face acute shortage of fresh water. It has been predicted that the roots of this water scarcity lie in the ongoing rapid deforestation. Since dense vegetations slow down the surface runoff and promote water percolation through the soil, deforestation has slowed down the rates of water recharge. Moreover, it has led to an ecological imbalance and an adverse impact on hydrological cycle. On the other hand, Rajasthan, despite of receiving very low annual rainfall, has no such problem of water scarcity. The reason has been the efficiency of watershed management in various parts of Rajasthan. At Bundi in Rajasthan from 1995 to 2004, the average rise in ground water and irrigated area were 5.7 m and 136 ha, respectively, as per Wani et al. (2009). An increase in annual rainfall has also been observed in almost every part of Rajasthan. However, the people started over-abstraction of ground water. This indicates that the success of any water management strategy depends on the awareness among the residents.

As discussed by Wani et al. (2009), watersheds have also been a great success at other places like Lalatora, Kothapally, and Rajasamadhiyala in India. Moreover the use of rain water harvesting structures and well-planned drainage systems with commendable success has been found in the excavations of many ancient civilizations.

Today the concept of social pumping or water markets (as mentioned by Manjunatha et al. 2011) may be adopted. This can prevent over-abstraction of ground water along with maintaining the quality of potable water. Being economically favorable over individual ones, social pumping is suited for most of the regions of the world.

Integrated urban water management (IUWM) has been an overall success and a good example of how a large metropolitan area with moderate income disparity can adequately operate and maintain quality water supply to its 3.3 million citizens. It is the practice of managing fresh water, waste water and storm water as components of a basin-wide management strategy. Sound urban water management is carried out by a set of technically strong institutions with financial independence without any political intervention.

The concept of national water reserve has been a grand success in South Africa. This represents the quantity of water resources needed to satisfy basic human needs and protect aquatic ecosystems. The concept ensures the integration of demand and sustainability aspects of hydrologic system.

However, the reclamation of waste water has been a complete failure in South Africa as compared to that in Israel. At present South Africa reclaims about 3 % of its waste water compared with Israel's 84 % reclamation of waste water. High reclamation rate lowers down the abstraction rate of ground water along with quality control of potable water.

The "Dublin Principles" established by the International Conference on Water and the Environment in January 1992 recognized that water resources are finite and vulnerable and that sustainability should be a management objective. Sustainable water use supports the human ability to endure and flourish in future without undermining the integrity of the hydrological cycle of the ecological systems.

4 Conclusion

With rapid growth of population, the demand for ground water has soared high. High rates of urbanization and industrialisation have resulted into over-abstraction of ground water. This has led to fall in water tables in many regions of the world. The quality of ground water has also been degraded by the rapidly flourishing industries and changing lifestyles. Degradation of ground water has affected the environment, health of individuals and economy of nations. The Earth's life support system is turning fragile.

Afforestation, water harvesting and minimum use of chemicals are among the efficient methods to curb the menace. The notion of social pumping is an economical way to prevent the qualitative and quantitative degradation of ground water.

There is an urgent need for community monitoring and bringing up the best suited water management strategies. Access to ground water can be a major engine for food security, poverty alleviation, and economic development. The effective management and utilization of ground water not only for consumption, but also as a source of surface water flows, wetlands, and wildlife habitats calls for an increased attention. Therefore, the development activities should aim at achieving a sustainable utilization of ground water resources. Sustainable ground water development can bring major benefits such as food security, safe drinking water, high levels of production, development of quality of human resources, and an economic boom.

Acknowledgements We sincerely acknowledge the help of residents of village Malawan in India for their valuable help in estimation of benefits of social pumping over individual ones. We also gratefully acknowledge the residents of Cherrapunji in India for the valuable information shared by them about the present deforestation and climate changes. Words are inadequate in offering our thanks to all those people who encouraged us and wished for the successful completion of the work.

References

Foster SSD (2001) The interdependence of groundwater and urbanisation in rapidly developing cities. J Urban Water 3:185–192

Manjunatha AV, Speelman S, Chandrakanth MG, Huylenbroeck GV (2011) Impact of groundwater markets in peninsular India on water use efficiency: a data envelopment analysis approach. J Environ Manag 92(11):2924–2929

Stone AW (2000) Education: ground water protection and management strategy in developing countries. In: International association of hydrogeologists XXX congress

Stone AW (2001) Technology: a bridge to the middle ground of ground water sustainability. In: Groundwater foundation annual conference

Stone AW (2005) Science as the basis for local and regional water policy. In: Biennial ground water conference

Wani SP, Sudi R, Pathak P (2009) Sustainable groundwater development through integrated watershed management for food security. International Crops Research Institute for the Semi Arid Tropics

Water Supply System Planning by Artificial Groundwater Recharge from Rooftop Rainwater Harvesting

Sirajul Islam and Bipul Talukdar

Abstract Rainwater harvesting could be an effective means of supplementing the growing demand of potable water in urban areas in India, especially in high rainfall areas such as the north-eastern part of the country. Although rainwater harvesting has been practiced in India since a long time, but it has never been seriously thought of as a component for integrated management of water resources for urban water supply. Among the various rainwater harvesting techniques, artificial recharge of groundwater from rooftop rainwater harvesting is the most effective means of collecting a sizeable quantity of the rainwater for future use. As per the estimation of the Central Ground Water Board (2000), about 0.008 m^3 of rainwater can be obtained from the rooftops of buildings per square metre area for 1 cm of rainfall. A mathematical model for simulation of the three-dimensional transient groundwater flow process is developed for a confined aquifer using finite difference method. The model results are validated by comparing with the results obtained from MODFLOW simulation. This model is then applied in a hypothetical urban water supply system to provide drinking water to a population of about 75,000 considering pumping from a set of discharge wells along with a set of recharge wells for rainwater harvesting. The recharge to the aquifer takes place from eight injection wells within the study area; each collecting rainwater from the roof tops of a network of buildings located within 200 m radius. The recharge to the aquifer is considered to be lumped for each well in quarterly time steps. The recharge wells are also used for pumping water to the supply system. The operation policy of the wells are designed in such a way that the recharge wells shall be used for pumping water to the supply system only for 6 months during the period from October to March. This will permit the recharge water from monsoon storm to settle down in the aquifer and will lead to substantial water quality improvement. However, the discharge wells shall be operated throughout the year. An optimization problem is

S. Islam (✉) · B. Talukdar
Civil Engineering Department, Assam Engineering College,
Gauhati University, Guwahati, Assam, India
e-mail: sirajulislam09@gmail.com

B. Talukdar
e-mail: bipulaec@gmail.com

A.K. Sarma et al. (eds.), *Urban Hydrology, Watershed Management and Socio-Economic Aspects*, Water Science and Technology Library 73,
DOI 10.1007/978-3-319-40195-9_12

formulated with the objective of minimizing the squared difference of supply and demand. This optimization model is externally linked to the groundwater flow simulation model to observe the tradeoff between withdrawal from aquifer and recharge from rainwater. The constraints to the optimization model are related to groundwater depletion and temporal and spatial operation of the two sets of wells. The optimization problem turns out to be nonlinear due to incorporation of the simulation constraints. A simple genetic algorithm is used to solve the problem and the results are obtained in the form of head distribution and drawdown after every successive time steps.

Keywords Groundwater flow · Rainwater harvesting · Artificial recharge · Genetic algorithm

1 Introduction

Migration of rural population to urban areas for economic and social reasons has led to rapid growth of urban population in India. The trend of urbanization has thrown up a serious challenge to the civic authorities to provide potable water to the urban mass both qualitatively and quantitatively. Nonavailability or distant location of perennial source of water, deterioration of aesthetic quality of water due to over treatment are some of the problems encountered while planning an urban water supply system with surface water sources. On the other hand, indiscriminate use of groundwater may lead to irreversible depletion of the aquifer system. Lack of a proper pricing policy has given freedom to the end users to misuse the precious water. Hence, if groundwater is the means of community water supply, it is necessary to plan the water supply system in an integrated manner considering recharge to the aquifer system. Natural recharge is a very slow process which is a function of the infiltration capacity of soil. Artificial recharge may therefore be considered as the best alternative to augment the depleting aquifer.

Rainwater harvesting could be a very effective way of augmenting the water table in an aquifer in areas with high rainfall such as the north-eastern Region of India. It has been shown that up to 40 % of onsite rainfall can be recharged back to the aquifer using appropriate techniques (Stout et al. 2015). Among the various rainwater harvesting techniques, collecting rainwater from the rooftops of buildings could be the best option in urban areas due to the presence of large structures such as commercial buildings, institutions and multi-family dwellings, etc. On an average, about 0.008 m^3 of rainwater is available for recharge from a rooftop area of 1 m^2 for 1 cm of rainfall (Central Groundwater Board 2000). Aquifer recharge from rainwater not only augments the water level in the aquifer but also improves the groundwater quality as rainwater is free of any impurity and hardness. Moreover, the problem of clogging of injection well is addressed to certain extent as they deal with a pure form of water (Central Groundwater Board 2007).

This study aims at developing a simulation-optimization model for integrated management of ground water with artificial recharge from rooftop rainwater harvesting for an urban water supply system. Model results are presented to show how rainwater harvesting could improve the sustainability of a medium scale water supply scheme.

2 Materials and Method

While making a long-term plan for extracting groundwater for any water supply scheme, it is very important to understand the aquifer characteristics, the hydraulic properties of the porous media, and the limits on drawdown and recharge characteristics.

A three-dimensional groundwater flow simulation model is developed for a confined aquifer using finite difference method. This model is applied to hypothetical urban water supply scheme to supply potable water to a population of about 75,000. A set of eight discharge wells are defined at specific locations. Another set of eight recharge wells are used to collect rainwater from rooftops of buildings located within a radius of 200 m. An optimization model is formulated with the objective of minimizing the squared difference of supply and demand. The groundwater flow model is externally linked to the optimization model to incorporate the simulation constraints viz., the limits on drawdown at each well. Discharge wells are operated throughout the year for supplying water to the distribution network, while the recharge wells are used only for pumping in water to the aquifer during the period from April to September assuming that most of the rainfall occurs during this period. These wells are also used for extracting water during the period from October to March. Such an operation policy of the wells permits the rainwater collected during rainy season to settle down in the aquifer and blend with ground water, resulting in water quality improvement. Further, the cost-effectiveness is improved by utilizing these wells for extracting water (which otherwise would remain idle) during the dry season. A simple binary-coded genetic algorithm is used to solve this nonlinear optimization problem in four quarterly time steps.

2.1 Ground Water Flow Simulation Model

The major components of a ground water flow model are the partial differential equation governing the flow process, the initial conditions and boundary conditions. The solution of such a model refers to the calculation of head values at each point of the system (Anderson and Woessner 1992).

2.1.1 Ground Water Flow Equation

The three-dimensional transient ground water flow through a porous medium can be represented by the following partial differential equation (McDonald and Harbaugh 1988).

$$\frac{\partial}{\partial x}\left(K_{xx}\frac{\partial h}{\partial x}\right)+\frac{\partial}{\partial y}\left(K_{yy}\frac{\partial h}{\partial y}\right)+\frac{\partial}{\partial z}\left(K_{zz}\frac{\partial h}{\partial z}\right)-W=S_S\frac{\partial h}{\partial t} \quad (1)$$

where h is the hydraulic head (L); K_{xx}, K_{yy} and K_{zz} are values of hydraulic conductivity along the x, y, and z coordinate axes, which are assumed to be parallel to the major axes of hydraulic conductivity (L/T); W is a volumetric flow rate per unit volume and represents sources and/or sinks of water (T^{-1}); S_S is the specific storage of the porous medium (L^{-1}); and t is time (T).

2.1.2 Finite Difference Discretization

The finite difference discretization of the partial differential equation representing ground water flow converts the continuous variables into discrete variables that are defined at the grid points. Finite difference grids can be either block centred, where the nodes are located in the centre of each grid cell or mesh centred, where the nodes are located at the intersection of grid cells (Wang and Anderson 1982). In this study, block-centred grids as illustrated in Fig. 1 are used. ΔX, ΔY and ΔZ represent the size of grids in x, y and z directions, respectively.

The discretization of the ground water flow equation results in a set of simultaneous linear equations. The finite difference approximation of cell (i, j, k) can be represented by Eq. 2.

$$\begin{aligned}&C1h^t_{i-1,j,k}+C2h^t_{i,j-1,k}+C3h^t_{i,j,k-1}+C4h^t_{i+1,j,k}+C5h_{i,j+1,k}+C6h_{i,j,k+1}+C7h_{i,j,k}+Q_{i,j,k}\\&\quad=S_{i,j,k}(\Delta X\Delta Y\Delta Z)\frac{h^t_{i,j,k}-h^{t-1}_{i,j,k}}{\Delta t}\end{aligned} \quad (2)$$

The space derivatives are discretized using central difference scheme, while the time derivatives are discretized using backward difference scheme.

2.1.3 Initial and Boundary Conditions

While simulating the ground water flow equation, the initial and boundary conditions representing an aquifer system must be specified. The initial conditions are specified in the form of hydraulic head distribution for the aquifer domain at the initial time.

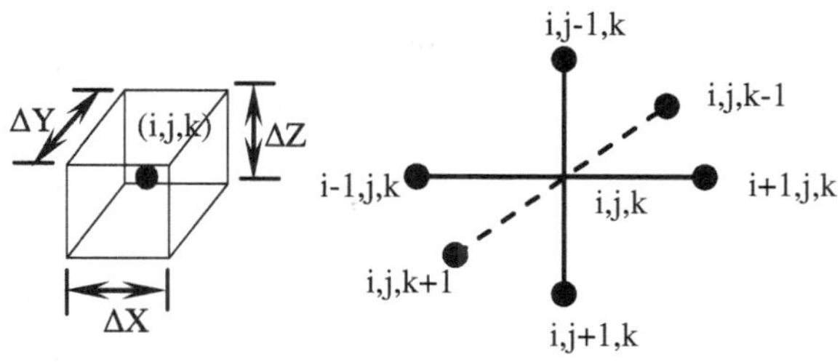

Fig. 1 *Block* centred grid system with associated indices

Mathematically,

$$h(x, y, z) = \Phi(x, y) \tag{3}$$

where Φ is a known function.

Boundary conditions refer to the physical features that act as hydrologic boundaries in an actual groundwater system. Generally, three types of boundary conditions are encountered in an aquifer system (Reilly 2001).

(a) Specified head (Dirichlet) boundary conditions mathematically represented as:

$$h(x, y, z) = \text{constant} \tag{4}$$

(b) Specified flow (Neumann) conditions mathematically represented as:

$$\frac{\mathrm{d}h(x, y, z, t)}{\mathrm{d}} n = \text{Constant} \tag{5}$$

(c) Head-dependent flow (Cauchy) conditions mathematically represented as

$$\frac{\mathrm{d}}{h} \mathrm{d}n + \mathrm{ch} = \text{Constant} \tag{6}$$

The first two types of boundary conditions are applied in this model in terms of constant head and no flow boundaries. In finite difference approach, the aquifer domain is discretized into a system of rectangular grids. So, the aquifer boundaries are supposed to be along a straight line. But in most real life situations, the boundaries are of irregular nature, necessitating their proper representation in the mathematical model. This model represents such irregular boundaries as shown in Fig. 2.

2.1.4 Study Area and Aquifer Parameters

A hypothetical aquifer system is defined for simulating the ground water flow process which shall be used for optimal management of an urban water supply

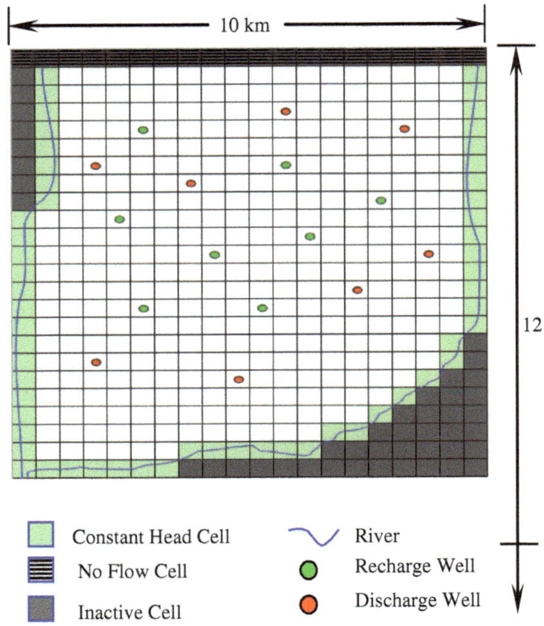

Fig. 2 Schematic diagram of hypothetical study area showing boundary conditions and well locations

system. Figure 2 shows the discretized study area, aquifer boundaries and pumping locations. The aquifer area of 10 km × 12 km is discretized into rectangular grids of 500 × 500 m size.

The eastern, western and southern boundaries of the aquifer are designated as constant head boundaries (river boundary) while the northern side is considered to be bounded by a hydraulic divide line representing no flow conditions across it. The hydraulic properties such as Horizontal hydraulic conductivity, storage coefficient are specified for each cell. However, vertical hydraulic conductivity is not taken into account as this model considers only a single confined layer of the aquifer. Also for simplicity, the aquifer is assumed to be homogeneous and isotropic in nature. Pumping locations are shown in Fig. 2. These locations are chosen in such a way that the radii of influence of any two wells do not intersect. Minimum distance of 500 m is maintained between two adjacent wells. The hydraulic gradient is considered to be in north–south and east–west directions. The head values at constant head boundaries are assumed to decrease linearly at a rate of 0.1 m. The constant head at the western boundary starts at 75 m and ends at 72.8 m. Similarly, a head value of 76.2 m is assumed at the starting point of eastern boundary, which extends through the southern part of the aquifer and merges with the end point of western boundary.

2.1.5 Method of Solution

The discretization of the flow equation results in a set of simultaneous linear equations of the form represented by Eq. 2. There are various methods for solving simultaneous linear equations. A computer code is written in MATLAB which solves this set of linear equations using matrix inversion method. This program returns the head values at different grid points as a function of discharge rates at different pumping locations and resulting draw down at these locations. MATLAB uses backlash operator to solve simultaneous equations.

$$[A]\{h\} = [B] \tag{7}$$

$$h = B/A$$

where A is the coefficient of flow equation, B is the constant on right-hand side and h is the peizometric head.

2.1.6 Model Validation

There are several numerical models for simulating the three-dimensional ground water flow process. In this study, MODFLOW, developed by the U.S. Geological Survey (McDonald and Harbaugh 1988), is used. MODFLOW is a complete ground water flow and transport model that uses block centred finite difference grid system. It consists of a main program and sub-routines called modules. The original MODFLOW program has a difficult interface, due to which different user friendly Graphical User Interfaces have been developed by software companies. In this study, PMWIN (Processing MODFLOW for WINDOWS), developed by Chiang and Kinzelbach (1995) is used for simulation. Identical flow parameters and discharge rates as presented in Table 1 and boundary conditions as discussed in Sect. 2.1.4 are used for simulation of both MODFLOW and this model. The results of simulation are shown in Fig. 3 as comparative plots of contours of head values (m) after each time step of simulation. The contours do not seem to deviate even after three continuous stress periods with high discharge rates. The numerical values differ in the order of 10^{-3} in decimal fractions.

2.2 Rainwater Harvesting Model

This study considers rooftop rainwater harvesting scheme for groundwater recharge in a confined aquifer through injection wells which would also be used for groundwater extraction. It consists of a hypothetical system of eight rainwater harvesting stations each collects rainwater from the rooftops of a network of buildings. These buildings include govenment establishments, institutions,

Table 1 Flow parameters and pumping rates for comparative simulation of flow model and MODFLOW

Flow parameter	Value	Discharge rates (m^3/day)				
		Well location	Steady state	t = 1	t = 2	t = 3
Horizontal hydraulic conductivity						
Saturated thickness of aquifer		(8,13)	–	6500	1500	4500
Storage coefficient	0.0001 m/s	(12,5)	–	2000	5000	3000
Time period		(17,11)	–	3000	4500	1000
	40 m					
	0.0001					
	30 days					

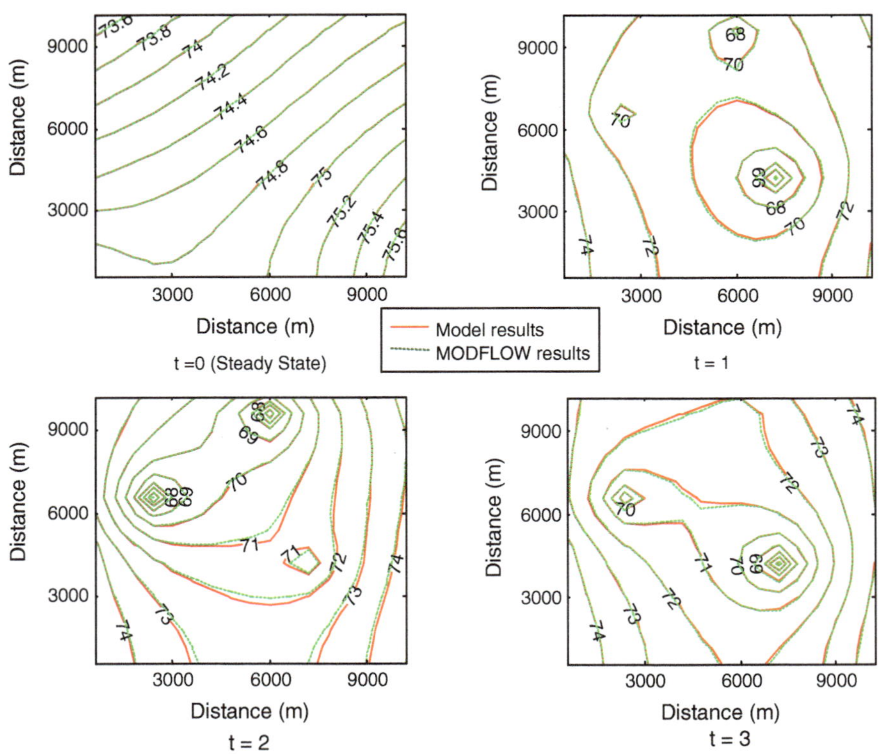

Fig. 3 Comparative contours of head values obtained from simulation of flow model and MODFLOW

Table 2 Rooftop area and quantity of rainwater collected at different recharge stations

Recharge station	no. of structures	Total roof area (m^2)	Quantity of rainwater collected (m^3)			
			Apr–June	July–Sept	Oct–Dec	Jan–Mar
RWS1	6	2119	1830.60	2237.40	305.10	339.00
RWS2	5	2073	1790.64	2188.56	298.44	331.60
RSW3	4	1543	1332.72	1628.88	222.12	246.80
RSW4	6	2713	2343.60	2864.40	390.60	434.00
RSW5	5	2102	1816.56	2220.24	302.76	336.40
RSW6	4	1079	932.04	1139.16	155.34	172.60
RSW7	5	1933	1669.68	2040.72	278.28	309.20
RSW8	6	2254	1947.24	2379.96	324.54	360.60

commercial buildings or multi-family dwellings. A set of data for each station showing rooftop areas and quantity of rainwater collected, is presented in Table 2 depicting a real life situation. The rainfall data is considered in four time steps of three months each viz. Apr–June ($t = 1$), July–Sept ($t = 2$), Oct–Dec ($t = 3$) and Jan–Mar ($t = 4$) with cumulative rainfall of 108, 132, 18 and 20 cm, respectively. For convenience in simulation, each of these time periods is taken as 90 days.

2.3 *Optimization Model*

The basic objective of this study is to provide drinking water facilities to a population of about 75,000 living in an urban area. The per capita water supply norm in India as specified by Central Public Health and Environmental Engineering Organization for towns with piped water supply without sewerage system is 70 L/day (Central Pollution Control Board 2010). As such a water demand of 5250 m^3 is the target to be fulfilled per day.

2.3.1 Objective Function

The objective function is formulated as the squared difference of supply and demand.

$$\text{Minimize:} f(x) = \left(Q_{\text{d}}^{t} - \sum_{n=1}^{N} Q_{\text{p}_n}^{t} \right)^2 \tag{8}$$

where Q_{d} is the water demand (m^3), Q_{p} is the quantity of water to be pumped (m^3), n is the pumping location index, N the number of wells and t is the time index.

The decision variables in this optimization problem are the discharge rates of different pumping wells which may be expressed as $\boldsymbol{q}_{\mathbf{p}}$ (m^3/day)

$$F(x) = \frac{1}{1 + P(x)}.$$

2.3.2 Constraints

The groundwater flow model described in Sect. 2.1 is externally linked to the optimization model. An upper limit is specified for the drawdown at different pumping locations at any instance of pumping. This constraint function can be expressed as

$$g(x) = S_{\mathrm{s}} - S_n^t \geq 0 \tag{9}$$

where s_n^t is the drawdown at pumping location n at any time period t and s_{s} is specified value of drawdown (m). Apart from the drawdown constraint, upper and lower limits are fixed for the decision variables, i.e. the discharge rates of different pumping wells. This may be expressed as:

$$q^u \geq q_{\mathrm{p}} \geq q^l \tag{10}$$

where q^u is specified upper limit on discharge (m^3/day), q^l is the specified lower limit on discharge (m^3/day).

Due to incorporation of the simulation constraints, the optimization problem turns out be nonlinear in nature. A binary-coded genetic algorithm (GA) is used to solve this nonlinear minimization problem. The details of GA can be found in Goldberg (1989). Due to their evolutionary nature based on survival of fittest principle, GAs are naturally suitable for solving maximization problems. However, minimization problems can also be efficiently solved with appropriate conversions. The following conversions are made in this problem.

First, the constrained minimization problem is converted to an unconstrained problem using penalty parameter approach as given below:

$$p(x) = f(x) + \sum_{n=1}^{N} R_n \langle g_{n(x)} \rangle^2 \tag{11}$$

where $P(x)$ is the penalty function, $g(x)$ is the constraint function and R is the penalty coefficient.

This unconstrained minimization problem is then converted to an equivalent maximization problem by introduction of fitness function.

3 Results and Discussion

The optimization model is externally linked to the groundwater flow simulation model. The decision variables are the discharge rates of the wells as specified in Fig. 2. During first two time steps, i.e. during the periods April–June and July–September, only eight discharge wells are used for pumping, while the recharge wells are used only for pumping in rainwater to the aquifer. But during the other two time steps, the recharge wells are also used for pumping, thus constituting 16 decision variables altogether.

The optimization results (Scenario-I) are presented in Fig. 4 as the contours of drawdown after each time step. An upper limit of 10 m is fixed for drawdown after every time step. Table 3 shows the discharge rates of wells. The upper bound on discharge rate is set at 1000 m^3/day.

With same limits on drawdown and discharge rates, optimization is carried out in the same manner as above, but without considering recharge from rainwater harvesting (Scenario-II). Results show a different trend with drawdown level reaching the maximum limit and shortage in the supply level during the last time step (t = 4) (Fig. 5, Table 4).

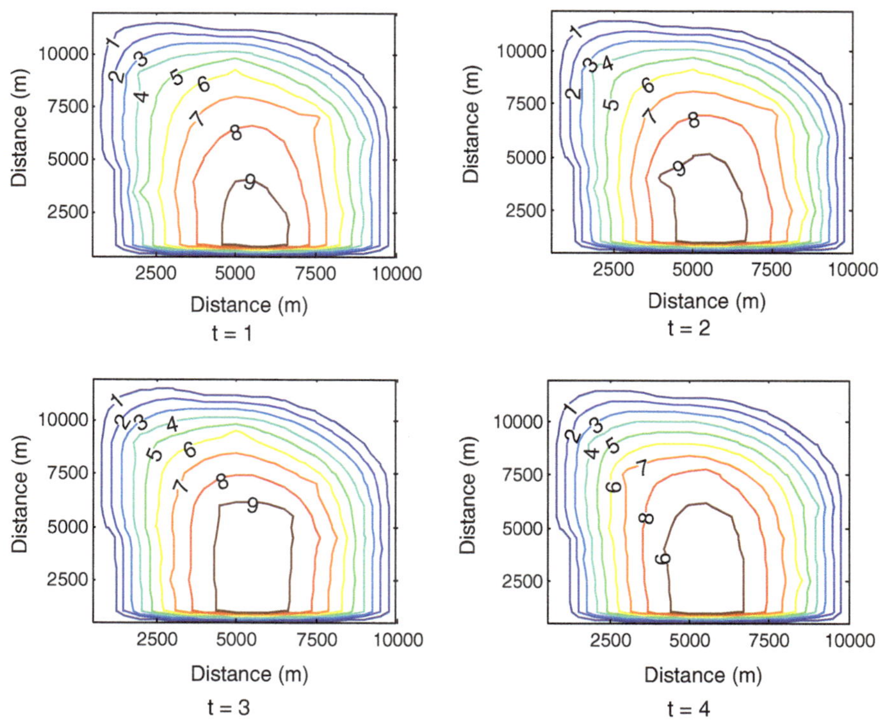

Fig. 4 Post-optimization contours of drawdown during different time steps (Scenario-I)

Table 3 Discharge rates of different pumping wells obtained from optimization using Genetic Algorithm during different time steps (Scenario-I)

Cell no.	Discharge rate (m^3/day)			
	t = 1	t = 2	t = 3	t = 4
(5,6)	0	0	357.25	153.43
(7,12)	0	0	54.64	135.08
(10,5)	0	0	728.32	245.23
(9,16)	0	0	252.75	310.53
(11,13)	0	0	538.11	210.65
12,9)	0	0	256.48	66.53
15,6)	0	0	202.83	758.29
15,11)	0	0	46.66	612.09
(4,12)	622.21	123.52	102.87	94.05
(5,17)	429.53	891.97	715.28	548.98
(7,4)	973.95	552.85	48.98	344.75
(8,8)	288.55	579.13	67.55	162.24
(12,18)	603.84	711.31	243.23	626.89
(14,15)	944.85	895.81	460.73	253.56
(18,4)	614.65	912.26	830.20	692.73
(19,10)	841.64	583.40	844.12	34.97

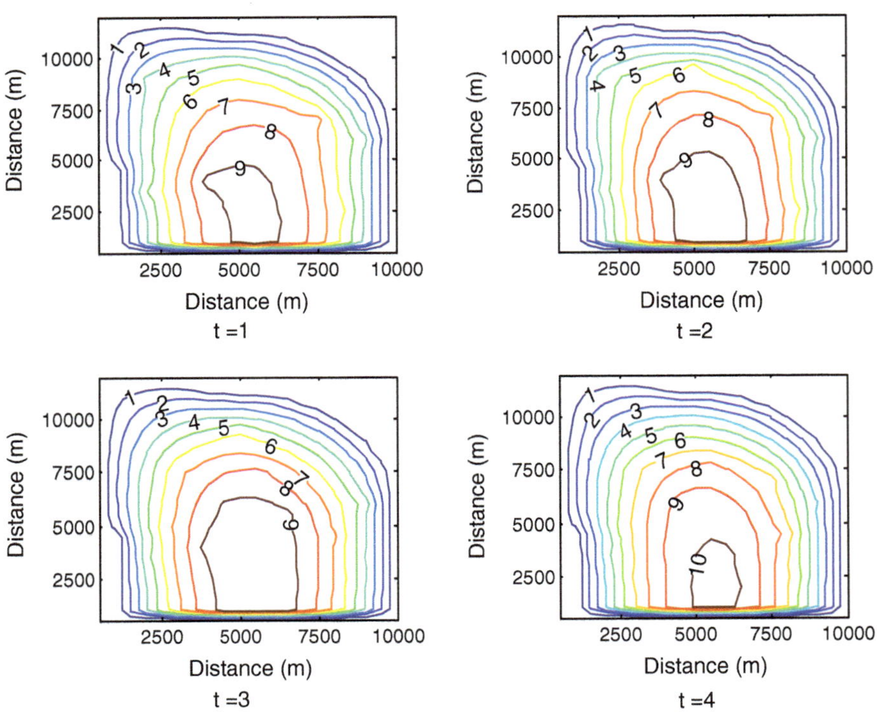

Fig. 5 Post-optimization contours of drawdown during different time steps (Scenario-II)

Table 4 Discharge rates of different pumping wells obtained from optimization using Genetic Algorithm during different time steps (Scenario-II)

Cell no.	Discharge rate (m^3/day)			
	$i = 1$	$t = 2$	$t = 3$	$t = 4$
(5,6)	0	0	27.70	168.93
(7,12)	0	0	55.12	749.57
(10,5)	0	0	369.33	99.90
(9,16)	0	0	279.26	367.20
(11,13)	0	0	388.50	596.78
12,9)	0	0	215.14	209.75
15,6)	0	0	403.32	327.82
15,11)	0	0	377.27	34.76
(4,12)	153.02	279.35	234.62	253.33
(5,17)	499.03	521.37	226.05	388.69
(7,4)	776.92	494.61	513.70	400.75
(8,8)	906.69	625.93	598.00	489.92
(12,18)	749.67	779.03	873.95	182.46
(14,15)	975.43	748.55	31.86	95.71
(18,4)	569.27	900.57	72.50	629.73
(19,10)	619.96	900.60	583.68	254.61

Table 5 Parameters of Genetic Algorithm used in optimization

Sl. no.	G.A. parameter	Specification
1	Population size	100
2	Selection operator	Binary tournament selection
3	Cross over probability	0.75
4	Mutation probability	0.005
5	No. of generations	40

Using the GA parameters presented in Table 5, the GA is run for 40 generations in each time step for both Scenario-I and II. Generation versus Fitness curves are drawn for both Scenario-I and II as shown in Fig. 6a, b.

These curves show different trends. In scenario-I, the maximum fitness is reached within 20 generation except for time period, $t = 4$, where the same is obtained after 35 generations. Contrary to this, scenario-II needs more number of generations to reach near the maximum fitness level. Due to imposition of high penalty for violating drawdown constraints ($R = 1000$), constraints are not violated in any case as evident from contours of drawdown (Figs. 4 and 5). During Time step, $t = 4$ of scenario-II, the drawdown level reaches the maximum level of 10 m and maximum fitness in this case is less than the desired value indicating shortage in water supply level.

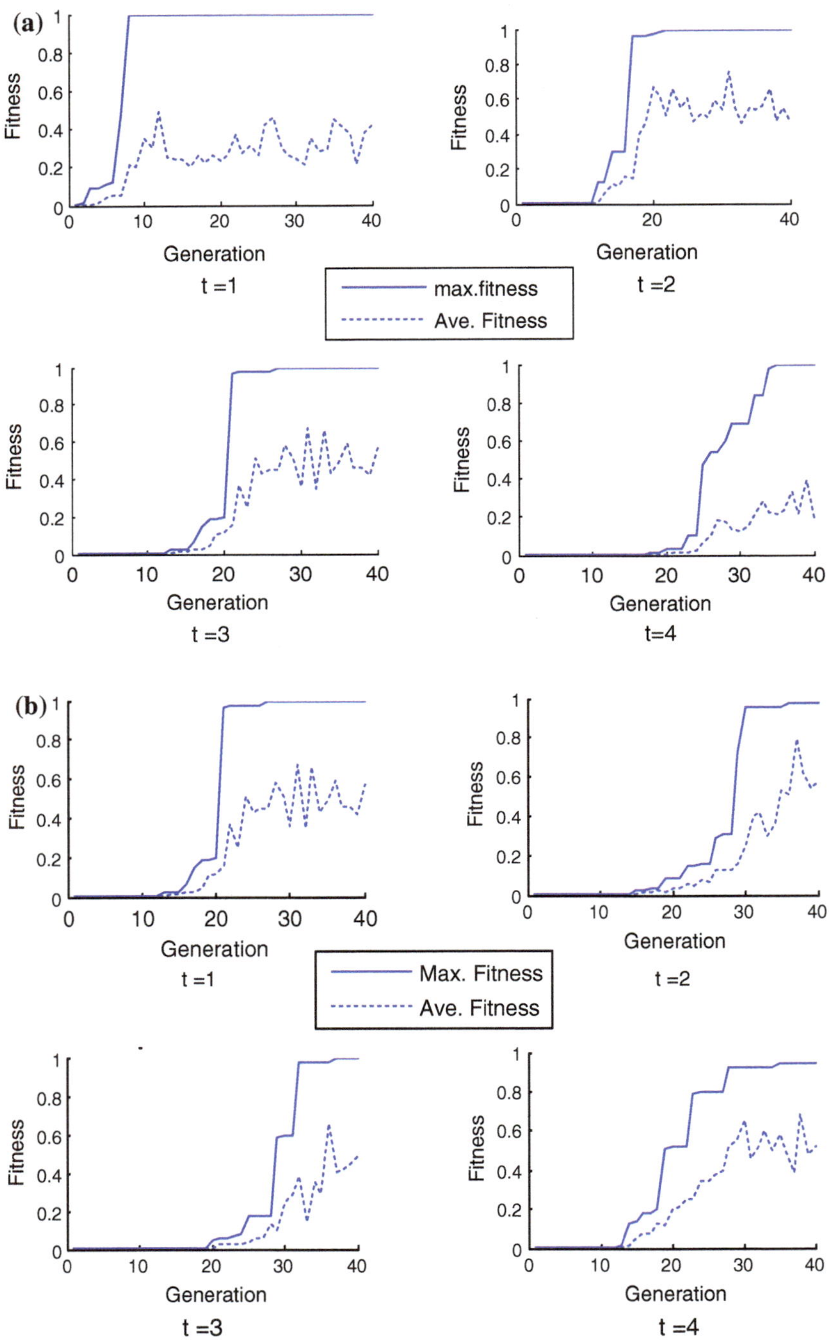

Fig. 6 **a** Fitness versus generation curves (Scenario-I). **b** Fitness versus generation curves (scenario-II)

4 Summary and Conclusion

A groundwater flow simulation model is developed for a confined aquifer using finite difference method considering artificial recharge from rainwater harvesting. This model is applied to a hypothetical urban water supply system to provide drinking water to a population of about 75,000. Optimization is done with the objective of minimizing the difference between supply and demand of water in quarterly time steps. Optimization is carried out under two different scenarios, first with aquifer recharge from rainwater and then without considering the recharge. Different trends in groundwater response are observed under the two scenarios of optimization. Although the water requirements are fulfilled under scenario-I (with rainwater recharge) during all the time steps, but scenario-II (without recharge) shows shortage in supply level towards the end with the same limits on drawdown.

This study considers only eight recharge stations with limited number of roof structures for collecting rainwater. But with more number of recharge points and a good number of buildings connected through piped network, rainwater harvesting could be a very effective component of urban water supply system. The water quality issues can also be addressed to a great extent by rainwater recharge, if the same is properly planned.

References

Anderson M, Woessner W (1992) Applied groundwater modeling: simulation of flow and advective transport. Academic Press, Inc, San Diego, p 381

Central Groundwater Board (2000) Guide on artificial recharge to groundwater. Ministry of Water Resources, New Delhi, India

Central Groundwater Board (2007) Manual on artificial recharge of groundwater. Ministry of Water Resources, New Delhi, India

Central Pollution Control Board (2010) Status of water treatment plants in India. Status Report. Ministry of Environment and Forests, New Delhi, India

Chiang WH, Kinzelbach W (1995) Processing MODFLOW. Hamburg, Zurich

Goldberg DE (1989) Genetic Algorithms. In: Search, optimization and machine learning. Addison-Wesley, New York

McDonald MG, Harbaugh AW (1988) A modular three dimensional groundwater flow model (Chap. A1, Book 6). Techniques of Water Resources Investigations of the United States Geological Survey. Denver, Colorado

Reilly TE (2001) System and boundary conceptualization in groundwater flow simulation (Book 3, Chap. B8). Techniques of water-resources investigations of the United States geological survey

Stout DT, Walsh TC, Burian SJ (2015) Ecosystem services from rainwater harvesting in India. Urban Water J. doi:10.1080/1573062X.2015.1049280 (Tailor & Francis, Online)

Wang H, Anderson MP (1982) Introduction to groundwater modeling. Finite difference and finite element methods. Academic Press, Inc., San Diego, 237 pp

Part IV
Application of Computational and Numerical Models in Urban Hydrology

Two-Dimensional Numerical Model for Urban Drainage System

Sudarshan Patowary and Arup K. Sarma

Abstract Modeling of urban drainage system is carried out for understanding and predicting the flow behavior in the drain, so that an adequate design of the drainage system can be achieved. However, many a time urban drainage models are developed without considering some important aspects like erosion and deposition processes in the drain. Though the erosion process can be neglected in the lined canal, deposition process has a major role in the urban drainage system. Therefore, study of sediment deposition in the urban drainage system is an essential aspect for minimization of artificial flood in the urban area. Sediment deposition in the drain depends upon various factors like particle size, settling velocity, critical shear stress, and bed shear stress. Therefore, in this paper an attempt has been made to develop a numerical model, which can identify the critical section of the urban drain where sediment starts to deposit. Two-dimensional continuity and momentum equations of unsteady free surface flow are solved by second order finite difference implicit scheme. To find the settling velocity of the particle Newton's law and Stock's law are used. After calculating the settling velocity of the particle, the position where the particle starts to settle down is determined. Critical sections of the drain from sediment deposition point of view are identified by comparing bed shear stress and critical shear stress computed by the Shields' equation for incipient motion. No deposition occurs if bed shear stress is greater than critical shear stress. Again, result obtained from the sensitivity analysis of average particles size has shown that with the increases of sediment size, problem of sediment deposition aggravates.

Keywords Settling velocity · Critical section · Unsteady flow · Critical shear stress

---

S. Patowary (✉) · A.K. Sarma
Department of Civil Engineering, Indian Institute of Technology Guwahati, Guwahati, India
e-mail: s.patowary@iitg.ernet.in

A.K. Sarma
e-mail: aks@iitg.ernet.in

A.K. Sarma et al. (eds.), *Urban Hydrology, Watershed Management and Socio-Economic Aspects*, Water Science and Technology Library 73,
DOI 10.1007/978-3-319-40195-9_13

1 Introduction

With the development of efficient numerical algorithm and the advent of high speed digital computer, solution of unsteady one-dimensional and two-dimensional hydrodynamic equation is becoming quite common for addressing various unsteady flow problem. Solution of unsteady flow in river is quite common. Though the basic equations for developing river model and urban drainage model are same but the river model cannot be applied directly to urban drainage model. In this paper, an attempt has been made to study the unsteady flow pattern considering some important aspect like sediment deposition in the urban drainage system. Keskin et al. (1997) developed a one-dimensional flood routing model using Saint-Venant equation. Hashemi et al. (2007) developed unsteady flow model and they had used differential quadrature method for simulation of the unsteady flow. Zhang and Shen (2007) developed steady and unsteady flow model for channel networks. Fang et al. (2008) developed one-dimensional numerical simulation of nonuniform sediment transport under unsteady flows using Saint-Venant equation. To minimize the artificial flood problem, critical section of the drain where the sediment starts to deposit is determined and if we can remove the sediment from that location, then it is possible to abatement the flood problem in urban area. Critical section of the drain depends upon various factors like particle size, settling velocity, critical shear stress, bed shear stress, and incipient motion of the particle. Sheilds (1936) had been the pioneer to establish a functional relationship between critical shear stress and boundary Reynolds number. Shields diagram has extensively used for determination of incipient conditions for sediment movement problems. Beheshti and Ataie-Ashtiani (2008) compared most widely used threshold curves presented to Shields diagram, a method based on the concept of probability of sediment movement and empirical methods based on movability number for predicting incipient motion. If the bed shear stress is greater than critical shear stress then incipient motion occurs. Many researchers developed some empirical formula to find the critical shear stress (Lane 1995; Highway Research Board 1970).

2 Mathematical Model

The proposed model has two components: (1) 2-D unsteady flow model for free surface flow computation, (2) Identify the critical section of the urban drain where sediment starts to deposit.

2.1 2-D Unsteady Flow Model

Two-dimensional continuity equation and fully dynamic form of the momentum equation in non-conservation form has been used as a governing equation for solution of unsteady flow in a hypothetical non-prismatic rectangular river reach considered in this study. The continuity and momentum equation are given in Eq. (1)

$$U_t + E_x + F_y + S = 0, \tag{1}$$

where

$$s_{f_y} = \frac{n^2 v\sqrt{u^2+v^2}}{1.49h^{1.33}} \quad E = \begin{pmatrix} uh \\ u^2h + \frac{1}{2}gh^2 \\ uvh \end{pmatrix} \quad F = \begin{pmatrix} vh \\ uvh \\ v^2h + \frac{1}{2}gh^2 \end{pmatrix}$$

$$S = \begin{pmatrix} 0 \\ -gh\left(S_{o_x} - S_{f_x}\right) \\ -gh\left(S_{o_y} - S_{f_y}\right) \end{pmatrix} \quad s_{f_x} = \frac{n^2 u\sqrt{u^2+v^2}}{1.49h^{1.33}} \quad s_{f_y} = \frac{n^2 v\sqrt{u^2+v^2}}{1.49h^{1.33}},$$

where h is the water depth, n is the manning's coefficient, u is velocity in the x direction, v is velocity in y direction, g is acceleration due to gravity, S_{o_x} and S_{o_y} are bed slope in x and y direction, s_{f_x} and s_{f_y} are friction slope in x and y direction.

2.2 Sediment Deposition Model

Sediment deposition in drain depends upon various factors like particle size, settling velocity, critical shear stress and bed shear stress. To find the settling velocity of the particle Newton's law and Stock's law are used.

2.2.1 Newton's Law for Settling Velocity

Newton's law of settling velocity is given by Eq. (2)

$$v_s = \sqrt{\frac{4g}{3c_D}(s_s - 1)d}, \tag{2}$$

where v_s is settling velocity of the particle, s_s is specific gravity of the particle, c_D is the drag coefficient and d is the diameter of the particle.

Again drag coefficient c_D is related to Reynolds number R by following observational relationships.

For R between 0.5 and 10^4

$$c_D = \frac{24}{R} + \frac{3}{\sqrt{R}} + 0.34$$

For high Reynolds number ($R > 10^3$–10^4) $c_D = 0.4$.

For low Reynolds number $c_D = \frac{24}{R}$.

2.2.2 Stoke's Law

If the diameter of the particle less than 0.1 mm, involving value of R less than 1, Stoke's law is applied for finding the settling velocity of the particle.

$$v_s = \frac{g}{18\upsilon}(s_s - 1)d^2, \tag{3}$$

where v_s is settling velocity of the particle, s_s is specific gravity of the particle, d is the diameter of the particle, υ kinematic viscosity of water in centistokes.

After calculating the settling velocity of the particle, critical section of the drain where particle starts to deposit is calculated and then at that particular section critical shear stress and bed shear stress is determined.

Shields (1936) established a functional relationship based on experimental data

$$\frac{\tau_c}{(\gamma_s - \gamma)d} = F\left(\frac{U_{*c}d}{\upsilon}\right), \tag{4}$$

where $U_{*c} = \left(\frac{\tau_c}{\rho}\right)^{1/2}$ is the critical friction velocity. The left hand side of Eq. (4) is the dimensionless critical shear stress and the right hand side is called critical boundary Reynolds number.

The Shields diagram contains the critical shear stress as an implicit variable that cannot be obtained directly. To overcome this difficulty, the ASCE Sediment Manual (1975) utilizes a third dimensionless parameter

$$\frac{d}{\upsilon}\left[0.1\left(\frac{\gamma_s}{\gamma} - 1\right)gd\right]^{1/2} \tag{5}$$

From the value of the third parameter, the value of the critical Shields stress is obtained at an intersection with the Shields curve from which critical shear stress τ_c can be calculated.

3 Numerical Formulation

The Saint-Venant equation is not amenable to analytical solution except for a few special cases.

They are partial differential equation that, in general, must be solved using numerical methods. To approximate Saint-Venant equation many numerical schemes have been developed. Beam and Warming scheme, which basically uses implicit finite difference method has been implemented for solving the governing equation. The central finite difference method is derivative.

The final expression using Beam and Warming finite difference scheme is (Bean and Warming 1976)

$$\left[I + \Delta t \frac{\theta}{1+\xi}\left(\frac{\partial}{\partial x}A^k + \frac{\partial}{\partial y}B^k + Q^k\right)\right]\Delta_t U^{k+1} = -\Delta t \frac{1}{1+\xi}\left(\frac{\partial E}{\partial x} + \frac{\partial F}{\partial y} + S\right)^k + \frac{1}{1+\xi}\Delta_t U^k \tag{6}$$

For Euler Implicit scheme $\theta = 1$ and $\xi = 0$.

The final expression after replacing $\theta = 1$ and $\xi = 0$ is as follows:

$$\left[I + \Delta t\left(\frac{\partial}{\partial x}A^k + \frac{\partial}{\partial y}B^k + Q^k\right)\right]\Delta_t U^{k+1} = -\Delta t\left(\frac{\partial E}{\partial x} + \frac{\partial F}{\partial y} + S\right)^k + \Delta_t U^k \tag{7}$$

4 Initial Conditions and Bathymetry of the River

A hypothetical drain of 2 km length and 2 m width was considered for simulation of the model. Total number of grids in longitudinal and transverse direction of the drain is 200 and 10 with Δx = 10 m and Δy = 0.2 m, respectively. The longitudinal slope of the bed is 0.0005. Figure 1 shows the initial condition and bathymetry of the drain.

5 Boundary Condition

Numerical form of the Saint-Venant equation is used in the interior grid points to compute the unsteady flow and velocity. At the boundaries, however, we cannot use this equation, since there is no grid point outside flow domain. Therefore, for

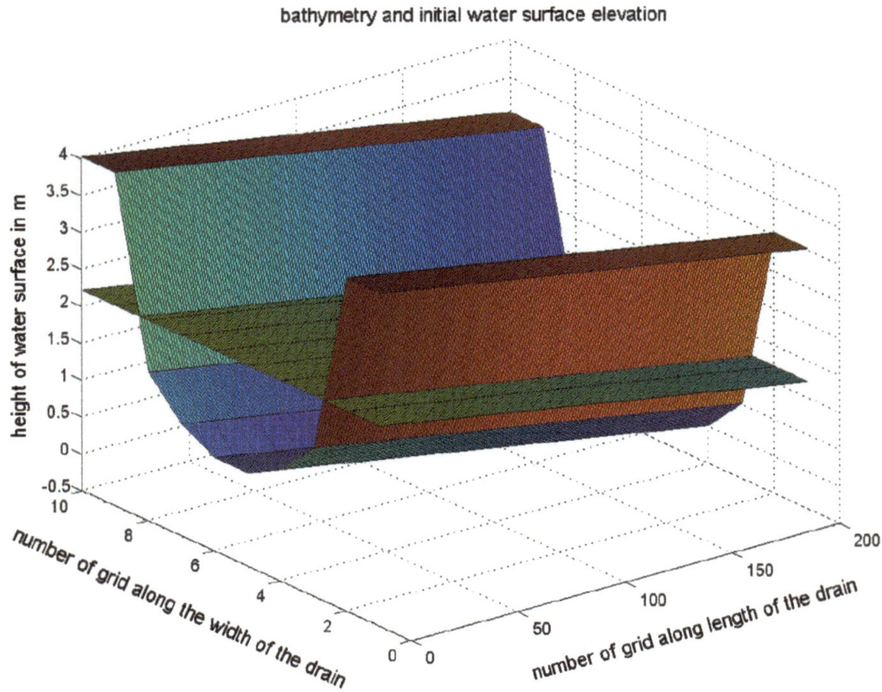

Fig. 1 Bathymetry and initial condition

two-dimensional unsteady flow, we need two boundary conditions. One is upstream boundary condition and the other is downstream boundary condition.

5.1 Upstream Boundary Condition

The discharge hydrograph as shown in Fig. 2 has been taken as the upstream boundary condition.

Fig. 2 Upstream boundary condition

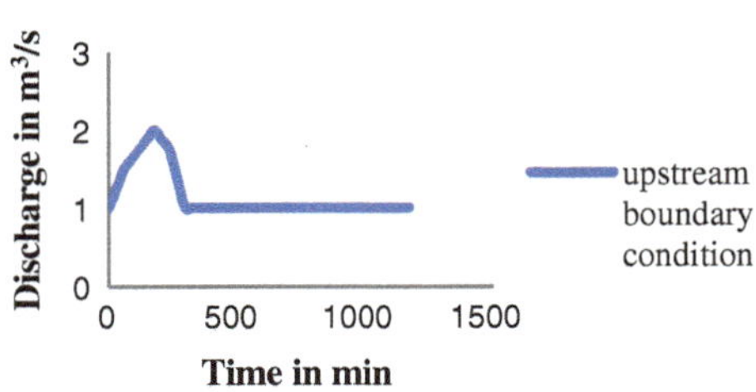

5.2 Downstream Boundary Condition

To calculate the flow parameter at the downstream boundary we have used two equations, positive characteristic equation [Eq. (8)], and Manning's equation [Eq. (9)]

$$u_{i,j-1}^{k} + 2c_{i,j-1}^{k} = u_{i,j}^{k+1} + 2c_{i,j}^{k+1} \tag{8}$$

For

$$\frac{dx}{dt} = u + c$$

$$u = \frac{1}{n} y^{\frac{2}{3}} s_f^{\frac{1}{2}} \tag{9}$$

Celerity c of shallow water wave is given by $c = \sqrt{gh}$.

Reflective boundary condition technique was applied as the solid boundary. For this dummy cells are created outside the bank line and the flow variables at those cells are made equal to the inner line to make the bank behave like a wall. The normal velocity component at the dummy cell is just opposite to the velocity at the inner line of bank. The values of the variables at the dummy cells for left bank are as follows:

$$u_{i,j}^{k} = u_{i+2,j}^{k}$$
$$h_{i,j}^{k} = h_{i+2,j}^{k}$$
$$v_{i,j}^{k} = -v_{i+2,j}^{k}$$

6 Result and Discussion

The elevation at different time, discharge hydrograph and the depth hydrograph at downstream section are presented in this section. The data used in the model are hypothetical. After running the unsteady flow model, the critical section of the sediment deposition is calculated.

6.1 Elevation

The elevation of water surface profile at different time step is shown in Fig. 3. From this figure it has been observed that with the increases of time the water surface

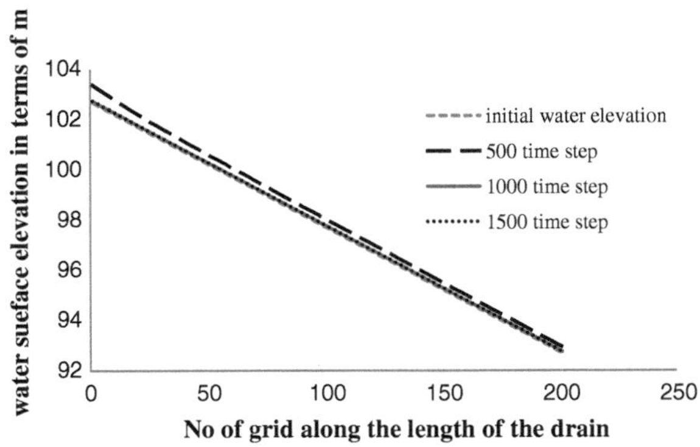

Fig. 3 Water surface elevation at different time step

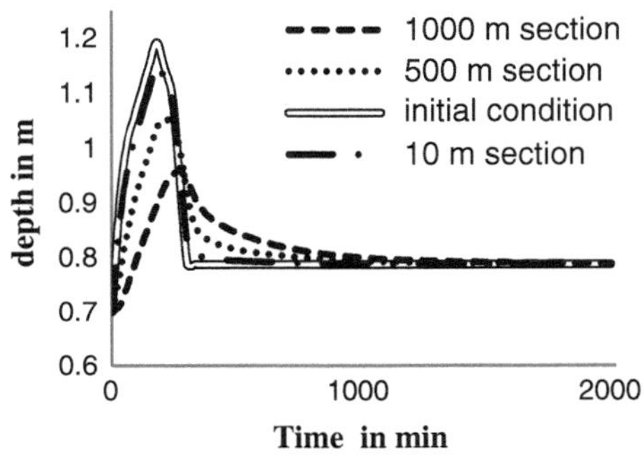

Fig. 4 Depth hydrograph at different section

elevation is increases up to a certain time step (500 time step). After that water surface elevation again starts to decreases with time.

6.2 Depth Hydrograph

The depth hydrograph at different section of the drain is shown in Fig. 4. Depth hydrograph is considered at section 10, 500, 1000, and 1500 m of the drain. From this result, it has been observed that as the distance of the observed section from upstream boundary of the drain increases, the peak of the depth hydrograph decreases and time required to attain peak increases.

6.3 Settling Velocity and Sediment Deposition

Figure 5 shows the bed shear stress at different time. From this figure it has been observed that bed shear stress increases with time up to a certain limit (198 min),

Fig. 5 Bed shear stress at different time

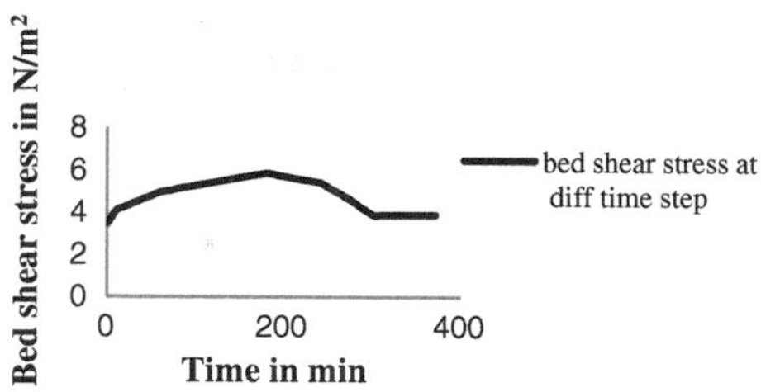

after it decreases and then it become constant. This implies that with the increases of water depth, bed shear stress increases.

6.4 *Critical Section*

Figure 6 represents the critical section of different particles at different time. Critical section of 0.0001, 0.0002, 0.001 m diameter particle are considered. From this figure, it has been seen that as the diameter of the particle increases the location of critical section from the upstream boundary decreases.

Table 1 shows different parameters of the particle like settling velocity, critical shear stress, critical section of drain, and bed shear stress. From this table it has been observed that with the increases of particle diameter settling velocity of the particle increases and larger sizes particles settle down at a nearest location from the upstream. If the bed shear stress of the particle is more than critical shear stress of the particle, incipient motion of the particle occurs. From the table below it has been observed that critical shear stress of 0.008 and 0.01 m particles are more than bed shear stress of the drain. Therefore, these particles start to deposit in their respective critical section of the drain. Smaller particles (<0.008 m) are washed away from the drain because critical shear stress of those particle are smaller than bed shear stress.

Fig. 6 Critical section of different particle with time

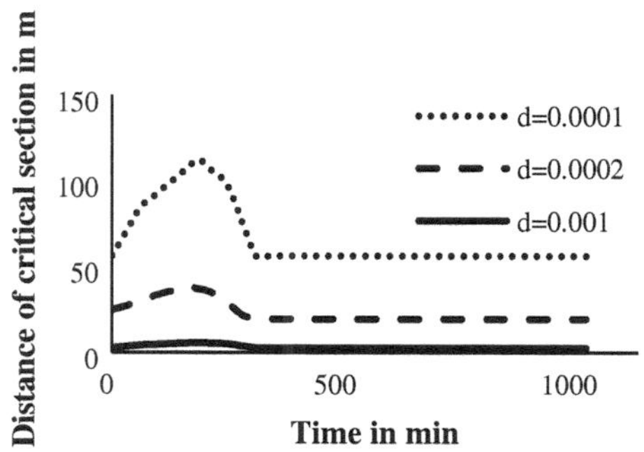

Table 1 Different parameters of the particle

Diameter of the particle (m)	Settling velocity (m/s)	Critical section of the drain (m)	Critical shear stress (N/m^2)	Bed shear stress (N/m^2)	Remarks
0.0001	0.0089	112–56	0.49	3.4–5.8	No sediment deposition
0.0002	0.0265	24–37	0.32	3.4–5.8	No sediment deposition
0.001	0.175	2.85–5.7	0.61	3.4–5.8	No sediment deposition
0.005	0.45	1.5–3.7	3.89	3.4–5.8	No sediment deposition
0.008	0.66	1.3–3	7.7	3.4–5.8	Sediment starts to deposition
0.01	0.75	1–2.5	9.7	3.4–5.8	Sediment starts to deposition

7 Conclusion

A two-dimensional numerical model considering the sediment deposition in the drain is developed. From the above results it has been observed that, particle size diameter has a great influence in determining the settling velocity of the particle and critical section of sediment deposition. In the proposed model particle sizes between 0.008 and 0.01 m are started to deposit in their respective critical location of the drain. Therefore, to minimize the artificial flood, it is necessary to remove the deposited sediment from the critical section of the drain.

References

Beam RM, Warming RF (1976) An implicit finite-difference algorithm for hyperbolic systems in conservation-law form. J Comp Phys 22:87–110

Beheshti AA, Ataie-Ashtiani B (2008) Analysis of threshold and incipient conditions for sediment movement. J Coastal Eng 55:423–430

Fang H, Chen M, Chen Q (2008) One-dimensional numerical simulation of non-uniform sediment transport under unsteady flows. J Sed Res 23:315–328

Hashemi MR, Abedini MJ, Malekzadeh P (2007) A differential quadrature analysis of unsteady open channel flow. J Appl Math Model 31:1594–1608

Highway Research Board (1970) Tentative design procedure for riprap-lined channels. National Academy of Science, National Cooperative Highway Research Program, Report 108

Keskin ME, Agiralioglu, Necati (1997) A simplified dynamic model for flood routing in rectangular channels. J Hydrol 202:302–314

Lane EW (1995) Design of stable channels. Trans ASCE 2776:1234–1279

Sheilds A (1936) Anwendung der ahnlichkeitsmechanik under turbulenzforschung auf geschiebebewegung. Mitteilungen der PrevssischenVersuchsanstalt fur Wasserbau und Schiffbau, Berlin, Germany (trans: Ott WP, van Uchelen JC, into English) California Institute of Technology, Pasadena, California

Zhang M-l, Shen Y-m (2007) Study and application of steady flow and unsteady flow mathematical model for channel networks. J Hydrodynamics 19:572–578

Need of Two-Dimensional Consideration for Modelling Urban Drainage

Hriday Mani Kalita and Arup K. Sarma

Abstract The computation of unsteady free surface flow is essential in urban drainage systems to estimate values of various flow parameters like depth and velocity of flow in the different parts of a drain at different times of flow. Generally, 1-D governing equations of unsteady free surface flow (Saint-Venant equations) are solved for simulating storm water flow in drains. This gives acceptable result for turbulent flow in a channel of small width. However, for relatively larger channel width, flow parameters may vary significantly within a channel cross section and therefore, for proper analysis, 2-D simulation becomes essential. Need of using 2-D simulation for analyzing flow characteristic in a typical city drain is studied and presented in this paper. Comparison of 1-D and 2-D results has revealed that though both the models give same surface elevations along the drain at different times, flow velocity varies significantly across the channel in 2-D simulation. 2-D result shows higher velocity at middle of the channel and lower flow velocity near the bank, whereas, because of inherent limitation, the variation of flow velocity across the channel cross section never gets reflected in the result computed by 1-D analysis. Thus, the fact that a channel having low velocity near the bank may experience sediment deposition, might go unnoticed if it is analyzed by a 1-D model.

Keywords Unsteady flow · Urban drainage · 2-D simulation

1 Introduction

The computation of unsteady free surface flow is essential in urban drainage systems to estimate values of various flow parameters like depth and velocity of flow in the different parts of a drain at different times of flow. Generally, 1-D governing

H.M. Kalita (✉) · A.K. Sarma
Department of Civil Engineering, Indian Institute of Technology, Guwahati, India
e-mail: hriday@iith.ernet.in

A.K. Sarma
e-mail: aks@iitg.ernet.in

A.K. Sarma et al. (eds.), *Urban Hydrology, Watershed Management and Socio-Economic Aspects*, Water Science and Technology Library 73,
DOI 10.1007/978-3-319-40195-9_14

equations of unsteady free surface flow (Saint-Venant equations) are solved for simulating storm water flow in drains. Many researchers (Jingxiang and Charles 1985; Khan 2000; Ramesh et al. 2000; Patricia and Raimundo 2005; Akbari and Firoozi 2010; Yong 2010) have been using the 1-D Saint-Venant's equation for modelling channels. Though one-dimensional numerical model have been used extensively in this regards, in many situations, however, one-dimensional solution is not sufficient to describe the actual flow scenario. For example, to study the changes in different flow variables at different places of a same cross section in a channel, the computation of two-dimensional unsteady free surface flows becomes necessary. Due to this criterion, some investigators (Fennema and Chaudhry 1990; Weiming 2004; Schwanenberg and Harms 2004) have applied the 2-D Saint-Venant's equations for modelling channels. From the survey of literature, it has been observed that though numerous models have been developed for simulation of water flow in channels in 1-D and 2-D, it appears that there is a scope of doing comparative study between 1-D and 2-D simulations of drains, for selecting the efficient one. In this paper, details of the study done on comparison of 1-D and 2-D unsteady open channel flow simulation for different shaped drains are presented.

2 Governing Equation

Saint-Venant equations describing unsteady free surface flows are composed of continuity and momentum equations. For prismatic channels having no lateral inflow or outflow the Saint-Venant equations in 1-D (Jingxiang and Charles 1985) are defined as

$$\frac{\partial h}{\partial t} + D\frac{\partial u}{\partial x} + u\frac{\partial h}{\partial x} = 0 \tag{1}$$

$$\frac{\partial u}{\partial t} + u\frac{\partial u}{\partial x} + g\frac{\partial h}{\partial x} = g(S_O - S_\mathrm{f}) \tag{2}$$

In Eqs. (1) and (2), h is flow depth; u is flow velocity; $D = A/B$ is the hydraulic depth; A is flow area; B is top water surface width; S_O is channel bottom slope; $S_f = (u^2 \times n^2)/h^{4/3}$ is friction slope; x is distance along the channel length; t is time and g is the acceleration due to gravity; n is Manning's coefficient.

In 2-D form, the Saint-Venant equations with no lateral inflow and outflow in matrix form Fennema and Chaudhry (1990) are as follows:

$$U_t + E_x + F_y + S = 0 \tag{3}$$

In Eq. (3),

$$U = \begin{pmatrix} h \\ hu \\ hv \end{pmatrix} \quad E = \begin{pmatrix} hu \\ hu^2 + gh^2/2 \\ huv \end{pmatrix}$$

$$F = \begin{pmatrix} hv \\ huv \\ hv^2 + gh^2/2 \end{pmatrix} \quad S = \begin{pmatrix} 0 \\ -gh(S_{ox} - S_{fx}) \\ -gh(S_{oy} - S_{fy}) \end{pmatrix}$$

where h is flow depth; u is flow velocity in x direction; v is the flow velocity in y direction; S_{ox} is channel bottom slope in x direction; $S_{fx} = n^2u(u^2 + v^2)^{1/2}/h^{4/3}$ is the friction slope in x direction and x is the distance along the channel length; S_{oy} is channel bottom slope in y direction; $S_{fy} = n^2v(u^2 + v^2)^{1/2}/h^{4/3}$ is friction slope in y direction; y is distance across the channel length; n is Manning's coefficient; t is the time and g is the acceleration due to gravity.

3 Finite Difference Method

The governing equations are nonlinear first-order, hyperbolic partial differential equations for which closed-form solutions are not available. Therefore, these equations are solved numerically. One of the common numerical methods applied for the solution of these types of equation is finite difference method.

For 1-D case, the x direction is designated by the subscript i and the t direction is by subscript k. The time interval where the flow variable is known is denoted by subscript k and the unknown time level is by subscript $k + 1$. The increment in x direction is designated by Δx and Δt is the increment in the time direction.

Again for 2-D case, the x direction is designated by the subscript j, the y direction, by the subscript i and the t direction by the subscript k. The time interval where the flow variable is known is denoted by subscript k and the unknown time level is shown by subscript $k + 1$. Δx is the increment in the x direction, Δy is the increment in the y direction and Δt is the increment in the t direction.

4 Lax Diffusive Scheme

When a direct computation of the dependent variables can be made in terms of known quantities, the computation is said to be explicit. Lax diffusive finite difference scheme (Chaudhry 2008) is an explicit scheme. It is first-order accurate in time and second order accurate in space. Flow variables are known at time level k and their values are to be determined at time level $k + 1$.

The finite difference forms of the governing equations in one dimension are as follows,

$$h_i^{k+1} = \frac{1}{2}\left(h_{i-1}^k + h_{i+1}^k\right) - \frac{\Delta t}{2\Delta x} D_i^*\left(u_{i+1}^k - u_{i-1}^k\right) - \frac{\Delta t}{2\Delta x} u_i^*\left(h_{i+1}^k - h_{i-1}^k\right) \quad (4)$$

$$\begin{aligned} u_i^{k+1} &= \frac{1}{2}\left(u_{i-1}^k + u_{i+1}^k\right) - \frac{\Delta t}{2\Delta x} g\left(y_{i+1}^k - y_{i-1}^k\right) \\ &\quad - \frac{\Delta t}{2\Delta x} u_i^*\left(u_{i+1}^k - u_{i-1}^k\right) + g\Delta t\left(S_o - S_f^*\right) \end{aligned} \quad (5)$$

The general formulation of this scheme in two dimensions is as follows:

$$\begin{aligned} U_{i,j}^{k+1} &= \frac{1}{2}\left(U_{i,j+1}^k - U_{i,j-1}^k\right) - \frac{\Delta t}{2\Delta x}\left(E_{i,j+1}^k - E_{i,j-1}^k\right) - \frac{\Delta t}{2\Delta y}\left(F_{i+1,j}^k - F_{i-1,j}^k\right) \\ &\quad - \Delta t S_{i,j}^* \end{aligned} \quad (6)$$

From Eqs. (4)–(6), the dependent flow variables are computed at unknown time step.

5 Application of the Scheme

The above numerical schemes were then applied to solve the governing equations for studying the unsteady flow behaviour in different shaped drains by routing hydrographs through the drains. The details of the tests done for all the drains in 1-D and in 2-D and their results are presented below.

5.1 Drain Size Consideration

Test drain used for simulation was 200 m in length and 15 m in top width. Two different cross sections were taken under consideration. They are as follows:

- Trapezoidal shaped drain.
- Rectangular shaped drain.

The bottom widths of both the drains were 3 m. The total length of the drain was divided into 40 grids with $\Delta x = 5$ m for 1-D and 2-D considerations, whereas, for 2-D consideration the total width of the drain was divided into 5 grids with $\Delta y = 3$ m. The roughness of the drain was introduced by applying manning's *n*. A classical value of 0.031 was applied over the whole computational domain for all the cases. The longitudinal bed slope of the drain was assumed to be 1:1000.

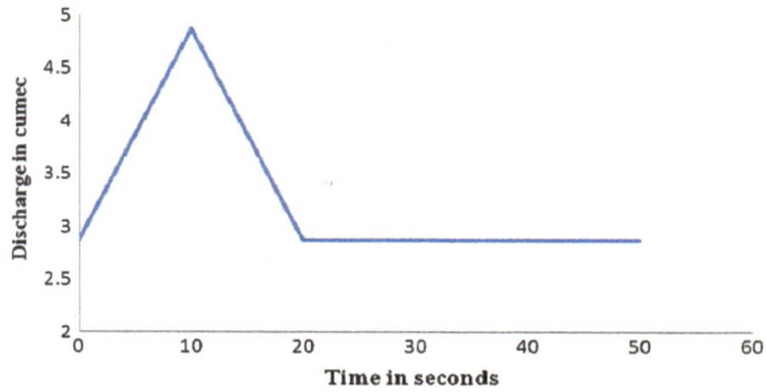

Fig. 1 Upstream flow hydrograph

5.2 *Initial and Boundary Condition*

Initial values of all the dependent variables were applied on the drain as initial condition. Flow hydrographs were applied at the upstream as upstream boundary conditions. For the downstream boundary, all the dependent variables were extrapolated from the interior domain. For two-dimensional simulations the solid boundary was simulated by Reflective boundary condition technique (Fennema and chaudhry 1990). Figure 1 shows the upstream flow hydrograph.

5.3 *Stability of the Models*

The Lax diffusive scheme is an explicit finite difference scheme and hence the model will be stable if Courant–Friedrichs–Lewy (CFL) criterion is satisfied. In one dimension according to CFL condition,

$$\frac{((u+\sqrt{gh})\times\Delta t)}{\Delta x}\leq 1$$

And for two-dimensional cases the condition is

$$\frac{((u+\sqrt{gh})\times\Delta t)}{\Delta x}\leq 1 \text{ and} \frac{((v+\sqrt{gh})\times\Delta t)}{\Delta y}\leq 1$$

5.4 *Results and Discussions*

Using the above initial and boundary conditions, the governing equations of free surface flow (Saint-Venant's equation) were solved by Lax diffusive scheme both in 1-D and 2-D for all the drains. To compare the performances of 1-D and 2-D simulations, surface elevations along the channel at different time steps were

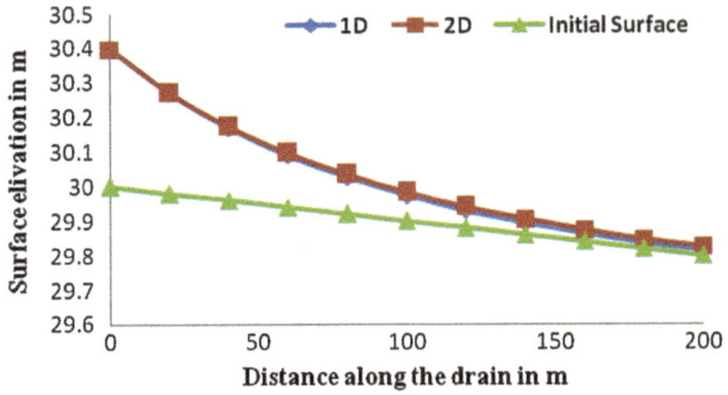

Fig. 2 Water surface profiles after 500 time steps for trapezoidal drain

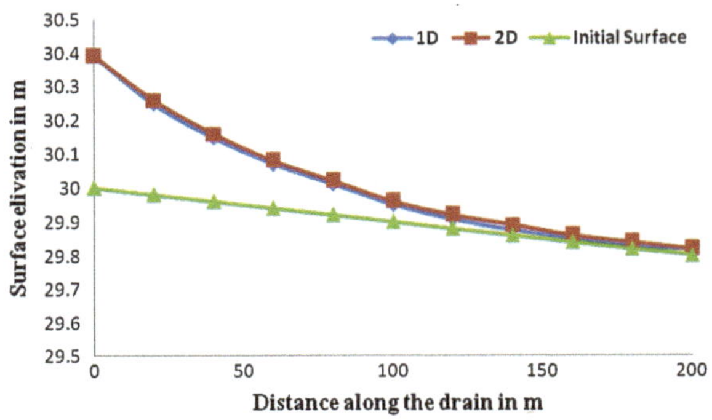

Fig. 3 Water surface profiles after 500 time steps for rectangular drain

plotted. From the plots, same surface elevations along the channels between 1-D and 2-D results were observed, for both the drains. Figure 2 shows the water surface profiles after elapsing of 500 time steps from beginning for the trapezoidal drain. Figure 3 shows same for the rectangular drain. From the results of surface elevation profiles, the potential of 1-D modelling regarding computation of water depth in drains same as that of 2-D modelling can be evaluated.

The water velocities obtained from 1-D and 2-D models were also compared for both the drains to have an idea about the performances of the models in calculating water velocities. From the result of 2-D velocities, it was observed that velocities near the bank line are very lesser than the centreline velocities for both the drains, whereas in 1-D modelling since the values are averaged across the cross sections this difference velocity pattern cannot be observed. The lower velocity pattern near the banks establishes the possibilities of sedimentation on those less velocity areas. Figure 4 shows the comparison in velocities along the channel for one-dimensional and two-dimensional simulations for rectangular drain. Figure 5 shows the same for trapezoidal drain. The lower velocity pattern near the banks can also be observed from the velocity vector plotting of the drains in 2-D. Figure 6 shows the velocity vector plot for the trapezoidal drain.

Fig. 4 Water velocities after 500 time step for rectangular channel

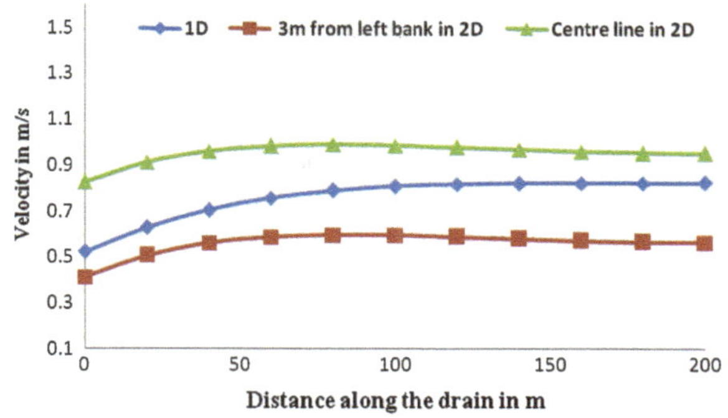

Fig. 5 Water velocities after 500 time step for trapezoidal channel

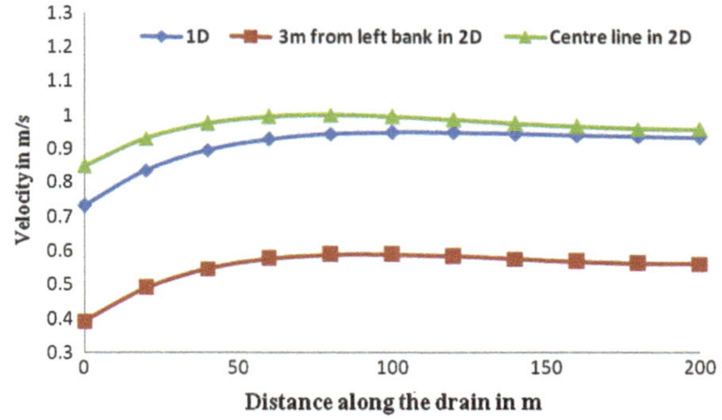

Fig. 6 Velocity vector for trapezoidal drain

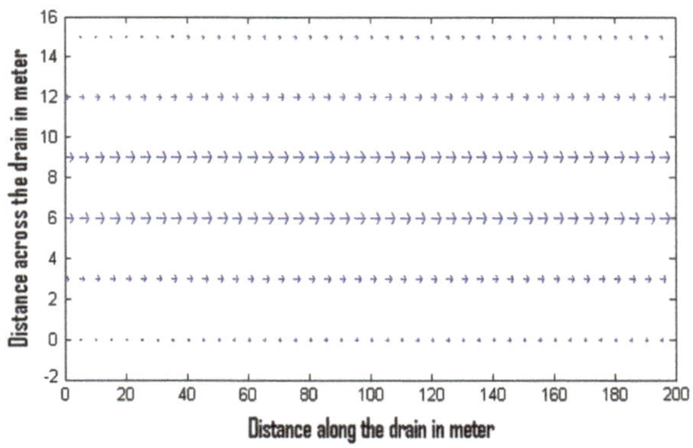

6 Conclusions

Lax diffusive finite difference explicit scheme has been applied for unsteady flow simulation in drains by solving the Saintt-Venant equation both in 1-D and 2-D forms. It has been found that 1-D modelling is sufficient for evaluating the correct water surface profiles. But, it has been observed that the flow velocity varies significantly across the channel in 2-D simulations for all the drains. Two-dimensional velocity results shows higher velocity at middle of the channel and lower flow velocity near the bank, whereas, because of inherent limitation, the variation of flow velocity across the channel cross section never gets reflected in the result computed by one-dimensional analysis. Thus, the fact that a channel having low velocity near the bank may experience sediment deposition, might go unnoticed if it is analyzed by a one-dimensional model. And hence, the present study facilitates the necessity of two-dimensional modelling in drain simulation problems.

References

Akbari G, Firoozi B (2010) Implicit and explicit numerical solution of Saint-Venant's equations for simulating flood wave in natural rivers. In: Proceedings of 5th national congress on civil engineering, May 4–6, Ferdowsi University of Mashhad, Mashhad, Iran

Chaudhry MH (2008) Open channel flow: Second edition. Springer Science+Business Media, LLC

Fennema RJ, Chaudhry MH (1990) Explicit methods for 2-D transient free surface flows. J Hydraul Eng Div ASCE 116(8):1013–1024

Jingxiang H, Charles CS (1985) Stability of dynamic flood routing schemes. J Hydraul Eng Div ASCE 111(12):1497–1505

Khan AA (2000) Modeling flow over an initially dry bed. J Hydraul Res 38(5)

Patricia C, Raimundo S (2005) Solution of Saint Venant's equation to study flood in rivers, through numerical methods. Hydrology days, Department of Environmental and Hydraulics engineering, Federal University of Ceara

Ramesh R, Datta B, Bhallamudi M, Narayana A (2000) Optimal estimation of roughness in open-channel flows. J Hydraul Eng Div ASCE 126(4):299–303

Schwanenberg D, Harms M (2004) Discontinuous Galerkin finite-element method for transcritical two-dimensional shallow water flows. J Hydraul Eng Div ASCE 130(5):412–421

Weiming W (2004) Depth-averaged two-dimensional numerical modeling of unsteady flow and no uniform sediment transport in open channels. J Hydraul Eng Div ASCE 130(10):1013–1024

Yong GL (2010) Two-dimensional depth-averaged flow modeling with an unstructured hybrid mesh. J Hydraul Eng Div ASCE 136(1):12–23

Modeling Well Level Fluctuations as Seismograph

Prerna Agrawal, Shweta Khemani, N.P. Dewangan and Amit Mishra

Abstract The development of Richter scale for earthquake magnitude in 1935 was a major breakthrough in seismology. The Richter magnitude is calculated from the maximum amplitude recorded by the seismograph. It is quite interesting and surprising to know that a well founded in confined aquifer exhibit the characteristics of a seismograph. Eaton and Takasaki (Seismol Soc Am Bull 49(3):227–245, 1958) extended this concept to measure seismically induced water level fluctuation of a well in confined aquifer. This measurement comprises of the hydroseismic magnitude, which is identical to the surface wave magnitude from the Richter scale and a factor which represent the influence of local geology, hydrology, well construction, and magnification on the water level fluctuation for a series of quakes recorded in the same wells as well as for a single quake recorded in the nearby and distant well. Each recorder-equipped well is a potential standby seismometer for the largest quakes because most seismographs to go off scale so maximum amplitude fails to record. The present paper is based on the project work for the B.E. degree of the authors in this paper, Vorhis method of calculating the hydroseismic magnitude has been presented so that it can be applied to the seismic water level fluctuation and help permanent installation of instruments in well so that it can act as standby seismograph and record micro water level fluctuations.

Keywords Richter magnitude · Hydroseismic magnitude · Nomogram · Stereographic chart

P. Agrawal (✉) · S. Khemani · N.P. Dewangan
Department of Civil Engineering, Government Engineering College, Raipur, Chhattisgarh, India
e-mail: agrawalprerna@gmail.com

A. Mishra
Department of Civil Engineering, Raipur Institute of Technology, Raipur, Chhattisgarh, India

A.K. Sarma et al. (eds.), *Urban Hydrology, Watershed Management and Socio-Economic Aspects*, Water Science and Technology Library 73,
DOI 10.1007/978-3-319-40195-9_15

1 Introduction

An earthquake is the result of a sudden release of energy in the earth's crust that creates seismic waves. The seismicity, seismic or seismic activity of an area refers to the frequency, type, and size of earthquakes experienced over a period of time. Earthquakes are measured using observations from seismometer. The moment magnitude is the most common scale on which earthquakes larger than approximately 5 are reported for the entire globe. The more numerous earthquakes smaller than magnitude 5 reported by national seismological observatories are measured mostly on the local magnitude scale, also referred to as the Richter scale. These two scales are numerically similar over their range of validity. Magnitude 3 or lower earthquakes are mostly almost imperceptible and magnitude 7 and over potentially causes serious damage over large areas, depending on their depth. The largest earthquakes in historic times have been of magnitude slightly over 9, although there is no limit to the possible magnitude. The most recent large earthquake, of magnitude 9.0 or larger, was the 9.0 magnitude earthquake in Japan in 2011 (as of March 2011), and it was the largest Japanese earthquake since records began.

Earthquake engineering deals with the effects of earthquakes on people and their environment with method of reducing those effects. It is very young discipline, many of its most important developments having occurred in the past 30–40 years. Earthquake engineering is a very broad field, drawing on aspect of geology, seismology, geotechnical engineering, structural engineering, risk analysis, and technical fields. Its practice also required consideration of social, economical, and political factors.

2 Earthquake and Richter Scale

2.1 Earthquake

2.1.1 Types

There are main three types of earthquake, *Tectonic earthquake, Volcanic earthquake, Explosion earthquake.*

2.1.2 Causes

There are two main causes of earthquakes. First, they can be linked to explosive volcanic eruptions; they are in fact very common in areas of volcanic activity where they either proceed or accompany eruptions. Second, they can be triggered by

Tectonic activity associated with plate margins and faults. The majority of earthquakes worldwide are of this type.

2.1.3 Faults

There are mainly three type of faults *Plate tectonics, Normal fault, Strike, and dip.*

2.1.4 Seismic Waves

Earthquakes are three-dimensional events, the waves move outwards from the focus, but can travel in both the horizontal and vertical plains. This produces three different of waves which have their own distinct characteristics and can only move through certain layers within the Earth.

$$M_{\circ} = \mu AD$$

μ Strength of material
A area
D avg. amount of slip

Wave is two types-Body waves and surface waves.

1. *Body wave*: Body wave is travel interior of the earth. Types of body wave are p wave and s wave
2. *Surface waves*: Wave is result from the interaction between body wave and S wave. Types of surface wave are Rayleigh and L wave.

2.1.5 Seismograph

Earthquakes generate seismic waves which can be detected with a sensitive instrument called a **seismograph**.

Advances in seismograph technology have increased our understanding of both earthquakes and the Earth itself.

A seismograph, or seismometer, is an instrument used to detect and record earthquakes. Generally, it consists of a mass attached to a fixed base. During an earthquake, the base moves and the mass do not. The motion of the base with respect to the mass is commonly transformed into an electrical voltage. The electrical voltage is recorded on paper, magnetic tape, or another recording medium. This record is proportional to the motion of the seismometer mass relative to the earth, but it can be mathematically converted to a record of the absolute motion of the ground. **Seismograph** generally refers to the seismometer and its recording device as a single unit.

2.1.6 Measurement

Earthquakes can be measured in several ways.

1. *Intensity*: Intensity is the measure, in terms of degrees, of damage to the surface and the effects on humans. Intensity records only observations of effects on the crust, not actual ground motion or wave amplitudes which can be recorded by instruments.
2. *Magnitude*: Magnitude is a measure of the amount of energy released during an earthquake.

2.2 Richter Scale

The Richter magnitude scale was developed in Richter (1935a, b) by Charles F. Richter of the California Institute of Technology as a mathematical device to compare the size of earthquakes. The magnitude of an earthquake is determined from the logarithm of the amplitude of waves recorded by seismographs. Adjustments are included for the variation in the distance between the various seismographs and the epicenter of the earthquakes. On the Richter scale, magnitude is expressed in whole numbers and decimal fractions. For example, a magnitude 5.3 might be computed for a moderate earthquake, and a strong earthquake might be rated as magnitude 6.3. Because of the logarithmic basis of the scale, each whole number increase in magnitude represents a tenfold increase in measured amplitude; as an estimate of energy, each whole number step in the magnitude scale corresponds to the release of about 31 times more energy than the amount associated with the preceding whole number value.

3 Vorhis Model and Its Verification

3.1 Introduction

The Robert C. Vorhis gives the second extension of the Richter scale in 1964 for a well in the crystalline rocks of north Georgia and he used the hydroseismic data for calculation of Earthquake magnitudes.

3.2 Purpose

Although data on earthquake-induced water level fluctuation in wells have been printed in many seismologic & hydrologic publications, they have been treated for the most part only as curiosities.

But the purpose of Vorhis was to stimulate evaluation of the accumulated published data, interpretation of newsworthy fluctuations, and permanent installation of instruments to record precise time and amplitudes of microscopic water level fluctuations.

3.3 Method and Formula

The seismologic method for calculating earthquake magnitude is adapted. Here the fluctuation is converted to M_s + log C or the hydroseismic magnitude plus a variable.

In this form rough comparisons can be made from quake to quake and from well to well.

A stereographic chart is included to calculate graphically distance from epicenter to well and a nomogram to estimate $\boldsymbol{M_s}$ **+ log** ***C***.

3.4 Calculation of M_s + log C

Hydroseismic magnitudes plus log *C* can be computed in three steps:

1. Determination of arc distance from the epicenter to the well in degrees.
2. Measure the maximum single amplitude of seismic fluctuation recorded in the well.
3. Draw a line in NOMOGRAM connecting the arc distance to single amplitude and read at the intersection with the center line.

Fluctuation and double amplitude are considered as synonyms. For those fluctuations where rise is equal to fall, single amplitude is any of three values: (1) the amount of rise, (2) the amount of fall, (3) one-half the fluctuation.

3.4.1 Calculation of Arc Distance

1. *By Stereographic chart*: The chart is a stereographic projection of a hemisphere on which the meridians and parallels are drawn at intervals of 2°. The significantly different features of the projection are four different steps by which degrees are numbered on it. The latitude north and south are given around the outside and are strictly conventional, north being at top and south at the bottom. Two sets of degrees are numbered along the equator; one to measure longitude west of the right-hand bounding meridian, so this set thus uses minus signs; the other to measure longitudes east of the left-hand bounding meridian. Along the vertical axis the latitude lines are numbered from 0° to 180° upward from the South Pole, and it is from this set that the arc distance is read.
2. *By trigonometric formulas*: Three formulas given below.

If a is the arc distance between two points that lie on latitudes λ_1 and λ_2 and meridian separated by $\Delta\Gamma$, then

$$\mathrm{hav\,a} = \mathrm{hav}(\lambda_1 - \lambda_2) + \cos\lambda_1 \cos\lambda_2 \,\mathrm{hav}\,\Delta\Gamma \tag{1}$$

$$\sin^2 a/2 = \sin^2(\lambda_1 - \lambda_2) + \cos\lambda_1 \cos\lambda_2 \sin^2 \Delta\Gamma/2 \tag{2}$$

$$\cos a = \cos(\lambda_1 - \lambda_2) - \cos\lambda_1 \cos\lambda_2 + \cos\lambda_1 \cos\lambda_2 \cos\Delta\Gamma \tag{3}$$

Using any of above equation arc distance can be calculated.

4 Result and Discussion

The Richter magnitude is calculated from the maximum amplitude recorded by the seismograph. After extension of Eaton and Takasaki (1958), this measurement comprises by Vorhis, as the hydroseismic magnitude, which is identical to the surface wave magnitude from the Richter scale (M_s) and a factor (log C) which depends upon parameters of well aquifer system as local geology, hydrology, well construction, and magnification on the water level fluctuation. The value of M_s + log C can be calculated rather easily but its use is limited. The variable log C is constant for a same well and vary for different wells. As the same well is considered in this study for all the regions of Table 1, the properties of representative well situated at 37.137N/80.4386W is considered that will remain constant for all regions in the analysis (Figs. 1, 2 and 3).

Table 1 Calculation of hydroseismic magnitude by Vorhis model

S. no.	Region	Latitude/longitude	Arc distance	Fluctuation	Single amplitude (in feet)	M_s + log C	log C	M_s
1	M9.0, Honshu Japan, 03/11/11	38.297N/142.372E	65.1°	4.00	2.00	10.70	2.1	8.60
2	M7.5, Nicobar Island, India region 21/06/10	7.848N/91.917E	31.1°	1.08	0.54	9.50	2.1	7.40
3	M7.6, Southern 30/09/09	0.724S/99.856E	40.6°	1.08	0.54	9.80	2.1	7.70
4	M7.6, Andaman Island India, 10/08/09	14N/92.9E	25.7°	1.33	0.67	9.50	2.1	7.40
5	M7.7, Sea of Okhotsk, 05/07/08	53.888N/142.869E	78.1°	0.42	0.21	9.80	2.1	7.70
6	M7.9, Eastern Sichuan, China, 12/05/08	30.67N/104.07E	112.1°	5.00	2.50	11.20	2.1	8.10
7	M8.0, Coast of Central Peru, 15/8/07	13.3545S/76.5092W	81.43°	2.92	1.46	10.70	2.1	8.60
8	M7.2, Vanuatu, 01/08/07	17.561S/168.028E	95.6°	0.08	0.04	9.30	2.1	7.20
9	M8.2, East of the Kuril Islands, 13/01/07	46.272N/154.455E	83.2°	2.58	1.29	10.70	2.1	8.60
10	M7.9, Tonga, 03/05/06	20.088S/174.219W	80.91°	1.67	0.84	10.50	2.1	8.4

Worldwide Earthquakes Recorded in well 27F 2 SOW 019 located in Christiansburg, Virginia (USA) at Latitude and Longitude of Well: 37.137N/80.4386W

5 Conclusion

Vorhis method of calculating the hydroseismic magnitude can be applied to the seismic water level fluctuation and help permanent installation of instruments in well so that it can act as standby seismograph and record micro water level fluctuations. Results obtain from Vorhis method, i.e., hydroseismic magnitude is almost same as result from Richter scale.

6 Brief on Kyoshin Net

The strong motion data for the present work has been acquired from the kyoshin net. The K-net is a Japanese earthquake record system, which sends strong motion data on the internet, which are obtained from 1000 observatories, deployed all Japan. The avg. station to station distance is about 25 km. Each station has a digital strong motion seismograph with white dynamic characteristics and frequency band, having a maximum measurable acceleration of 2000 Ga. The records obtained from

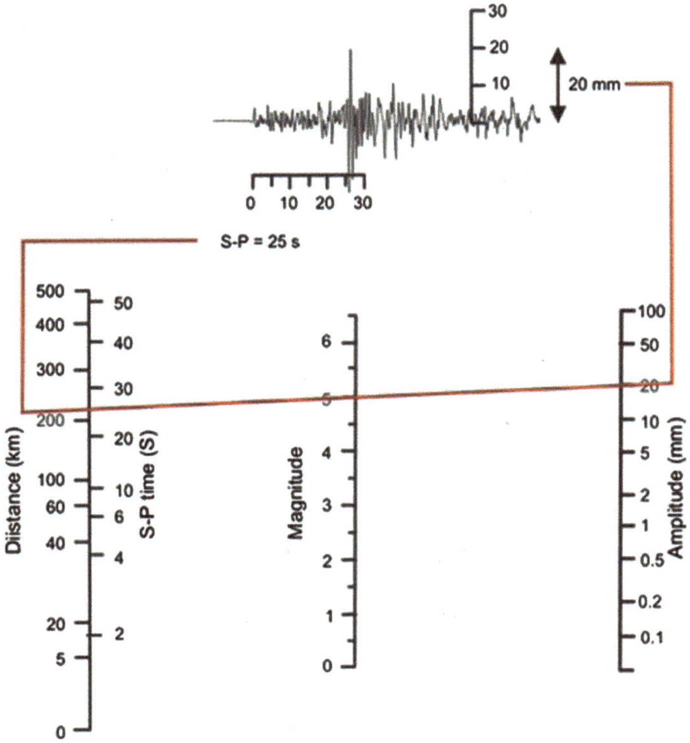

Fig. 1 Richter scale

these seismographs are acquired at a controlled center in Tsukuba by telemetry. At each site, the soil condition, and the P&S-waves velocity profiles, has been obtained by the down-hole measurement. The controlled center compiles these strong motion records, including the distribution map of maximum acceleration and makes it available on the internet.

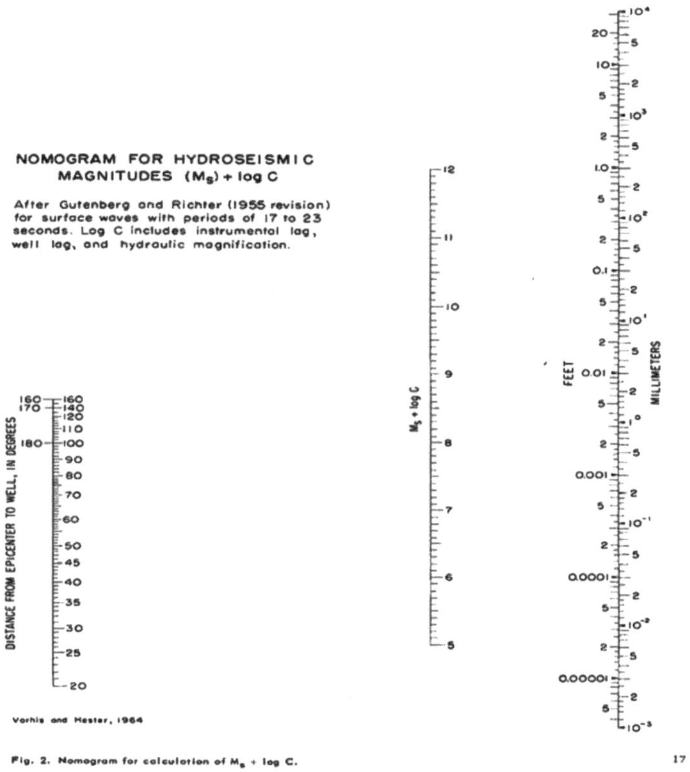

Fig. 2 Nomogram for calculation of M_s + log C (Vorhis 1964)

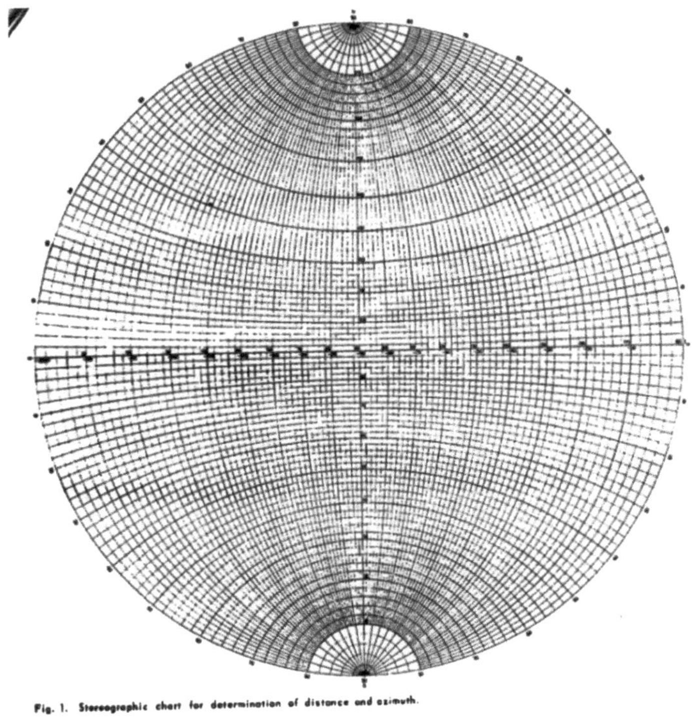

Fig. 3 Stereographic chart for determination of distance and azimuth (Vorhis 1964)

Acknowledgments The progress of our paper MODELING OF WELL AS A SEISMOGRAPH - A CASE STUDY OF VERIFICATION OF VOHRIS MODEL, could not have been achieved without the kind support of our guide Mr. N.P. DEWANGAN. We would like to thank all those persons who have directly or indirectly helped us by sharing their knowledge and experience as it might become an impossible task to complete our paper without their help guidance. A special thank to Kyoshin Net Strong Motion Network of Japan for providing an excellent earthquake database for conducting this research. This support is gratefully acknowledged.

References

Eaton JP, Takasaki KJ (1958) Seismological interpretation of earthquake induced water-level fluctuation in wells. Seismol Soc Am Bull 49(3):227–245

Richter CF (1935a) An instrumental earthquake magnitude scale. Seismol Soc Am Bull 25:1–32

Richter CF (1935b) Elementary seismology. W.A. Freeman & Co., San Francisco

Vorhis RC (1964) Earthquake-induced water-level fluctuation from a well in Dawson Country. Georgia Seismol Soc Am Bull 54:1023–1033

Flow Analysis in Compound Channel Considering Momentum Transfer Mechanism

Thappeta Suresh Kumar and Arup K. Sarma

Abstract Many engineering practices analysis of flow characteristic in compound channel becomes essential for adequate design. Several studies have been carried out since last three decades, where results of conventional methods are not well defined. It happens because of neglecting the concept of "Momentum transfer mechanism", which occurs at interface of main channel and flood plain. Therefore, an effort is made to develop a method considering "Momentum transfer mechanism" for steady flow conditions in case of both symmetrical and unsymmetrical multi-stage compound channels, i.e., for generalized condition of compound channel. A computer program has been developed for this purpose. Observations of stage-discharge diagrams using different equivalent roughness formulae have been presented. Apparent shear stress, velocities, and discharges were compared with previous studies for symmetrical compound channel. Furthermore, stage-discharge curves are presented for a four stage compound channel.

Keywords Discharge estimation · Compound channel · Momentum transfer

1 Introduction

Analysis of open channel flow is important for planning, design, and operation of various water related projects, such as city drainage, navigation, irrigation, etc. For analysis of flow in natural channel like river, for simplifying mathematical computation, many a times it is approximated as a simple rectangular, trapezoidal, or parabolic channel, though such channels are generally compound channel with different surface roughness in different subsections. To avoid flooding in rivers levees are provided in many river reaches, which also make the channel a compound channel. Similarly, for artificial channel like city drains having varying discharges in different seasons, it is advisable to go for a compound channel to

T.S. Kumar (✉) · A.K. Sarma
Department of Civil Engineering, IIT Guwahati, Guwahati, India
e-mail: sthappeta@gmail.com

A.K. Sarma et al. (eds.), *Urban Hydrology, Watershed Management and Socio-Economic Aspects*, Water Science and Technology Library 73,
DOI 10.1007/978-3-319-40195-9_16

make it self-cleansing in lean period and to have sufficiently carrying capacity to carry large volume in the flood period. Thus for many engineering practices, we need to design compound channel. Approximate design by considering the channel as rectangular may lead to oversize or undersize depending on the case. For design of a perfect compound channel, one must consider momentum transfer effect and effect of composite nature of the channel bed. If it is undersized flooding may occur and if it is oversized, problems like channel instabilities, erosion, sediment problems will arise. Improper design may also effect natural development of biodiversity in the river bank. Therefore accurate design of compound channel is quite important. Failure of traditional methods to predict the accurate discharge capacity of compound river channels is due to the presence of a momentum transfer mechanism between the fast flowing main channel and the slow moving flow in the flood plains. Some investigators attempted to improve the accuracy of the traditional methods. Conventionally compound channels are designed by considering the flow in different subsections separately. Sellin (1964) was the first person to investigate this mechanism. Dracos and Hardegger (1987) presented a modification of the single-channel method, which involved the development of an apparent roughness coefficient that can take account of the momentum transfer. Posey (1967) and others suggested modifications to the way in which the zones of flow may be identified in the divided channel method. In addition, Myers (1978), Wormleaton et al. (1982) and Knight and Demetriou (1983) showed that the apparent shear stress on the main channel/flood plain interface is many times greater than the boundary stress in main channel/flood plain particularly at low depths. In this study different available methods of flow analysis are compared and asystematic procedure for analyzing flow in a symmetrical and asymmetrical multi-stage compound channel has been attempted. A mathematical model for analysis of steady state has been developed.

2 Observations of Stage-Discharge Curves Using Different Equivalent Roughness Equations

The following formulae are used for calculating equivalent roughness value in compound channels

(i) Pavlovskii (1931): The assumption under this method is that the total resistance force is equal to the sum of subarea resistance forces.

$$n_e = \left(\frac{1}{p} \sum_{i=1}^{N} p_i n_i^2 \right)^{0.5}$$

(ii) Einstein and Banks (1950): In this method, it is assumed that the friction slope is same for all subsections and velocities are equal in all subsections.

$$n_e = \left(\frac{1}{p}\sum_{i=1}^{N} p_i n_i^{3/2}\right)^{2/3}$$

(iii) Lotter (1933): This is based on the assumptions that the friction slope is the same for all subsections and velocities are equal in all subsections and the total discharge equals to the sum of the subsection discharges.

$$n_e = pR^{(5/3)}\left(\sum_{i=1}^{N}\frac{p_i R_i^{5/3}}{n_i}\right)^{-1}$$

(iv) Yen (1991): This is based on the assumption that the total shear velocity is equal to sum of subareas shear velocities

$$n_e = \left(\frac{1}{pR^{\left(\frac{5}{3}\right)}}\right)\left(\sum_{i=1}^{N}\frac{p_i R_i^{5/3}}{1}\right)$$

(v) General considerations: Here general consideration means that, without considering any equivalent roughness formulas for calculating discharge. Thus in this conventional approach, for computing flow through the subareas and main channel their respective roughness coefficient values were used in Manning's equation. The plots of stage-discharge diagram for Fig. 1 cross section is shown in Fig. 2.

Where

h = Relative depth of water; H = Total depth of water in compound channel; Q_g = Discharge at depth of H; MCC = Maximum discharge carrying capacity of deep channel.

Observation of the above diagrams reveals that, in case of Pavlovskii, Einstein, and Bank formulae, maximum discharge carrying capacity of entire compound

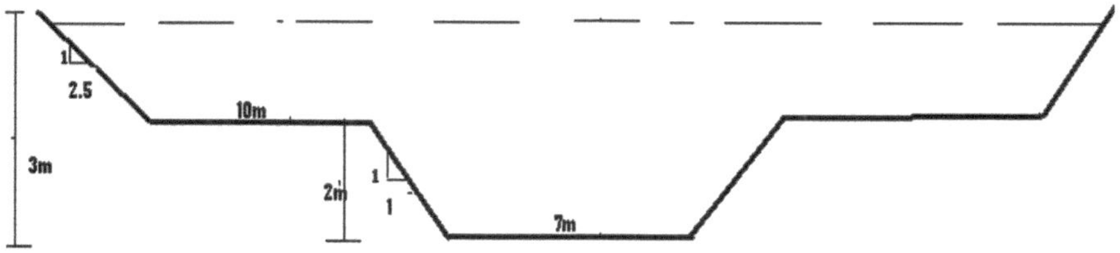

Fig. 1 Trapezoidal symmetrical channel

Fig. 2 Stage-discharge diagram

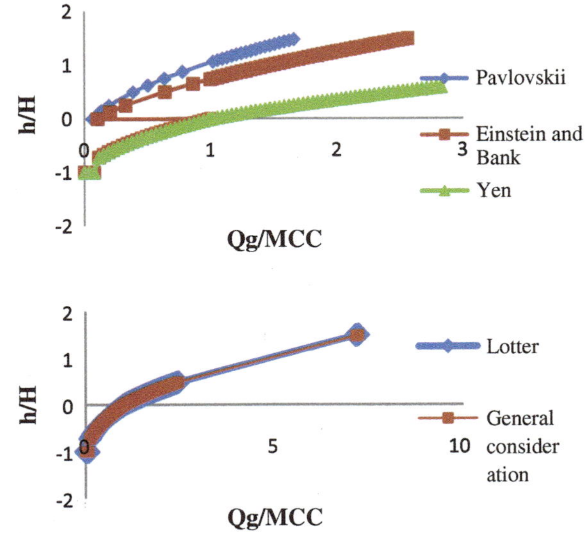

channel is getting less than that of maximum discharge carrying capacity of deep channel up to certain small depth of water, when flow occurs in flood plain. But in case of Lotter, Yen and general considerations, this condition is not occurring.

3 Momentum Transfer Mechanism (MTM)

The momentum transfer coefficient is an important parameter for determining the apparent shear stress at the vertical interface between the main channel and flood plains, the cross-sectional mean velocity and the discharge capacity in compound channels.

3.1 Momentum Transfer Coefficient (α)

Expression for momentum transfer coefficient is shown below.

$$\alpha = \frac{2\tau_a}{\rho(v_m^2 - v_f^2)} \tag{1}$$

where α = Momentum transfer coefficient, τ_a = Apparent shear stress (N/m^2) at interface of flood plain and main channel, v_m = Velocity of water in main channel, v_f = Velocity of water in flood plain.

3.2 *For Symmetrical Channel*

3.2.1 Derivation of Apparent Shear

By applying force balance concept in left side of the flood plain, the expression for apparent shear stress can be obtained. Therefore, for unit length of compound channel,

$$\begin{aligned} &\text{Resolved weight force} - \text{Boundary shear force} \\ &+ \text{Apparent shear force at the interface} = 0 \end{aligned} \tag{2}$$

Resolved weight force = $w \sin \theta = \gamma A_{\mathrm{FL}} S_{\mathrm{o}}$; Boundary shear force = $\int \tau_{\mathrm{b}} \mathrm{d}p$; Apparent shear force at interface = $\tau_{\mathrm{a}} H_{\mathrm{FL}}$.

After substituting all above equations in Eq. (2), we can obtain following expression for apparent shear stress at interface

$$\tau_{\mathrm{a}} = \left(\int \tau_{\mathrm{b}} \mathrm{d}p - \gamma A_{\mathrm{FL}} S_{\mathrm{o}} \right) \Big/ H_{\mathrm{FL}} \tag{3}$$

where τ_{a} = Apparent shear stress at interface, τ_{b} = Boundary shear stress of flood plain (N/m^2), dp = Elemental wetted perimeter, A_{FL} = Area of left or right flood plain for symmetrical channel (m^2), S_{o} = Bed slope of flood plain, γ = Density of water = 9810 N/m^3, H_{FL} = Height of water in left or right flood plain for symmetrical section (Figs. 3 and 4).

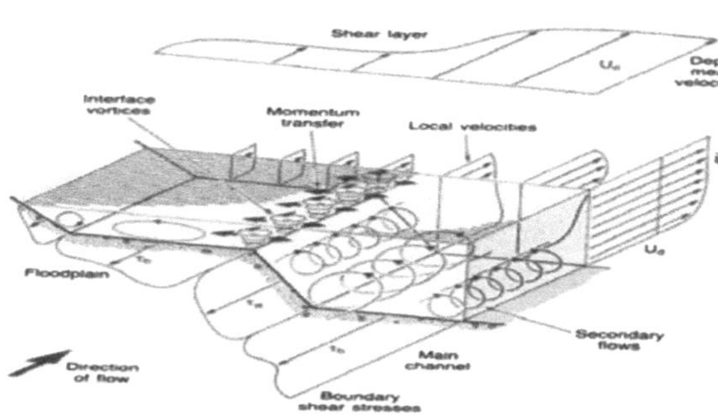

Fig. 3 Mechanism of over bank flow in compound channel

Fig. 4 Vortices formed at interface of main channel and flood plain

3.2.2 Derivation of Expression for Boundary Shear Force for Flood Plain (Fig. 5)

$$\text{Boundary shear force in flood plain} = \tau_{\text{side}}\left(\sqrt{(H_{\text{FL}}^2 z_1^2) + H_{\text{FL}}^2}\right) + (\tau_{\text{bed}} b_{\text{FL1}}) \quad (4)$$

τ_{side} = Shear stress on side of flood plain = $\frac{w}{a}\cos\theta\tan\emptyset\sqrt{1-\left(\frac{\tan\theta\tan\theta}{\tan\emptyset\tan\emptyset}\right)}$

$\tau_{\text{bed}} = \frac{w}{a}\tan\emptyset = 5503.41d$

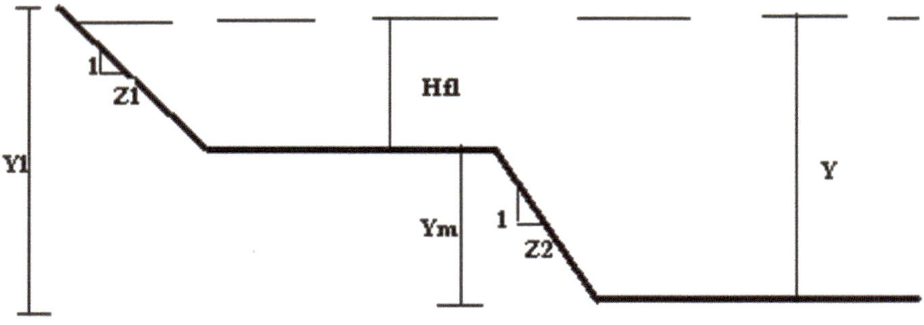

Fig. 5 Symmetrical compound channel

w = Submerged weight of sediment partial = $C_2(\gamma_s - \gamma)d^3 = \left(\frac{\pi}{6}\right)(\gamma_s - \gamma)d$; a = Area of sediment particle = $\left(\frac{\pi}{4}\right)(d^2)\theta$ = Side slope of channel; $\emptyset$ = Angle of repose = 27°; d = Diameter of particle (m).

Finally Eq. (4) is as below

$$\text{Boundary shear force in flood plain} = \int \tau_b \mathrm{d}p$$

$$= \left[10{,}791(d)\cos\theta(0.5)\sqrt{1-\left(\frac{\tan\theta\tan\theta}{\tan\emptyset\tan\emptyset}\right)}\sqrt{(H_{FL}^2 z_1^2)+H_{FL}^2}\right] + [5503.41\mathrm{d}b_{FL}]$$

3.2.3 Derivation of Velocities in Main Channel and Flood Plain Considering Apparent Shear Stress

Applying force balance concept in flood plain, (Resolved weight force) − (Boundary shear force) + (Apparent shear force at the interface) = 0

$$(\gamma A_{FL} S_o) - \left(\int \tau_b \mathrm{d}p_{FL}\right) + (\tau_a H_{FL}) = 0$$

$$\tau_b = \frac{(\gamma A_{FL} S_o) + \tau_a H_{FL})}{\int \mathrm{d}p_{FL}} \tag{5}$$

But for uniform flow,

$$\text{Boundary shear stress} = \frac{\gamma V_{FL}^2}{C_{FL}^2} \tag{6}$$

After Substituting (4), (6) equations in (5) we can obtain below equation

$$V_{FL}^2\left[\frac{1}{C_{FL}^2} + \frac{\propto H_{FL}}{2g\sum \mathrm{d}p_{FL}}\right] - \left[\frac{\propto H_{FL} V_m^2}{2g\sum \mathrm{d}p_{FL}}\right] - R_{FL} S_o = 0 \tag{7}$$

Applying force balance concept in main channel

$$(\gamma A_m S_o) - \left(\int \tau_b \mathrm{d}p_m\right) - 2(\tau_a H_{FL}) = 0$$

$$\tau_b = \frac{(\gamma A_m S_o) - 2\tau_a H_{FL})}{\int \mathrm{d}p_m} \tag{8}$$

Similarly for uniform flow,

$$\text{Boundary shear stress} = \frac{\gamma V_{\mathrm{m}}^2}{C_{\mathrm{m}}^2} \tag{9}$$

After substituting Eqs. (4), (9) in Eq. (8) we can obtain below equation

$$V_{\mathrm{m}}^2\left[\frac{1}{C_{\mathrm{m}}^2} + \frac{\propto H_{\mathrm{FL}}}{g\sum \mathrm{d}p_{\mathrm{m}}}\right] - \left[\frac{\propto H_{\mathrm{FL}} V_{\mathrm{FL}}^2}{g\sum \mathrm{d}p_{\mathrm{m}}}\right] - R_{\mathrm{m}}S_{\mathrm{o}} = 0 \tag{10}$$

By solving Eqs. (7) and (10) we can get velocity expressions as below.

Velocity of water in main channel considering apparent shear stress $V_{\mathrm{m}} = \sqrt{\frac{B}{A}}$. Where

$$\mathrm{B} = R_{\mathrm{FL}}S_{\mathrm{o}} + \left[\frac{\sum \mathrm{d}p_{\mathrm{m}}R_{\mathrm{m}}S_{\mathrm{o}}H_{\mathrm{FL}}}{2H_{\mathrm{FL}}\sum \mathrm{d}p_{\mathrm{m}}}\right] + \left[\frac{\sum \mathrm{d}p_{\mathrm{m}}gR_{\mathrm{m}}S_{\mathrm{o}}}{\alpha H_{\mathrm{FL}}C_{\mathrm{FL}}^2}\right]$$

$$\mathrm{A} = -\frac{\propto H_{\mathrm{FL}}}{2g\sum \mathrm{d}p_{\mathrm{FL}}} + \left[\frac{\sum g\mathrm{d}p_{\mathrm{m}}}{\alpha H_{\mathrm{FL}}C_{\mathrm{m}}^2} + 1\right]\left[\frac{\propto H_{\mathrm{FL}}}{2g\sum \mathrm{d}p_{\mathrm{FL}}}\right] + \left[\frac{\sum \mathrm{d}p_{\mathrm{m}}g}{\alpha H_{\mathrm{FL}}C_{\mathrm{m}}^2} + 1\right]\frac{1}{C_{\mathrm{FL}}^2}$$

Velocity of water in flood plain considering apparent shear stress from Eq. (7)

$$\mathrm{V}_{\mathrm{FL}} = \sqrt{\left[\frac{-\sum \mathrm{d}p_{\mathrm{m}}gR_{\mathrm{m}}S_{\mathrm{o}}}{\alpha H_{\mathrm{FL}}} + \left[\frac{\sum \mathrm{d}p_{\mathrm{m}}g}{\alpha H_{\mathrm{FL}}C_{\mathrm{m}}^2} + 1\right]\mathrm{V}_{\mathrm{m}}^2\right]}$$

where

$\sum \mathrm{d}p_{\mathrm{FL}} = ((y - y_{\mathrm{m}})^2 + z_1^2(y - y_{\mathrm{m}})^2)^{0.5} + B_{\mathrm{FL}}$ = wetted perimeter of side slope;
$\sum \mathrm{d}p_{\mathrm{m}} = 2(y_{\mathrm{m}}^2 + z_2^2 y_{\mathrm{m}}^2)^{0.5} + B_{\mathrm{m}}$ = wetted perimeter of main channel;
$C_{\mathrm{FL}}^2 = \frac{R_{\mathrm{FL}}^{1.666}}{n_{\mathrm{FL}}}$ = Chezy's coefficient of flood plain;
$C_{\mathrm{m}}^2 = \frac{R_{\mathrm{m}}^{1.666}}{n_{\mathrm{m}}}$ = Chezys coefficient of main channel $H_{\mathrm{FL}} = -y_{\mathrm{m}}$;
y = Depth of water in entire compound channel;
y_{m} = Maximum depth of water that can occur in deep channel;
$z_1 = z_2$ = Side slopes of flood plain;
$n_{\mathrm{FL}}, n_{\mathrm{m}}$ = mannings roughness coefficients of flood plain and main channel respectively;
B_{FL}, B_{m} are widths of flood plain and main channel respectively.

3.2.4 Discharge Calculation

For symmetrical channel

Discharge in left or right flood plain $Q_{\mathrm{FL\ or\ RF}} = A_{\mathrm{FL}}\ V_{\mathrm{FL}}$.

Then, total discharge in entire compound channel $Q = 2Q_{\mathrm{FL\ or\ RF}} + (A_{\mathrm{m}}V_{\mathrm{m}})$.

3.3 For Unsymmetrical Channel (Fig. 6)

Here, for unsymmetrical channel velocities in left and right floodplains are different so we have to apply force balance concept for left, right and main channels separately. After applying we can obtain three equations with three unknowns as below

$$V_{\mathrm{FL}}^2\left[\frac{1}{C_{\mathrm{FL}}^2}+\frac{L\,.\propto H_{\mathrm{FL}}}{2g\sum \mathrm{d}p_{\mathrm{FL}}}\right]-\left[\frac{L\,.\propto H_{\mathrm{FL}}V_{\mathrm{m}}^2}{2g\sum \mathrm{d}p_{\mathrm{FL}}}\right]-R_{\mathrm{FL}}S_{\mathrm{ol}}=0,\quad \text{For left flood plain} \tag{11}$$

$$V_{\mathrm{FR}}^2\left[\frac{1}{C_{\mathrm{FR}}^2}+\frac{R\,.\propto H_{\mathrm{FR}}}{2g\sum \mathrm{d}p_{\mathrm{FR}}}\right]-\left[\frac{R\,.\propto H_{\mathrm{FR}}V_{\mathrm{m}}^2}{2g\sum \mathrm{d}p_{\mathrm{FR}}}\right]-R_{\mathrm{FR}}S_{\mathrm{or}}=0,\quad \text{For right flood plain} \tag{12}$$

$$\begin{aligned}\frac{V_{\mathrm{m}}^2}{C_{\mathrm{m}}^2}=&-\left[\frac{R\,.\propto H_{\mathrm{FL}}(V_{\mathrm{m}}^2-V_{\mathrm{FL}}^2)}{2g\sum \mathrm{d}p_{\mathrm{m}}}\right]\\&-\left[\frac{L\,.\propto H_{\mathrm{FL}}(V_{\mathrm{m}}^2-V_{\mathrm{FL}}^2)}{2g\sum \mathrm{d}p_{\mathrm{m}}}\right]+R_{\mathrm{m}}S_{\mathrm{om}},\quad \text{For main channel}\end{aligned} \tag{13}$$

After solving Eqs. (11–13) assuming that

$$\begin{aligned}&E=\frac{1}{C_{\mathrm{FL}}^2}+\frac{L\,.\propto H_{\mathrm{FL}}}{2g\sum \mathrm{d}p_{\mathrm{FL}}};\quad K=\frac{R\,.\propto H_{\mathrm{FR}}V_{\mathrm{m}}^2}{2g\sum \mathrm{d}p_{\mathrm{FL}}};\quad N=R_{\mathrm{m}}S_{\mathrm{om}}\\&F=\frac{1}{C_{\mathrm{FR}}^2}+\frac{R\,.\propto H_{\mathrm{FR}}}{2g\sum \mathrm{d}p_{\mathrm{FR}}};\quad L=R_{\mathrm{FL}}S_{\mathrm{ol}};\quad O=\frac{L\,.\propto H_{\mathrm{FL}}}{2g\sum \mathrm{d}p_{\mathrm{m}}}\\&J=\frac{L;.\propto H_{\mathrm{FL}}V_{\mathrm{m}}^2}{2g\sum \mathrm{d}p_{\mathrm{FL}}};\quad MR_{\mathrm{FR}}S_{\mathrm{or}};\quad P=\frac{R\,.\propto H_{\mathrm{FR}}}{2g\sum \mathrm{d}p_{\mathrm{m}}}\\&Q=\left[\frac{1+(O+P)C_{\mathrm{m}}^2}{C_{\mathrm{m}}^2}\right]-\left[\frac{OJ}{E}\right]-\left[\frac{PK}{F}\right]\end{aligned}$$

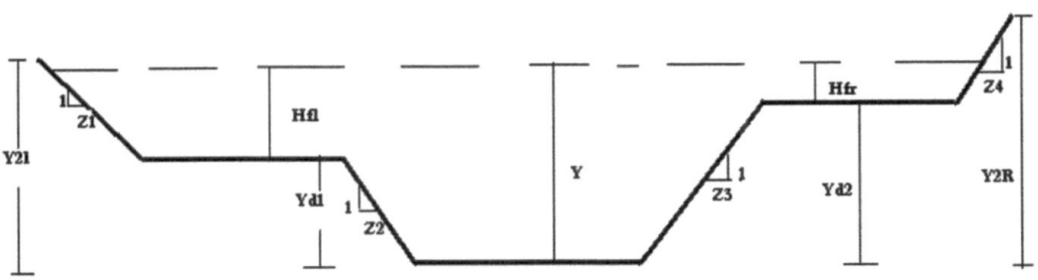

Fig. 6 Unsymmetrical cross section of compound channel

Velocity of water in main channel considering apparent shear stress

$$V_{\mathrm{m}} = \sqrt{PM/_{FQ} + OL/_{EQ} + N/_{Q}}$$

Velocity of water in left flood plain considering apparent shear stress

$$\mathrm{V}_{\mathrm{FL}} = \sqrt{L/_{E} + j\mathrm{v}_{\mathrm{m}}^2/_{E}}$$

Velocity of water in right flood plain considering apparent shear stress

$$V_{\mathrm{RL}} = \sqrt{M/_{F} + K\mathrm{v}_{\mathrm{m}}^2/_{F}}$$

where

$\sum \mathrm{d}p_{\mathrm{FL}} = \left((y - y_{\mathrm{m}})^2 + z_1^2(y - y_{\mathrm{m}})^2\right)^{0.5} + B_{\mathrm{FL}}$ = wetted perimeter of side slope;

$\sum \mathrm{d}p_{\mathrm{FR}} = \left((y - y_{\mathrm{m}})^2 + z4^2(y - y_{\mathrm{m}})^2\right)^{0.5} + B_{\mathrm{FR}}$ = wetted perimeter of side slope;

$\sum \mathrm{d}p_{\mathrm{m}} = \left(y_{\mathrm{m}}^2 + z_2^2 y_{\mathrm{m}}^2\right)^{0.5} + \left(y_{\mathrm{m}}^2 + z_3^2 y_{\mathrm{m}}^2\right)^{0.5} + B_{\mathrm{m}}$ = wetted perimeter of main channel;

$C_{\mathrm{FL}}^2 = \frac{R_{\mathrm{FL}}^{1.666}}{n_{\mathrm{FL}}}$ = Chezy's coefficient of left flood plain, $C_{\mathrm{FR}}^2 = \frac{R_{\mathrm{FR}}^{1.666}}{n_{\mathrm{FR}}}$ = Chezy's coefficient of right flood plain, $C_{\mathrm{m}}^2 = \frac{R_{\mathrm{m}}^{1.666}}{n_{\mathrm{m}}}$ = Chezy's coefficient of main channel,

y = Depth of water in entire compound channel;

y_{d1} = Left side elevation in deep channel;

y_{d2} = Right side elevation in deep channel;

Z_2, Z_3 are deep channel slopes of flood plain;

Z_1, Z_4 are left, right flood plain side slopes;

n_{FR}, n_{FL}, n_{m} = Manning's roughness coefficients of right, left flood plains, and main channel, respectively;

B_{FR} B_{FL}, B_{m} are widths of left, right flood plain, and main channel, respectively.

3.3.1 Discharge Calculation

For unsymmetrical section

Discharge in left flood plain $Q_{\mathrm{FL}} = A_{\mathrm{FL}} V_{\mathrm{FL}}$

Discharge in right flood plain Q_{FR}. Then, total discharge in entire compound channel

$$Q = Q_{\mathrm{FL}} + Q_{\mathrm{FR}} + A_{\mathrm{m}} V_{\mathrm{m}}$$

If flow occurs in first stage, the results for symmetrical section are compared with SERC-FCF Series 01 experimental measured data given by UK-FCF got from journal paper.

3.4 *Momentum Transfer Mechanism (MTM) Calculations Done in Program*

$$V_{\mathrm{m}} = Q_{\mathrm{m}}/A_{\mathrm{m}}; \quad V_{\mathrm{fl}} = Q_{\mathrm{fl}}/A_{\mathrm{fl}}; \quad H_{\mathrm{fl}} = y - y_{\mathrm{m}};$$

$$I_1 = \frac{Z_1}{(1+Z_1^2)^{0.5}}$$

$$G1 = \sqrt{1 - \left(\frac{1/Z_1^2}{0.15838444}\right)}; \quad G1 = \sqrt{(H_{\mathrm{fl}}^2 Z_1^2) + H_{\mathrm{fl}}^2}$$

$$G = (5503.41 \times \mathrm{d}a \times G1 \times I_1 \times G2)(5503.41 \times \mathrm{d}a \times b_{\mathrm{fl}})$$

$$\tau_{\mathrm{a}} = \frac{G - (A_{\mathrm{fl}} \times 9810 \times S_{\mathrm{ol}})}{H_{\mathrm{fl}}}; \quad \alpha = \frac{2\tau_{\mathrm{a}}}{1000(V_{\mathrm{m}}^2 - V_{\mathrm{fl}}^2)}$$

Calculate P_{m}, P_{fl}, A_{fl}, A_{m}, R_{fl}, R_{fl}, $C_{\mathrm{f\,1}}$, C_{fr}, C_{m}, A, B

$$A = -\frac{\propto H_{\mathrm{fl}}}{2g\sum \mathrm{d}p_{\mathrm{fl}}} + \left[\frac{\sum g\mathrm{d}p_{\mathrm{m}}}{\alpha H_{\mathrm{fl}} C_{\mathrm{m}}^2} + 1\right]\left[\frac{\propto H_{\mathrm{fl}}}{2g\sum \mathrm{d}p_{\mathrm{fl}}}\right] + \left[\frac{\sum \mathrm{d}p_{\mathrm{m}} g}{\alpha H_{\mathrm{fl}} C_{\mathrm{m}}^2} + 1\right]\frac{1}{C_{\mathrm{fl}}^2}$$

$$B = R_{\mathrm{fl}} So_{\mathrm{fl}} + \left[\frac{\sum \mathrm{d}p_{\mathrm{m}} R_{\mathrm{m}} S_{\mathrm{om}} H_{\mathrm{fl}}}{2H_{\mathrm{fl}}\sum \mathrm{d}p_{\mathrm{m}}}\right] + \left[\frac{\sum \mathrm{d}p_{\mathrm{m}} g R_{\mathrm{m}} S_{\mathrm{om}}}{\alpha H_{\mathrm{fl}} C_{\mathrm{fl}}^2}\right]$$

$$MV_{\mathrm{m}} = \sqrt{\frac{B}{A}}$$

$$MV_{\mathrm{fl}} = \sqrt{\left[\frac{-\sum \mathrm{d}p_{\mathrm{m}} g R_{\mathrm{m}} S_{\mathrm{o}}}{\alpha H_{\mathrm{fl}}} + \left[\frac{\sum \mathrm{d}p_{\mathrm{m}} g}{\alpha H_{\mathrm{fl}} C_{\mathrm{m}}^2} + 1\right] MV_{\mathrm{m}}^2\right]}$$

4 Results

4.1 *Cross-Sectional Detail of SERC-FCF Series 01 (Fig. 7)*

B_{m} = Main channel width = 0.75 m; H = Total depth of water (m); B_{f} = Flood channel width = 4.1 m; h = Flood water depth (m); B_{s} = Side slope horizontal = 0; S = Side slope = 1; Bed slope So = 0.001027; $H - h$ = 0.15 m; n = 0.01 (Figs. 8, 9, 10 and 11).

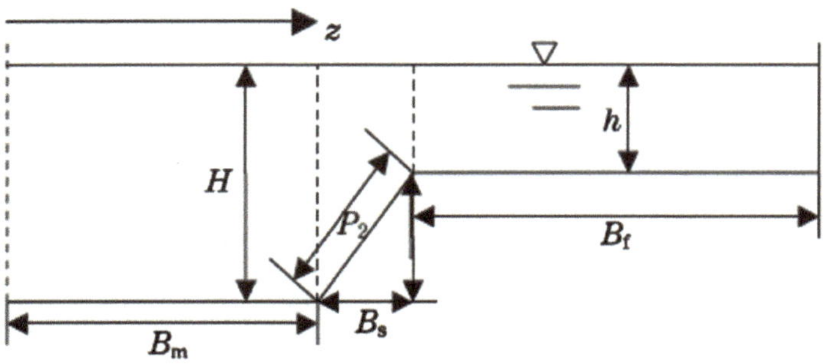

Fig. 7 Cross section of SERC-FCF Series 01

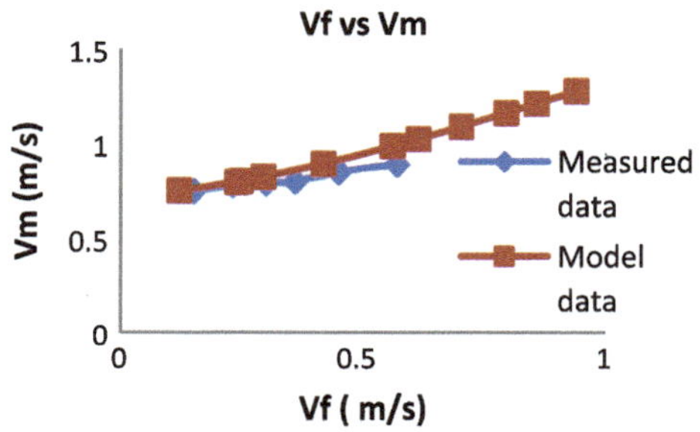

Fig. 8 Comparison of velocities in flood plain and in main channel with measured data

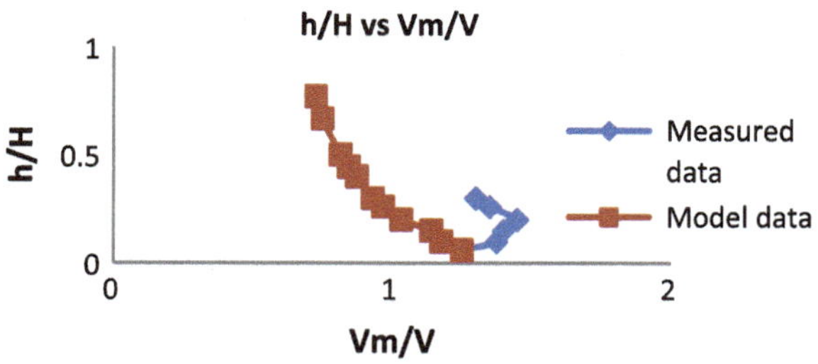

Fig. 9 Comparison of velocity ratios with measured data

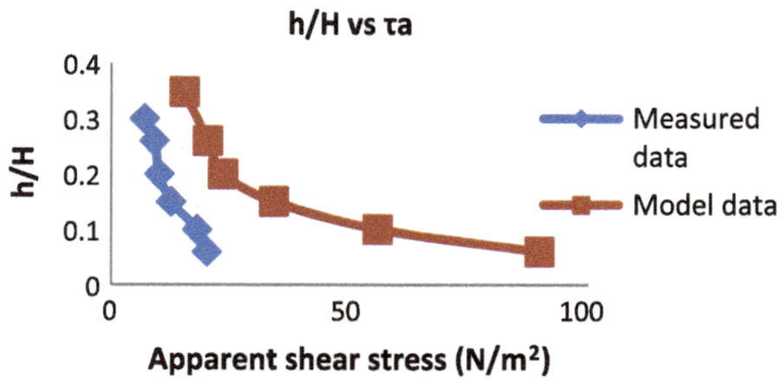

Fig. 10 Comparison of apparent shear stress with measured data

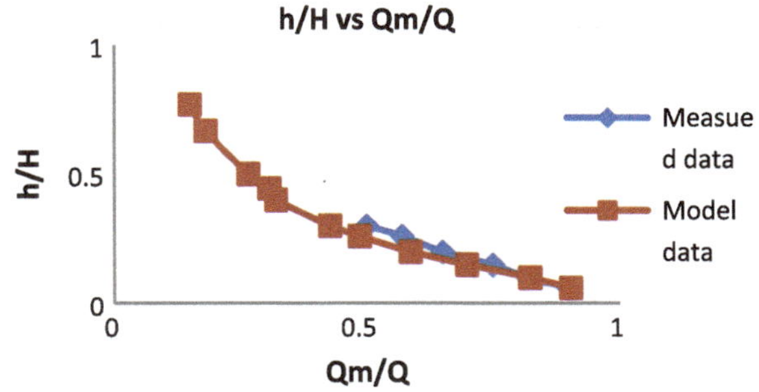

Fig. 11 Comparison of discharge ratios with measured data

4.2 Results for Four Stages Channel (Fig. 12)

Where

$B_m = 5$ m	$Z_m = 1$; $Z_1 = Z_2 = Z_3 = Z_4 = 1.5$
$B_{f1} = 10$ m	$n_m = 0.02$
$B_{f2} = 10$ m	$n_{f1} = n_{f2} = n_{f3} = n_{f4} = 0.1$
$B_{f3} = 10$ m	Bed slopes = 0.001
$B_{f4} = 10$ m	

4.3 Significant MTM Depths

Significant MTM (Momentum Transfer Mechanism) depths in different stages are as below (Fig. 13)

$$\text{1st stage} = 0.268\,\text{m}$$
$$\text{2nd stage} = 0.7\,\text{m}$$
$$\text{3rd stage} = 1.04\,\text{m}$$
$$\text{4th stage} = 1.23\,\text{m}$$

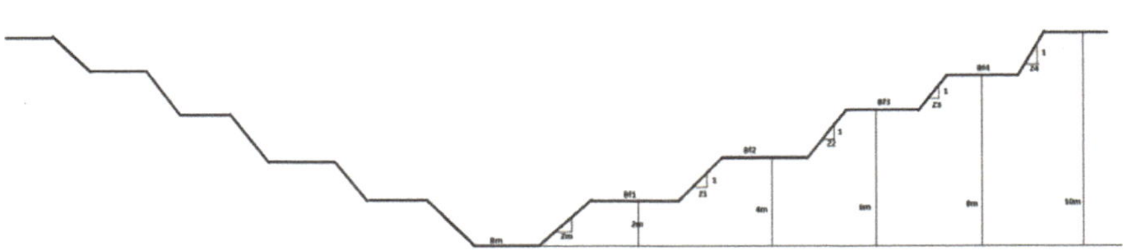

Fig. 12 Four stages symmetrical cross section

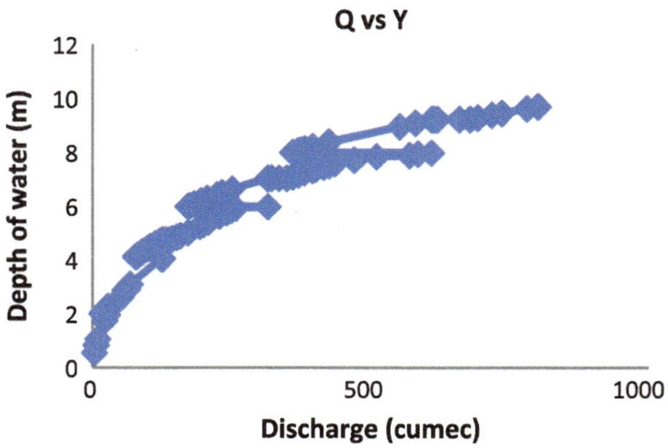

Fig. 13 Stage-discharge diagram

5 Conclusion

The two models are developed for symmetrical, unsymmetrical channels for one-dimensional (1-D) case only, but analysis of flow in compound channel in 1-D is not significant for practical purposes. Even though the results are obtained from symmetrical model correlated with measured data. Comparison of velocities in flood plain and in main channel with measured data and comparison of discharge ratios with measured data are well correlated.

Comparison of apparent shear stress with measured data is not well correlated; eventhough model apparent shear stress value is varying rapidly at lower depths, which is practically occurring. Moreover, all significant momentum transfer depths are increasing with increase of stage of compound channe, i.e., in first stage, it is very less and in second stage it is more.

6 Future Work

The results are not presented for unsymmetrical cross section, these results should be verified. Model should be developed for multi-stage for unsteady flow condition in two-dimensional analyses. There is another method for calculating of discharge considering momentum transfer mechanism (Liu and Dong method), so all results can further be verified using this method.

Notations used in program

Z_1, Z_2, Z_3, Z_4 are side slopes.
B_{fl}, B_{fr}, B_m are widths.
Y_1, Y_m are maximum depth of water that can occur in deep channel and main channel respectively.
S_{ol}, S_{or}, S_{om} are bed slopes of left, right flood plains, and main channel respectively.

n_{fl}, n_{fr}, n_{m} are Manning's roughness coefficients of left, right flood plains, and main channel respectively.
d_{a} is the diameter of sediment particle.
MCC, MCC1 are maximum discharge carrying capacities of deep channel and entire compound channel, respectively.
Q_{t} is "given total discharge".
Q_{g} is "given discharge": Y, H, H_1, H_2 are depth of water at that particular loop in program.
Q_{c} is "Entire compound channel computed discharge" with Y, H, H_1, H_2 values at that particular loop in program.
V_{fl}, V_{fr}, V_{m} velocities in left, right flood plains, and main channel without considering MTM respectively.
Q_1 is stored value of Q_{g} in loop.
A_{fl}, A_{fr}, A_{m} are area of left, right flood plains and main channel, respectively.
Q_{fl}, Q_{fr}, Q_{m} are discharges of left, right flood plains, and main channel, respectively.
P_{fl}, P_{fr}, P_{m}, are wetted perimeters of left, right flood plains, and main channel respectively.
H_{fl}, H_{fr} are depths of water in left, right flood plains, respectively.
C_{fl}, C_{fr}, C_{m} are Chezy's coefficients for left, right flood plains, respectively.
$\mathrm{MV}_{\mathrm{fl}}$, $\mathrm{MV}_{\mathrm{fr}}$, MV_{m} are velocities in left, right flood plains, and main channel considering MTM, respectively.

References

Dracos T, Hardegger P (1987) Steady uniform flow in prismatic channels with flood plains. J Hydr IAHR 25(2):169–185

Einstein HA, Banks RB (1950) Fluid resistance of composite roughness. Trans Am Geophys Union 314:603–610

Knight DW, Demetriou JD (1983) Flood plain and main channel flow interaction. J Hydraul Eng ASCE 109(8):1073–1092

Lotter GK (1933) Considerations on hydraulic design of channels with different roughness of walls. Transactions, All-Union Scientific Research Institute of Hydraulic Engineering, Leningrad 9:238–241

Myers WRC (1978) Momentum transfer in a compound channel. Hydr Res IAHR 16(2)139–150

Pavlovskii NN (1931) On a design formula for uniform flow in channels with non-homogeneous walls. Transactions, All-Union Scientific Research Institute of Hydraulic Engineering, Leningrad 3:157–164

Posey CJF (1967) Computations of discharge including overbank flow. Civil Eng ASCE 37(4): 62–63

Sellin RHJ (1964) A laboratory investigation into the interaction between the flow in the channel of a river and that over its flood plain, vol 7. La Houille Blanche, Paris, France, pp 793–802

Yen B (1991) Channel flow resistance: Centennial of manning's formula. Water Resource Publications, Highlands Ranch, Colorada

Wormleaton PR, Allen J, Hadjipanos P (1982) Discharge assessment in compoundchannel flow. J Hydraul Div 108(9):975–979

Zeng Y, Wang Y (2010) Hydrauliccalculation of steady uniform flows in trapezoidal compound open channels. Appl Math Mech 31(8):947–954

Optimal Reservoir Operation with Environmental Flows for Ranganadi Hydroelectric Project in Arunachal Pradesh

Mudo Puming and Ram Kailash Prasad

Abstract Optimal release policy for hydro power has been developed extensively in the past for several river basins in India and abroad. However, there is a lack of available literature which takes into account the optimal release along with the necessary environmental flow to the downstream side. As a consequence several hydropower projects are facing problem due to the environmental concerns. In Arunachal Pradesh, some of the hydropower projects are under critics basically due to environmental factors. In this study, a formulation has been developed to obtain the optimal release for hydropower for Ranganadi Hydro Electric Project (RHEP) in Arunachal Pradesh, considering the constraints of minimum environmental flow. Traditionally reservoirs used to be operated without considering the environmental flows. Whenever the release to the downstream side (or overflow) is to be considered, the flow exceeds the active storage capacity of the reservoir. In this study two models have been formulated, one considers the optimal release for hydropower without consideration of minimum environmental flow and the other with the consideration of environmental flows and have been solved using Lingo. Thus traditional optimal release policy has been compared with the condition of optimal release when minimum environmental flows are enforced at the downstream side of the reservoir. The minimum environmental flow has been obtained based on the practices being followed in India and the other parts of the world.

Keywords Optimal reservoir operation · Hydroelectric project · Environmental flows

---

M. Puming · R.K. Prasad (✉)
Department of Civil Engineering, NERIST, Nirjuli, Arunachal Pradesh, India
e-mail: rkp@nerist.ac.in

A.K. Sarma et al. (eds.), *Urban Hydrology, Watershed Management and Socio-Economic Aspects*, Water Science and Technology Library 73,
DOI 10.1007/978-3-319-40195-9_17

1 Introduction

In India, energy security is mostly linked to the renewable sources of energy such as hydropower generation. Arunachal Pradesh has the highest potential for hydropower development in India and among the northeastern states in particular. In order to develop this potential, it is necessary to utilize the available water resources effectively in a well-planned manner. Optimal release of water from the reservoir for the hydropower development is one of the ways to utilize available water resources for maximum power production. Hydropower development, in particular, is closely related with the economic and energy security of a country. Hydropower is considered as a green energy source which is generated from moving water. Hydropower is also known as a clean source of energy which does not produce solid, liquid, or gaseous pollutants. Hydropower is one renewable energy source which is being implemented to replace the fossil fuels, a commodity in the process of depletion. In India, there are plenty of water resources, but unevenly distributed from region to region or with time. It is also found that during hydropower generation flow on the downstream region is usually neglected. But to maintain an environmentally sustainable reservoir operation, flow on the downstream region should be checked. The environmental flow is the water that is left in a river ecosystem or released into it for the specific purpose of managing the condition of that ecosystem. Failure to maintain such flow may lead to decline in the health of water dependent system. Hence, it is necessary to maintain a minimum environmental flow during reservoir operations. The literature review (Kumar 1999; Mathur and Nikam 2009; Homa et al. 2005; Kumar et al. 2007; Vedula and Majumdar 2007) shows that many works are done in the past for reservoir operation. The theoretical aspects of influence of environment for river basins are also studied in the past. However, there is a lack of works which combines both environment and reservoir operation in a single framework. This study is an attempt to explore the optimal reservoir operation with environmental consideration. The study aims at the following objectives:

1. To study the flow pattern of river Ranganadi and the power production scenario from RHEP under the condition of different turbines in operation,
2. To develop a model for optimal release for hydropower from Ranganadi under existing condition of river flow,
3. To develop a model for optimal hydropower production from river Rangandi by considering the minimum environmental flow that is required to maintain the health of water dependent system. The minimum environmental flow considered here is required for quality habitat of flora and fauna on the downstream.

2 Methodolgy

The following two models have been formulated in this study.

Model-I: Optimal Release for hydropower without consideration of minimum environmental flow.

$$
\begin{aligned}
&\text{Maximize } Z \\
&\quad \text{where } Z = \sum_{t=1}^{12} R_t \\
&\text{Subject to} \\
&S_t \le S_{\max} && t = 1, 2, \ldots..12 \\
&R_t \le T_c && t = 1, 2, \ldots..12 \\
&S_{t+1} = S_1 + Q_1 - E_1 - R_1 - O_1 && t = 1, 2, \ldots..12 \\
&R_t \le D_t && t = 1, 2, \ldots..12 \\
&S_t \le K && t = 1, 2, \ldots..12 \\
&R_t \ge 0 && t = 1, 2, \ldots..12 \\
&S_t \ge 0 && t = 1, 2, \ldots..12 \\
&S_{T+1} = S_1 && t = 1, 2, \ldots..12 \\
&O_t \le X_t M && t = 1, 2, \ldots..12 \\
&X_t \le S_{t+1}/K && t = 1, 2, \ldots..12
\end{aligned}
$$

Model-II: Optimal Release for hydropower with consideration of minimum environmental flow.

$$
\begin{aligned}
&\text{Maximize } Z \\
&\quad \text{where } Z = \sum_{t=1}^{12} R_t \\
&\text{Subject to} \\
&S_t \le S_{\max} && t = 1, 2, \ldots..12 \\
&R_t \le T_c && t = 1, 2, \ldots..12 \\
&S_{t+1} = S_1 + Q_1 - E_1 - R_1 - O_1 - Rd_t && t = 1, 2, \ldots..12 \\
&R_t \le D_t && t = 1, 2, \ldots..12 \\
&S_t \le K && t = 1, 2, \ldots..12 \\
&R_t \ge 0 && t = 1, 2, \ldots..12 \\
&S_t \ge 0 && t = 1, 2, \ldots..12 \\
&S_{T+1} = S_1 && t = 1, 2, \ldots..12 \\
&O_t \le X_t M && t = 1, 2, \ldots..12 \\
&X_t \le S_{t+1}/K && t = 1, 2, \ldots..12
\end{aligned}
$$

Further,

$$
\begin{aligned}
&\text{If} \quad Rd_t \le Q_{\min} \quad \text{then} \quad Rd_t = Q_{\min} \quad t = 1, 2, \ldots..12 \\
&\text{If} \quad O_t > Q_{\min} \quad \text{then} \quad Rd_t = Q_{\min} \quad t = 1, 2, \ldots..12
\end{aligned}
$$

where, R_t is the release for hydropower production during period t; T is the last period in the year; S_t is the storage at the beginning of the period t; $S_{\max}$ is the maximum storage; T_c is the turbine capacity; S_{t+1} is the storage at the beginning of period $(t + 1)$ which is the same as the storage at the end of period t; D_t is the demand during the period t; Q_t is the inflow during the period t; K is the reservoir capacity during the period t; O_t is the Spill or overflow from reservoir; X_t is any integer; M is any arbitrary large constant; K is the reservoir storage; Rd_t is the actual downstream release, which includes O_t when O_t is greater than $Q_{\min}$; $Q_{\min}$ is the minimum environmental flow needed for maintaining sustainable environmental flow. Further, the power production can be calculated as

$$\begin{aligned} \text{Power P in kWH} &= 9.81 \times 0^6 R_t H_t/3600 \\ &= 2725\, R_t H_t \\ \text{Power P in MW} &= 2725\, R_t H_t \eta/(1000 * 30 * 24) \end{aligned}$$

where,

R_t Release into penstocks in Mm^3

H_t Design head in meters

η Plant efficiency

The above two models have been solved using LINGO (Sarker and Newton 2008) for the data of RHEP for year 2004–2008.

3 Results and Discussion

The following results are obtained from the present study.

3.1 *Optimal Release for Hydropower Without Considering Minimum Environmental Flows (MEF)*

3.1.1 Optimal Release for Two Turbines in Operation

Although, the RHEP has the installed capacity of three turbines each with a capacity of 135 MW, in the following discussions it is assumed that only two turbines are in operation for different duration.

Figures 1, 2, 3 and 4 depict the demand and optimal release pattern for two numbers of turbine running for different durations, such as 24, 12, 8, and 6 h. These figures can be used to study the effect of different hours of running of turbines on the demand and release patterns. To get an idea of general demand and release pattern only the data of the most recent year available, i.e., for the year 2008 has been plotted.

From Fig. 1 it is observed that when two numbers of turbines are operational for 24 h, the release satisfy the demand from June to October and for other months the demands are not satisfied.

From Fig. 2 it is observed that when two numbers of turbine is operational for 12 h, the release satisfies the demand from May to October.

From Fig. 3 it is observed that when two nos. of turbine is operational for 8 h release satisfy the demand from March to October.

From Fig. 4 it is observed that when two numbers of turbine is operational for 6 h release satisfy the demand almost from January to December. However, during the month of February due to decreased inflow the demand needed is not satisfied.

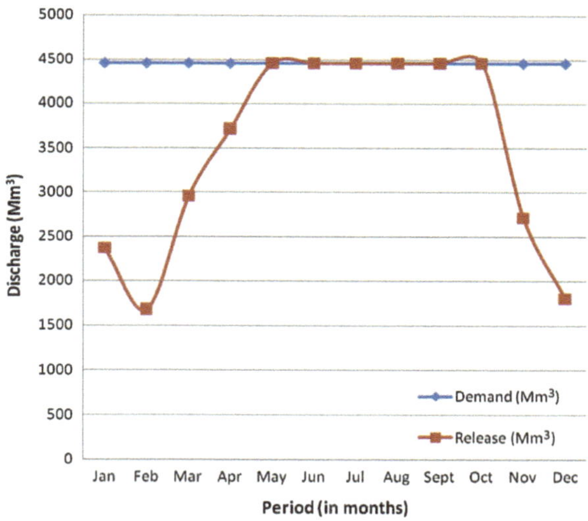

Fig. 1 Pattern of demand and optimal release when 2 nos. of turbine is running for 24 h for year 2008

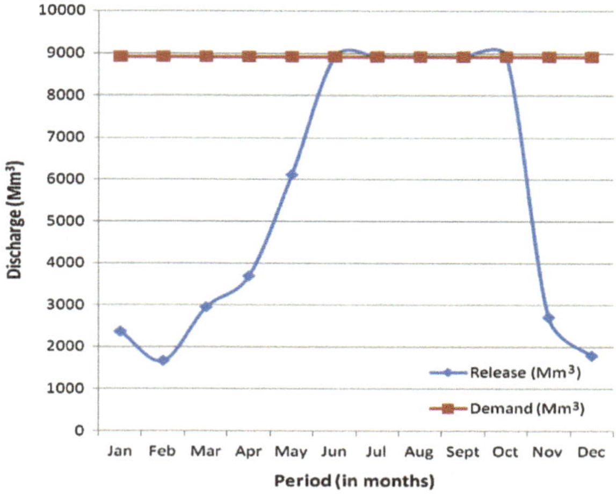

Fig. 2 Pattern of demand and optimal release when 2 nos. of turbine is running for 12 h for year 2008

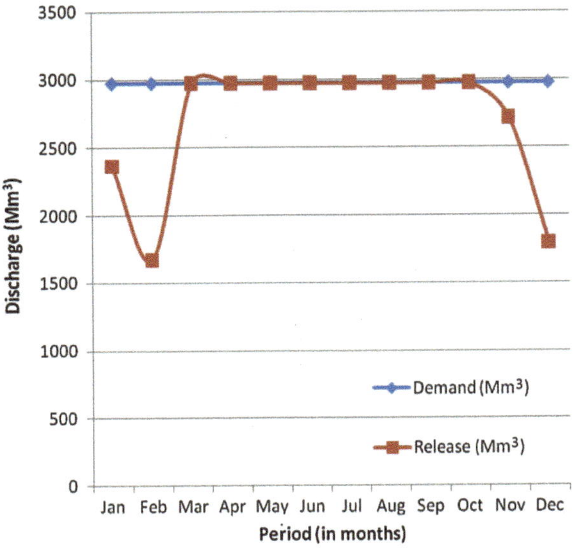

Fig. 3 Pattern of demand and optimal release when 2 nos. of turbine is running for 8 h for year 2008

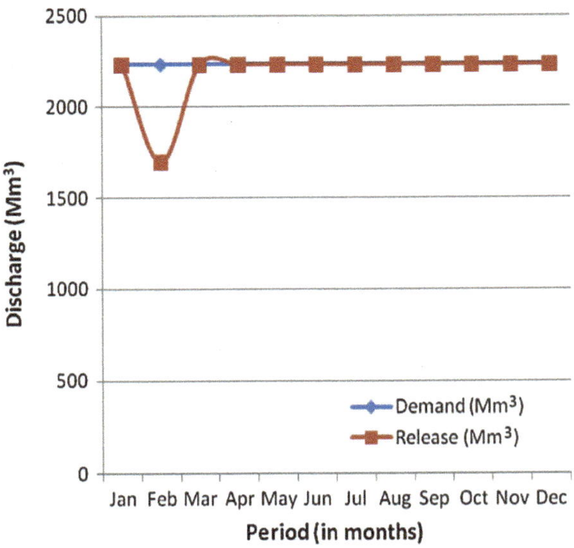

Fig. 4 Pattern of demand and optimal release when 2 nos. of turbine is running for 6 h for year 2008

Hence, it can be concluded that during period of lower inflows, the operational duration of two turbines can be decreased. This ensures that the optimal release would satisfy the demand. This study will also help in scheduling of turbines for optimal release and to identify the months during which the turbine can be made operational.

3.1.2 Optimal Release for Three Turbines in Operation

The RHEP has the capacity of three turbines each with an installed capacity of 135 MW, hence it is necessary to know whether the inflow available to the Ranganadi would meet the demand needed for three turbines to be operational simultaneously.

Figures 5, 6, 7 and 8 depict the demand and optimal release pattern for three numbers turbine running for different durations such as 24, 12, 8, and 6 h. These

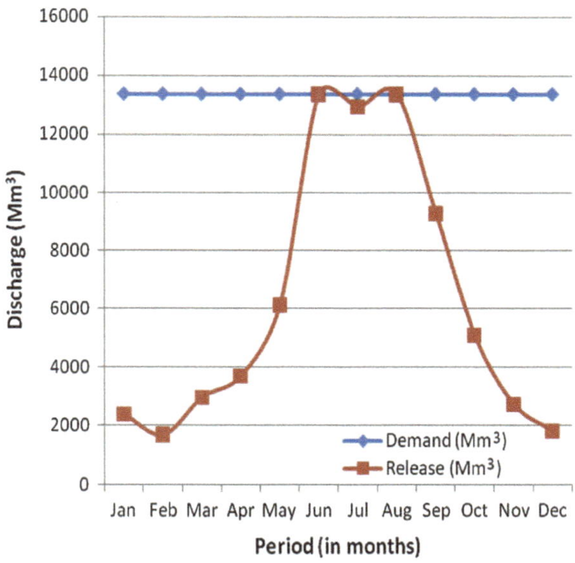

Fig. 5 Pattern of demand and release when 3 nos. of turbine is running for 24 h for year 2008

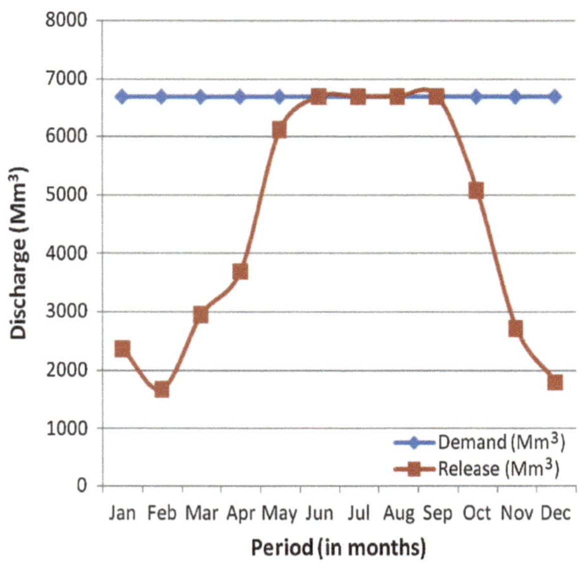

Fig. 6 Pattern of demand and release when 3 nos. of turbine is running for 12 h for year 2008

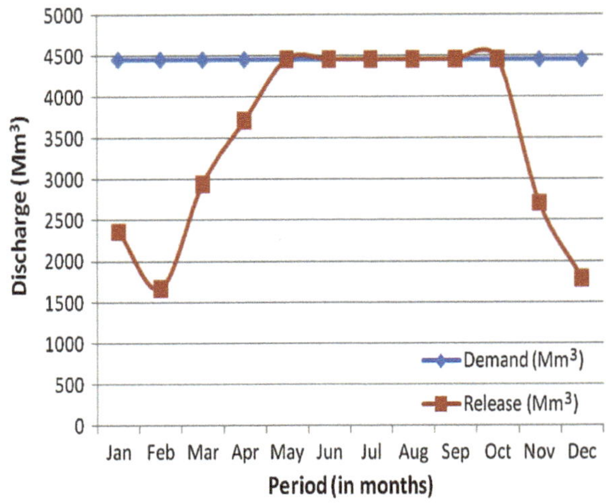

Fig. 7 Pattern of demand and release when 3 nos. of turbine is running for 8 h for year 2008

figures can be used to study the effect of different hours of running of turbines on the demand and release patterns.

From Fig. 5 it is observed that when three nos. of turbine is operational for 24 h release satisfies the demand from the month of June to August and other months demand is not satisfied.

From Fig. 6 it is observed that when three numbers of turbine is operational for 12 h release will satisfy the demand from the month of June to September.

From Fig. 7 it is observed that when three numbers of turbine is operational for 8 h release satisfy the demand from the month of May to October.

From Fig. 8 it is observed that when three numbers of turbine is operational for 6 h release satisfy the demand from the month of April to October. However,

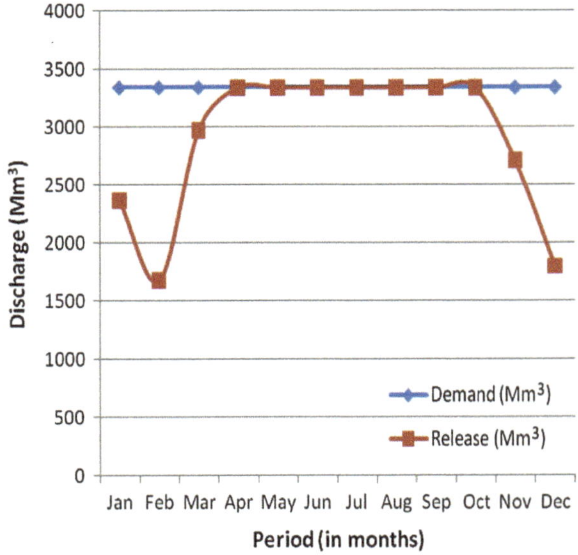

Fig. 8 Pattern of demand and release when 3 nos. of turbine is running for 6 h for year 2008

Table 1 Pattern of optimal release and power obtained for 2-nos. of turbine operational

Turbine operational period (h)	Optimal release (Mm^3)	Power (MW)
24	65,897.19	59,856.61
12	41,967.70	41,128.35
8	32,328.50	29,354.28
6	26,215.31	23,803.50

Table 2 Pattern of optimal release and power obtained for 3-nos. of turbine operational

Turbine operational period, t (h)	Release (Mm^3)	Power (MW)
24	75,356.69	68,423.87
12	53,140.32	4,825,141
8	41,967.58	38,106.56
6	34,929.34	31,715.84

during the months from January to March, and November to December the demands are not satisfied.

Hence it can be observed that during seven months (i.e., April–October) of the year, even three turbines can be made operational for 6 h only. However for 8 h of operation of turbine, the demands are satisfied for 6 months and for 12 h and 24 h demand are satisfied for 4 months and 3 months respectively. This further help in scheduling of three turbines.

Tables 1 and 2 summarizes the total optimal release for two numbers and three numbers of turbine in operation, respectively. The powers likely to be generated for these releases are also tabulated in Tables 1 and 2. It is evident that even though operation duration of two and three numbers of turbines is increased to fourfold the power generation increases to almost double only.

Comparing Table 1 with Table 2, it can be observed that during the period of sufficient inflow even three numbers of turbine can be made operational for smaller duration. However, for normal inflows running two turbines for relatively increased duration appears to be a better alternative than running three turbines for smaller duration, as this alternative produces more power.

Observing the optimal release which can be made for two and three turbine case, the analysis also supports the present strategy followed by RHEP for running two turbines and keeping one turbine as standby.

3.1.3 Optimal Release for Hydropower with Consideration of MEF

From the previous section, it is observed that the present condition of flow of the river Ranganadi only satisfy the demand needed for the operation of two turbines for most of the months of the year, and hence, the case of MEF has been studied for two turbines only. Further the inflow data of year 2004–2008 has been utilized to obtain optimal release with MEF.

3.1.4 Determination of Minimum Environmental Flow for River Ranganadi

Environmental flows are the water that is left in a river ecosystem or released into it for the specific purpose of managing the condition of that ecosystem. Failure to maintain such flows may lead to decline in the health of water dependent ecosystem.

The minimum environmental flows for river Ranganadi has been calculated taking into consideration of the available practices followed by different countries and agency.

The minimum environmental flow (Kumar et al. 2007) in U.K is fixed as $0.7 \times Q_{95}$ for least sensitive ecosystem. Where, Q_{95} is flow which is equaled or exceeded 95 % time. From Fig. 9 the value of Q_{95} is obtained as 1661.37 Mm^3 and the minimum environmental flows as 1162.96 Mm^3.

In USA minimum environmental flow is fixed as 10 % of mean flow for poor quality habitat of fish, and hence for river Ranganadi the mean flow was obtained for a period of 5 years that is from January 2004 to December 2008 and the value of 10 % of mean flow is obtained as 806.88 Mm^3.

Similarly, Himachal Pradesh state Environment Protection and Pollution Control Board has issued guidelines for minimum releases to be made from Nathpa dam as 15 % of the observed minimum flow in the lean season. After considering this guideline, for Ranganadi the value of lean flow was observed in the month of January 2007 and hence 15 % of the observed minimum flow in the lean season is obtained as 236.802 Mm^3.

Minimum environmental release as per prevailing practice has been calculated and is tabulated in Table 3.

From Figs. 10, 11 and 12 it can be observed that the release for generating hydropower is affected due to consideration for minimum environmental flows by considering different situations as MEF-I, MEF-II, and MEF-III. It can also be

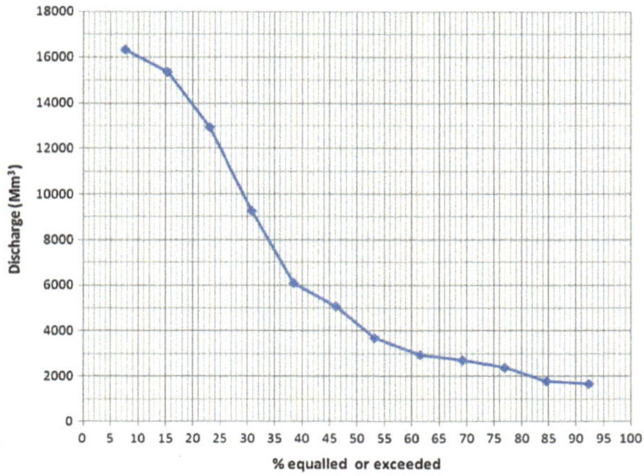

Fig. 9 Flow duration curve for river Ranganadi

Table 3 Minimum environmental flow as per available practices for river Ranganadi

Agency/country practice	Minimum flow requirement criteria	Minimum flow requirement d/s of Ranganadi dam (Mm^3)
U.K. (Environmental Agency 2002)	0.7 × Q_{95} for least sensitive ecosystem **(MEF-I)**	1162.96
USA (Montana method), Tenant 1976	10 % of mean flow for poor quality habitat of fish **(MEF-II)**	806.88
HP state environmental protection and pollution control board	15 % of the observed minimum flow in the lean season **(MEF-III)**	236.802

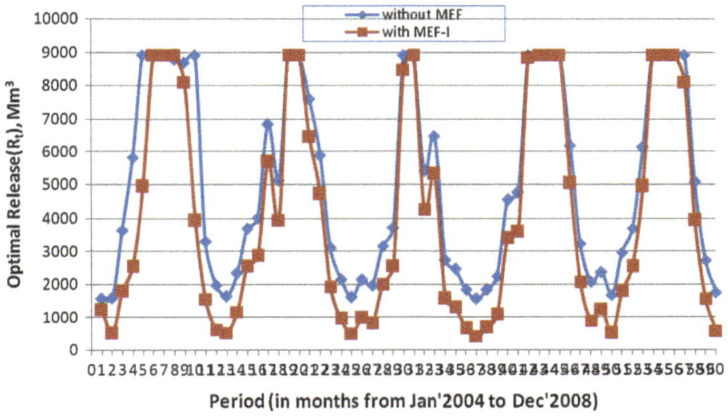

Fig. 10 Pattern of optimal release without considering MEF and by considering MEF-I when two nos. of turbine is running for 24 h for the period Jan 2004–Dec 2008

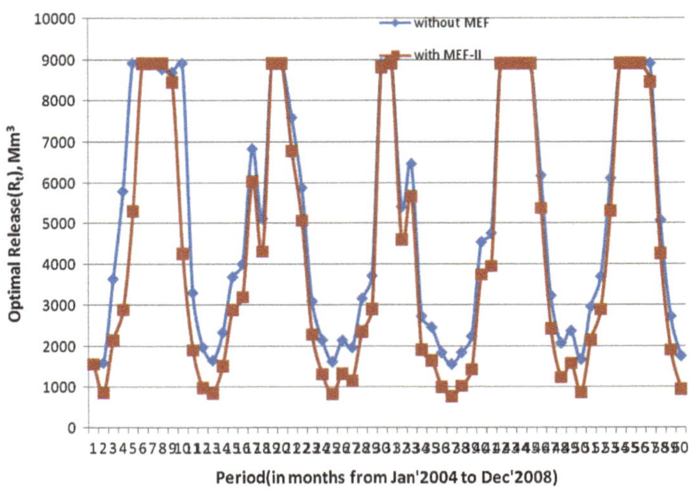

Fig. 11 Pattern of optimal release without considering MEF and by considering MEF-II when two nos. of turbine is running for 24 h for the period Jan 2004–Dec 2008

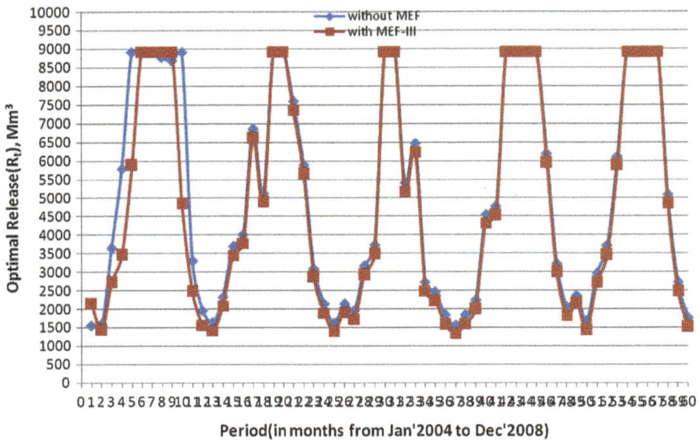

Fig. 12 Pattern of optimal release without considering MEF and by considering MEF-III when two nos. of turbine is running for 24 h for the period Jan 2004–Dec 2008

observed that with decreasing value of MEF from MEF-I to MEF-III, the hydropower generation increases without affecting environment conditions.

Hence, the minimum downstream release of approximately 236 Mm^3 may be enforced downstream inclusive of overflow which however, only occurs when flow exceeds the storage capacity of the reservoirs for RHEP.

Thus a minimum environmental flow policy similar to that of Himachal Pradesh Environment Protection Agency may be considered by Arunachal Pradesh state Government in consultation with Assam Government or Government of India to resolve the conflict to some extent. Hence, it can be concluded that MEF policy as practiced in different parts of country or states may lead to optimal release with a sustainable environment system. The policy may lead to small reduction in power prediction at the cost of preserving environment system downstream of dam. Thus, a concept of environment friendly optimal release developed in this study may be utilized in ongoing projects as well as proposed Hydropower Project in Arunachal Pradesh.

Figure 13 Shows the total power production under different environmental releases (MEF-I, MEF-II, MEF-III) for years 2004–2008. This figure also compares the power release without MEF.

It can be observed from the table that there is a sharp decline in power generation for the scenario MEF-I when compared with the situation of without MEF. However for MEF-III the power reduction appears to be sustainable with the environmental conditions. Figures 14, 15 and 16 compare the power generation without MEF with MEF-I, MEF-II, and MEF-III, respectively.

Fig. 13 Pattern of Power generation in MWh from year 2004 to 2008 under conditions of power generation without MEF and with MEF-I, MEF-II and MEF-III

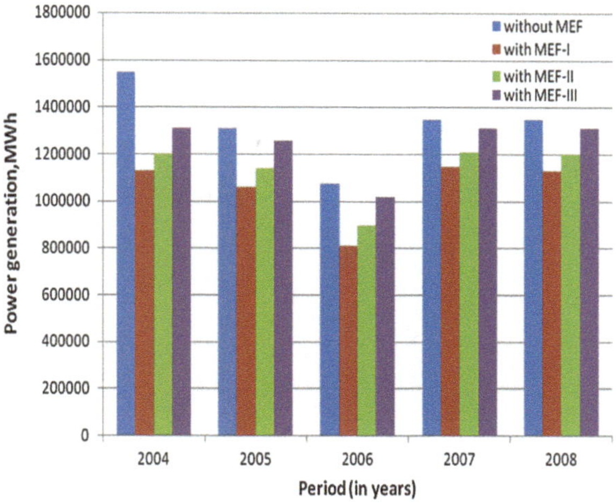

Fig. 14 Pattern of Power generation in MWh from year 2004 to 2008 under conditions of without MEF and with MEF-I

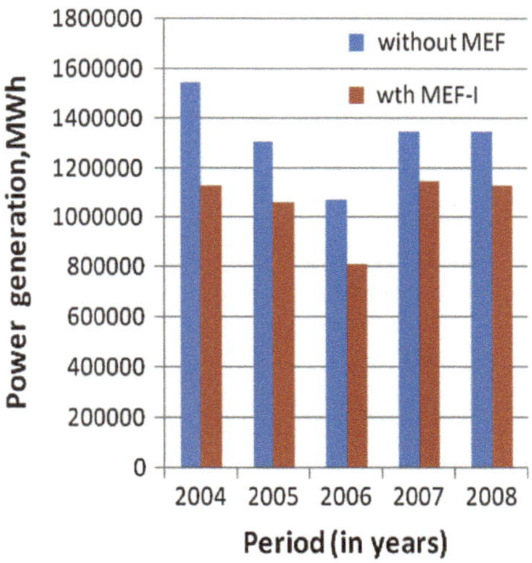

Fig. 15 Pattern of Power generation in MWh from year 2004 to 2008 under conditions of without MEF and with MEF-II

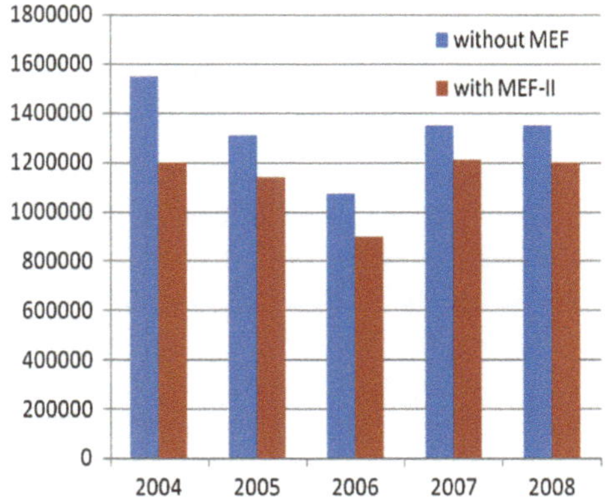

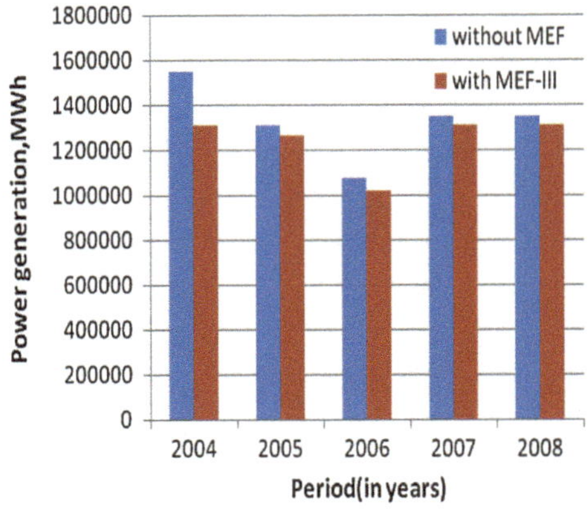

Fig. 16 Pattern of Power generation in MWh from year 2004 to 2008 under conditions of without MEF and with MEF-III

4 Conclusions

In this study, the optimal release from the Ranganadi Hydroelectric Project has been studied. The study explores the optimal release under different environmental conditions such as peak flow and lean flow in the river Ranganadi for a period of 5 years. The existing conditions of operation, along with the flow patterns available also have been studied. Based on the study the following conclusions can be drawn from this study.

The analysis of the discharge data shows that Ranganadi is perennial in nature with minimum flow during the lean period (December and January) and maximum flow during monsoon season (June to August). During the monsoon season the inflow is sufficient to run two turbines with the installed capacity, however during the lean period the inflow is insufficient to run two turbines with the installed capacity.

First, the optimal release without minimum environmental flows (MEF) has been considered with either three turbines or two turbines in operation. It is found that during the period of sufficient inflow three numbers of turbines can be made operational only for small duration. However, for normal inflows running of two turbines for less than 24 h appears to be a better alternative than running three turbines. This is in conformity with the present operating policy of RHEP of running two turbines and keeping one as standby. This study would further help in scheduling of turbines.

Second, the study of optimal release with consideration of minimum environmental flows has been discussed. Three cases of minimum environmental flows have been considered in this study and are observed that power generated by considering MEF-III is comparatively closer with the power generated without considering the MEF. Since the downstream side of river Ranganadi is not used for fish culture by the local people residing in the nearby area, while the downstream release is required only for maintaining the natural ecosystem such as flora and fauna habitat, etc. Hence, MEF-III with a MEF of 15 % of the observed minimum flow of the lean season may be considered to be justified for a environmentally sustainable hydropower project for Ranganadi. Also, the policy MEF-I, MEF-II, and MEF-III have been applied to the data from year 2004 to 2008 and it is found that during the lean period from January to May as well as from October to December, the policy MEF-I shows a sharp decline in power generation followed by MEF-II. However, in case of MEF-III, the power generation without MEF closely matches with MEF-III. This way any of the acceptable MEF policy can be implemented in hydro electric project.

References

Environmental Agency (2002) Managing water abstraction, the catchment abstraction management process, available on <http://www.environmentalagency.gov.uk/commondata/105385/mwaenglish.pdf>

Homa ES, Vogel RM, Smith MP, Apse CD, Huber-Lee A, Sieber J (2005) An optimization approach for balancing human and ecological flow needs. In: Proceedings of the EWRI world water and environmental resources congress. ASCE, Anchorage, Alaska

Kumar A (1999) Optimal reservoir operation using genetic algorithms. M.Tech Thesis, IIT Kharagpur

Kumar P, Umesh Chandra C, Surendra Kumar M (2007) Environmental flows for hydropower projects—a case study. In: International conference on small hydropower—hydro Sri Lanka

Mathur YP, Nikam SJ (2009) Optimal reservoir operation policies using genetic algorithm. Int J Eng Technol 1(2):184–187

Sarker RA, Newton CS (2008) Optimization modelling: a practical approach. CRC Press, New York

Vedula S, Majumdar PP (2007) Water resources systems modelling techniques and analysis. Graw-Hill Publishing Company Limited, Tata Mc

Part V
Soft Computing Techniques in Urban Hydrology

Analysis of Water Distribution Network Using Epanet and Vertex Method

P. Sivakumar and Ram Kailash Prasad

Abstract The analysis of hydraulic behavior of the water distribution network (WDN) is forefront part of the planning and augmentation of any water supply projects. The analysis of WDN determines the estimation of discharges, hydraulic gradient levels (HGL), nodal concentrations, etc., to fulfill the requirements of population. In the conventional approach of analysis, unique value of pipe discharges and hydraulic heads are obtained. The results so obtained may not give satisfactory performance in practice due to many uncertainties in nodal demands, pipe roughness, lengths, diameters of pipes, water levels in reservoirs, head-discharge characteristics of pumps, etc. In this study, the uncertainty in discharges and hydraulic heads are evaluated using two different pipe networks, the data for which is obtained from literature. Further, the membership function of pipe roughness has been used to calculate the membership function of discharges and hydraulic heads by incorporating EPANET with vertex method of fuzzy approach. The uncertainties in discharges and hydraulic heads at different α-cuts are evaluated considering the uncertainties in pipe roughness. The results of pipe discharges are found to vary between 15 and 30 % whereas the hydraulic heads at nodes vary between 0.3 and 3 m when the uncertainty is about 8 % in Hazen–Williams coefficient of pipe roughness in selected four- and five-pipe networks, respectively. Moreover, when the uncertainty of pipe roughness is combined with other kind of uncertainties as discussed above, would further aggravate the uncertainty in pipe discharges and nodal heads. As a result, the reliability of network would decrease in terms of either meeting the required discharges or the nodal heads to the consumers. This study would help to design the pipe network under the conditions of uncertainty in input parameters.

Keywords Water distribution network · EPANET 2 · Vertex method

P. Sivakumar (✉) · R.K. Prasad
Department of Civil Engineering, NERIST, Nirjuli, Arunachal Pradesh, India
e-mail: siva_nerist@yahoo.co.in

R.K. Prasad
e-mail: rkp@nerist.ac.in

A.K. Sarma et al. (eds.), *Urban Hydrology, Watershed Management and Socio-Economic Aspects*, Water Science and Technology Library 73,
DOI 10.1007/978-3-319-40195-9_18

1 Introduction

The analysis of water distribution network (WDN) determines the estimation of discharges, hydraulic gradient levels (HGL), nodal concentrations, etc., to fulfill the requirements of the population. In the conventional approach of analysis, unique value of the pipe discharges and hydraulic heads are obtained. The results so obtained may not give satisfactory performance due to many uncertainties in nodal demands, pipe roughness, lengths, diameters of pipes, water levels in reservoirs, head-discharge characteristics of pumps, etc. Due to the complex behavior of WDN, the reliable measurements is not normally possible in each and every node and links of the network. While augmenting the existing network, the length and diameters of pipe are assumed to be consistent even if the network has been used for many years. The diameter and friction coefficient of the pipe may vary due to scale formation on inside surface of pipes and aging process; length of pipe also vary due to introduction of joints or removal of a pipe line during the normal course of operation. However, the same have not been considered in the conventional method of analysis and hence do not give the expected result. Thus in the present study, pipe roughness is taken as an uncertainty parameter. The brief description of the EPANET, Fuzzy sets, and Vertex methods is given below.

EPANET (Rossman 2000) is a public domain software package that performs the extended period simulation of hydraulic and water quality behavior within the pressurized pipe networks which is based on the principle of gradient method (Todini and Pilati 1987). The output of the analysis of WDN such as pipe discharges, nodal heads, nodal concentrations, etc., can be readily calculated with available input parameters such as pipe roughness coefficient, length, diameter of pipe, etc.

The fuzzy set theory was introduced by Zadeh (1965), which is used for dealing with imprecision due to uncertainty and vagueness, which is essential for many engineering problems. It resembles human decision-making with its ability to work from approximate data and imprecise solutions. Since its inception, it has been used to describe imprecision in ground water (Dou et al. 1997a; Guan and Aral 2004; Prasad and Mathur 2007); and in the pipe network problems (Revelli and Ridolfi 2002; Bhave and Gupta 2004; Gupta and Bhave 2007). The imprecisely known parameters can be taken as fuzzy and it can be defined by membership function (Ross 1995; Kaufmann and Gupta 1991); however, it is computationally expensive for complex problems (Ross 1995; Guan and Aral 2004), because of large number of simulation run of the model.

Dong and Shah (1987) suggested the vertex method which is based on α level cut concept and interval analysis. In this method, the membership function is discretized, rather than discretizing the variable domain. The discretization of the membership domain is accomplished by dividing the membership domain into a series of equally spaced cuts, called α-cuts; α represents the possibility. For each α-cut, the maximum and minimum value of the fuzzy variable is selected. Further, the disadvantage of the vertex method is that the numbers of function evaluations are increased by the number of α-cuts raised to the power of the number of random variables.

Revelli and Ridolfi (2002) analyzed a pipe network by fuzzy approach through optimization considering uncertain parameters such as pipe roughness coefficient, nodal demands, and pizometric heads. Xu and Goulter (1999) and Bhave and Gupta (2004) optimized the water distribution networks with fuzzy demands using linear programming. Gupta and Bhave (2007) reanalyzed the network of Revelli and Ridolfi (2002) by direct method using impact table to obtain the dependent parameters of pipe discharges and nodal heads. They concluded that their proposed methodology requires less computational effort and time as compared to that of Revelli and Ridolfi (2002).

In the present study, two-pipe network problems are taken where the pipe roughness is taken as fuzzy and all other parameters are taken as crisp. Example 1 is a simple network problem taken from Bhave and Gupta (2006) to explain the proposed methodology. Example 2 is taken which has already been analyzed by Revelli and Ridolfi (2002) and Gupta and Bhave (2007). Revelli and Ridolfi (2002) used Strickler coefficient of pipe roughness as fuzzy whereas Bhave and Gupta (2006) have considered Hazen–Williams coefficient of pipe roughness as fuzzy. In this study, the Hazen–Williams coefficient of pipe roughness is taken as fuzzy for the simulation of WDN.

2 Fuzzy Sets and Vertex Method

Fuzzy set is a set of objects without clear boundaries or one without well-defined characteristics (Freissinet et al. 1999). The membership function establishes how much the element "belongs" to the set and is included in the interval [0, 1], the convention being that the closer it is to 1, the more the element belongs to the set and vice versa. Fuzzy sets, membership functions, and α level cuts are defined mathematically by Ross (1995). Thus, if X class of objects is denoted by x, then a fuzzy set in X is a set of ordered pairs $A = \{x, \mu_A(x) | x \in X\}$, where $\mu_A(x)$ represents the membership function for the fuzzy set A. The value representing the degree for an element x that belongs to the fuzzy set A is defined as the degree of membership for x, which is evaluated by the membership function. The α level cut of fuzzy set A is the set of those elements which have a membership value greater than or equal to α (Ross 1995). When $\alpha = 0$, the corresponding interval is called the "support" of the fuzzy member with extreme boundaries of the "minimum" and "maximum" values, respectively. Similarly, with a triangular function when $\alpha = 1$, the interval comes down to a crisp value, or the "most likely value" (Kaufmann and Gupta 1991).

Two most common types of membership function for fuzzy numbers are (1) Triangular; and (2) Trapezoidal. But, former one is preferred by some researchers (Dou et al. 1995, 1997a, b) because of its simple shape. The triangular membership function is a special case of trapezoidal membership function, because in this case, at $\alpha = 1$, there is a single point rather than a flat line as in the trapezoidal function. However, in this study triangular membership function has been used.

Moore (1979) has given a solution in interval computation, and found the basic problem as given a function $f(x_i, \ldots, x_n)$ and n intervals $[x_i^-, x_i^+]$ find the interval range of the variable $y = f(X)$ such that $x \in x_i\ [x_i^-, x_i^+]$. Yang et al. (1993) defined the goal of the interval computation is to find the minimum and the maximum of the function when the different possible values of the variables x_i range in their intervals $[x_i^-, x_i^+]$. Some methods are based on finding a finite set of points (called configurations or poles) on which this minimum and maximum is attained. For each α-cut, the minimum and maximum value of the fuzzy variable is selected as shown in Fig. 1. Prasad and Mathur (2007) compared the vertex method with the ANN-GA approach in their study of contaminant transport in groundwater flow. The formulation of vertex method (Dou et al. 1997b) is as follows. In the vertex method, fuzzy numbers $A_1, A_2, A_3, \ldots, A_N$ are defined on the real line L and the elements of A_1 are denoted by x_i, $i = 1, 2, \ldots, N$. If $x_1, x_2, \ldots, x_N$ are related to real number y by the mapping $y = f(x_1, x_2, \ldots, x_N)$ the solution of fuzzy number B in y that would correspond to fuzzy numbers A_1 in x_N can be obtained in the following procedure:

1. The range of membership [0, 1] is discretized into a finite number of values, called $\alpha_1, \alpha_2, \ldots, \alpha_M$. The refinement in discretization depends on the degree of accuracy desired.
2. For each membership value α_j, the corresponding intervals for A_i in x_i, $i = 1, 2, \ldots, N$ are determined. These are the supports of the α_j cuts of $A_1, A_2, A_3, \ldots, A_N$. The end points of these intervals are represented by $[a_1, b_1]$, $[a_2, b_2]$, …, $[a_N, b_N]$. Also a_i may be equal to b_i in which case the interval would reduce to a point.
3. Taking one end point from each of the intervals, the end points can be combined into 2^N distinct permutations, giving 2^N combinations for the vector $(x_1, x_2, \ldots, x_N)$. Thus for two-pipe network having the uncertain parameters as (a_1, b_1), (a_2, b_2) can be combined in ordered pair as follows: $[(a_1, a_2), (a_1, b_2), (b_1, a_2),$

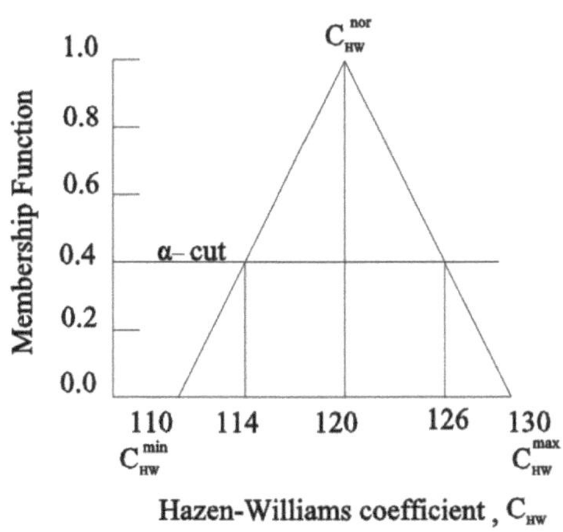

Fig. 1 Triangular membership function for Hazen–Williams coefficient

(b_1, b_2)]. Similarly, three-pipe network having the uncertain parameter as (a_1, b_1), (a_2, b_2), (a_3, b_3) can be combined as $[(a_1, a_2), (a_1, b_2), (b_1, a_2), (b_1, b_2)]$ $[a_3, b_3]$ and can be further combined as $[(a_3, a_1, a_2), (a_3, a_1, b_2), (a_3, b_1, a_2), (a_3, b_1, b_2), (b_3, a_1, a_2), (b_3, a_1, b_2), (b_3, b_1, a_2), (b_3, b_1, b_2)]$.

4. The function $f(x_1, x_2, \ldots, x_N)$ is evaluated for each of the 2^N combinations to obtain 2^N for y, denoted as $y_1, y_2, \ldots, y_N$. The desired interval for y is given by $[\wedge_k y_k, \vee_k y_k]$, which define support of the α_j-cut of B.
5. The process is repeated for other α-cuts to obtain additional α-cuts of B and the solution to the fuzzy number B.

3 Methodology

Hazen–Williams formula for head loss

The head loss due to pipe friction, h_f in meters (Bhave and Gupta 2006) is given by

$$h_f = \frac{10.68 \mathrm{LQ}^{1.85}}{C_{\mathrm{HW}}^{1.85} D^{4.87}} \tag{1}$$

where

- L length of the pipe in meters;
- D diameter of the pipe in meters;
- Q pipe discharge in cubic meters per second; and
- C_{HW} Hazen–Williams coefficient.

Fuzzy analysis through vertex method

Figure 1 shows Hazen–Williams coefficient, C_{HW} as a fuzzy parameter in a triangular membership function, having extreme boundary values $C_{\mathrm{HW}}^{\min}$ and $C_{\mathrm{HW}}^{\max}$ for each α-cuts (i.e., 0.0, 0.2, 0.4, 0.6, 0.8). Where $C_{\mathrm{HW}}^{\min}$ and $C_{\mathrm{HW}}^{\max}$ are the minimum and maximum values of C_{HW}. For $\alpha = 1$, it represents $C_{\mathrm{HW}}^{\mathrm{nor}}$, a single point value known as crisp value, i.e., most likely or normal value of C_{HW}. Hence for the triangular fuzzy number $A\left(C_{\mathrm{HW}}^{\min}, C_{\mathrm{HW}}^{\mathrm{nor}}, C_{\mathrm{HW}}^{\max}\right)$, the membership function μ_A is given by

$$\mu_A(C_{\mathrm{HW}}) = 0, \quad C_{\mathrm{HW}} \le C_{\mathrm{HW}}^{\min} \tag{2}$$

$$\mu_A(C_{\mathrm{HW}}) = \frac{C_{\mathrm{HW}} - C_{\mathrm{HW}}^{\min}}{C_{\mathrm{HW}}^{\mathrm{nor}} - C_{\mathrm{HW}}^{\min}}, \quad C_{\mathrm{HW}}^{\min} \le C_{\mathrm{HW}} \le C_{\mathrm{HW}}^{\mathrm{nor}} \tag{3}$$

$$\mu_A(C_{\mathrm{HW}}) = \frac{C_{\mathrm{HW}} - C_{\mathrm{HW}}^{\max}}{C_{\mathrm{HW}}^{\mathrm{nor}} - C_{\mathrm{HW}}^{\max}}, \quad C_{\mathrm{HW}}^{\mathrm{nor}} \le C_{\mathrm{HW}} \le C_{\mathrm{HW}}^{\max} \tag{4}$$

$$\mu_A(C_{\text{HW}}) = 0, \quad C_{\text{HW}} \geq C_{\text{HW}}^{\max} \tag{5}$$

Here α-cut is represented by α^*. If $\alpha^* = 0$, C_{HW} lies between 110 and 130; for $\alpha^* = 0.4$, C_{HW} lies between 114 and 126; $\alpha^* = 1$, normal value of C_{HW} is 120.

Analysis of Hydraulic Network

Consider a looped pipe network having M source nodes labeled $j = 1, 2, 3, \ldots, M$; N demand nodes labeled $j = M + 1, \ldots, M$; X pipes labeled $x = 1, 2, 3, \ldots, X$; C basic loops or circuits labeled $c = 1, 2, 3, \ldots, C$; and $M - 1$ pseudo loops labeled $c = C + 1, \ldots, C + M - 1$. Bhave and Gupta (2006) treated the Hazen–Williams coefficient of pipe roughness as fuzzy. For analyzing the networks, node flow continuity equations and head loss equations for loops have been used. Node flow continuity equation states that algebraic sum of flows at node is zero whereas loop head loss equation states that algebraic sum of head losses in pipes of a loop is zero, respectively. Hence

$$\underset{x\,\text{connected to}\,j}{\sum (Q_x)_{\alpha=\alpha*}} + q_j = 0, \quad j = 1, 2, 3, \ldots, J \tag{6}$$

and

$$\sum_{x \in c} \frac{10.68 L_x \left(Q_x^{1.85}\right)_{\alpha=\alpha^*}}{C_{\text{HW}}^{1.85} D_x^{4.87}} = 0, \quad c = 1, 2, 3, \ldots, C + M - 1 \tag{7}$$

are the node flow continuity and loop head loss equations, respectively.

In the vertex method, for n piped network, the order pair of the roughness coefficient of each membership would be 2^n. These roughness coefficients are formulated in ordered pair and output parameters like pipe discharges and nodal heads are obtained using EPANET 2. The output parameters can be grouped for pipe discharges $\left(Q_x^{\min}, Q_x^{\max}\right)$ and nodal heads $H_j^{\min}, H_j^{\max}$ for each membership function. Then the membership functions of discharges and nodal heads are plotted at each α level cut.

3.1 Example 1

Consider a one loop gravity network of Bhave and Gupta (2006) as shown in Fig. 2 has source node 1 with HGL of 100 m and nodes 2, 3, and 4 are demand nodes having zero elevations with demands of 0.04, 0.045, and 0.025 m^3/s, respectively. The pipes are labeled as (1), (2), (3), and (4). The length in meters and diameters in millimeters for different pipes are given in parentheses as pipe 1 (450 m, 300 mm); pipe 2 (500 m, 350 mm); pipe 3 (450 m, 250 mm); and pipe 4 (500 m, 200 mm), respectively.

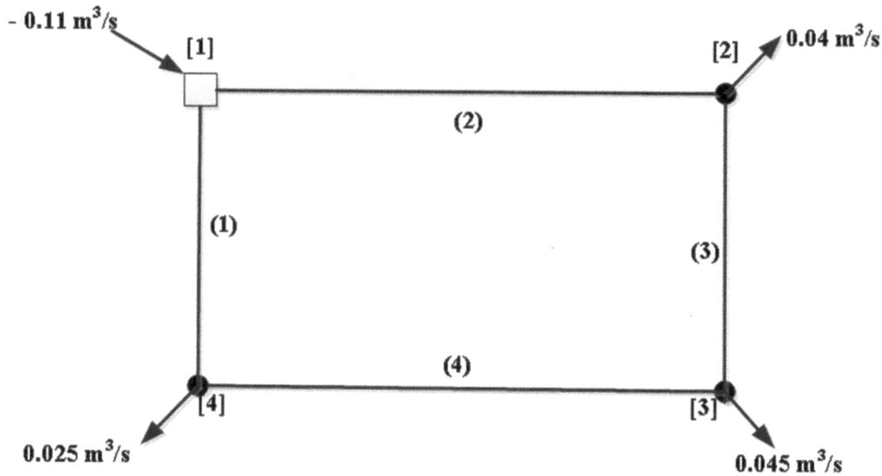

Fig. 2 A one-loop gravity network

As explained, the Hazen–Williams coefficient of pipe roughness (C_{HW}) is taken as uncertain and independent parameter to determine the dependent parameters like pipe discharges and nodal heads using EPANET 2. The results are shown in Table 1.

In the analysis of pipe network, taking all independent fuzzy parameters at their normal values ($\alpha = 1$) of pipe roughness for all pipes, i.e., $C_{\mathrm{HW1}} = C_{\mathrm{HW2}} = C_{\mathrm{HW3}} = C_{\mathrm{HW4}}$, the discharge value of pipe (1) is $Q_1 = 0.0418\ \mathrm{m^3/s}$ (Table 1).

For, $\alpha^* = 0$ the ordered pairs of C_{HW} are obtained using vertex method (Dong and Shah 1987).

[110,110,110,110]	[110,110,110,130]; [110,110,130,110]	[110,110,130,130]
[110,130,110,110]	[110,130,110,130]; [110,130,130,110]	[110,130,130,130]
[130,110,110,110]	[130,110,110,130]; [130,110,130,110]	[130,110,130,130]
[130,130,110,110]	[130,130,110,130]; [130,130,130,110]	[130,130,130,130]

The maximum value of pipe discharge; $Q_1^{\max} = 0.0442\ \mathrm{m^3/s}$ is obtained when $C_{\mathrm{HW1}} = C_{\mathrm{HW4}} = 130$ and $C_{\mathrm{HW2}} = C_{\mathrm{HW3}} = 110$. The minimum value of pipe discharge $Q_1^{\min} = 0.0394\ \mathrm{m^3/s}$ is obtained as when $C_{\mathrm{HW1}} = C_{\mathrm{HW4}} = 110$ and $C_{\mathrm{HW2}} = C_{\mathrm{HW3}} = 130$ in the membership function for $\alpha^* = 0.0$.

The maximum value of pipe discharge $Q_1^{\max} = 0.0437\ \mathrm{m^3/s}$ is obtained when $C_{\mathrm{HW1}} = C_{\mathrm{HW4}} = 128$ and $C_{\mathrm{HW2}} = C_{\mathrm{HW3}} = 112$. The minimum value of pipe discharge $Q_1^{\min} = 0.0399\ \mathrm{m^3/s}$ is obtained as when $C_{\mathrm{HW1}} = C_{\mathrm{HW4}} = 112$ and $C_{\mathrm{HW2}} = C_{\mathrm{HW3}} = 128$ in the membership function for $\alpha^* = 0.2$.

Table 1 Values of dependent parameters for different α-cuts for network of Fig. 2

Sl. No.	Pipe discharge (m^3/s)					Nodal HGL (m)		
	α cut value	Q_1	Q_2	Q_3	Q_4	H_2	H_3	H_4
(1)	(2)	(3)	(4)	(5)	(6)	(7)	(8)	(9)
1	$\alpha^* = 0.0$ Maximum	0.0442	0.0706	0.0306	0.0192	99.277	98.578	99.462
	Minimum	0.0394	0.0658	0.0258	0.0144	98.953	98.062	99.168
2	$\alpha^* = 0.2$ Maximum	0.0437	0.0701	0.0301	0.0187	99.251	98.536	99.439
	Minimum	0.0399	0.0663	0.0263	0.0149	98.992	98.125	99.205
3	$\alpha^* = 0.4$ Maximum	0.0432	0.0696	0.0296	0.0182	99.224	98.493	99.415
	Minimum	0.0404	0.0668	0.0268	0.0154	99.030	98.186	99.240
4	$\alpha^* = 0.6$ Maximum	0.0427	0.0692	0.0292	0.0177	99.195	98.448	99.389
	Minimum	0.0408	0.0673	0.0273	0.0158	99.066	98.243	99.273
5	$\alpha^* = 0.8$ Maximum	0.0423	0.0687	0.0287	0.0173	99.165	98.400	99.362
	Minimum	0.0413	0.0677	0.0277	0.0163	99.101	98.298	99.304
6	$\alpha^* = 1.0$ Normal	0.0418	0.0682	0.0282	0.0168	99.134	98.350	99.334

All the minimum and maximum values (except $\alpha^* = 1$) of pipe discharges and nodal heads are shown in the Table 1. The membership functions of the corresponding parameters are shown in Fig. 3.

3.2 Example 2

Another two-loop pipe network problem is taken from Bhave and Gupta (2006). The network has one source node, three demand nodes, and five pipes as shown in Fig. 4. Node 1 is a source node with HGL of 100 m and nodes 2, 3, and 4 are demand nodes having zero elevations with demands of 0.150, 0.300, and 0.230 m^3/s, respectively. The pipes are labeled as (1), (2), (3), (4), and (5) and its length in meters and diameters in millimeters for different pipes given in parentheses as pipe 1 (1200 m, 500 mm); pipe 2 (1100 m, 500 mm); pipe 3 (1500 m, 500 mm); pipe 4 (900 m, 350 mm); and pipe 5 (1000, 350), respectively.

In this analysis, the source HGL and nodal demands are precise, but uncertainty exists in Hazen–Williams coefficient, C_{HW} of all pipes.

The most likely value of C_{HW} for all pipes is 120 with minimum and maximum values of 110 and 130, respectively (Fig. 1).

In this network, there are five pipes with one uncertain parameter (i.e., Hazen–Williams coefficient, C_{HW}) for which the ordered pair becomes $2^5 = 32$ for each

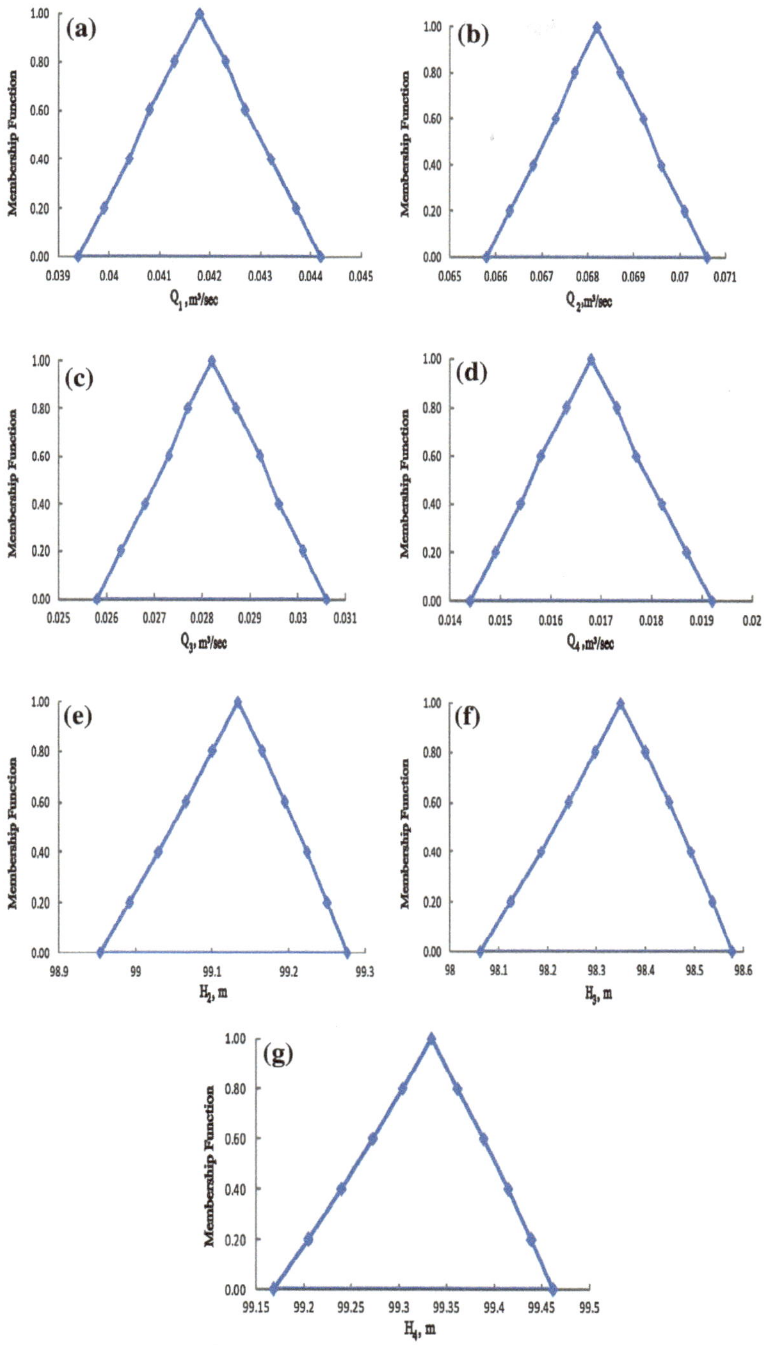

Fig. 3 Membership functions of unknown pipe discharges (m^3/s) and nodal HGL (m) for the network of Fig. 2

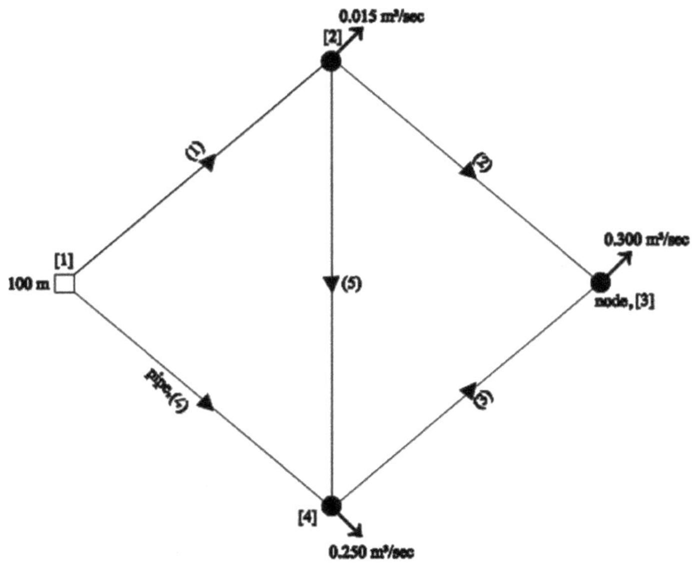

Fig. 4 A two-loop network

α-cut. Hence the EPANET 2 model has been executed 32 times and the minimum and maximum values of discharges and heads are obtained.

The results obtained by proposed method and those of Bhave and Gupta (2006) are identical up to three decimal places. Consequently the membership functions for vertex method and Bhave and Gupta (2006) are identical as shown in the Fig. 5.

4 Conclusions

Two water distribution network problems have been analyzed to determine the membership function of unknown pipe discharges and hydraulic heads by assuming pipe roughness of each pipe as uncertain. In this study, the membership function of pipe roughness is assumed to be triangular. Further, EPANET 2 is used to find the pipe discharges and hydraulic heads of each node by giving required input parameters. The resulted membership functions of output discharges and hydraulic heads are found to be triangular in nature. Further, the results obtained by proposed method and those of Bhave and Gupta (2006) are identical up to three decimal places. Consequently the membership functions for vertex method and Bhave and Gupta (2006) are identical. Thus, the proposed methodology using vertex method and EPANET 2 has been successful in quantifying uncertainty in pipe discharges and nodal HGLs. The results of pipe discharges are found to vary between 15 and 30 % whereas the hydraulic heads at nodes vary between 0.3 and 3 m when uncertainty of about 8 % in Hazen–Williams coefficient of pipe roughness is

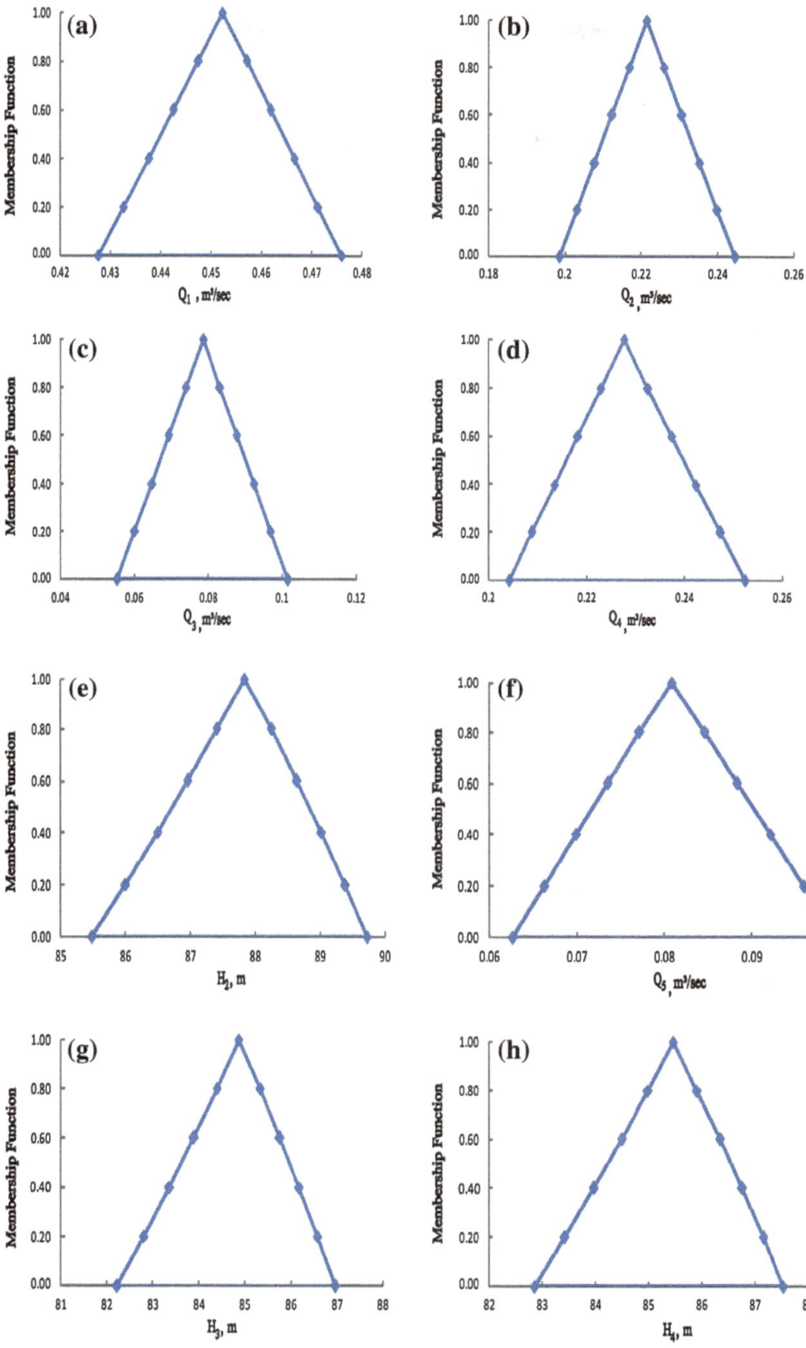

Fig. 5 Membership functions of unknown pipe discharges (m^3/s) and nodal HGL (m) for the network of Fig. 4

introduced in four- and five-pipe networks, respectively. The maximum uncertainty in discharges and hydraulic heads for both the networks are found to be at $\alpha^* = 0$. This may be due to maximum uncertainty in roughness coefficient of the pipe.

Furthermore, EPANET 2 required to run 16 and 32 times for four- and five-pipe network problem, respectively, at each α-cut level except $\alpha^* = 1$. Thus, the total number of EPANET 2 simulation needed are 65 and 129 for four- and five-pipe network problems, respectively. The process would be time consuming for large network even though there is a single uncertainty in input parameters. Furthermore, when this uncertainty in pipe roughness is combined with other kind of uncertainty like nodal demands, lengths, diameter of pipes, water levels in reservoirs, head-discharge characteristics of pumps, etc., would aggravate the uncertainty in discharges and nodal heads of the pipe. As a result, the reliability of network would decrease in terms of either meeting the required discharges or nodal heads to the consumers or both. This will lead the pipe network problem to be more complex in real life since the problem involves large number of pipes and different types of uncertainty in the pipes. As a result, there is a need to reduce this computational burden for the real-life problem in the proposed methodology.

References

Bhave PR, Gupta R (2004) Optimal design of water distribution networks for fuzzy demands. Civil Eng Environ Syst 21(4):229–245

Bhave PR, Gupta R (2006) Analysis of water distribution networks. Narosa Publishing Pvt. Ltd., New Delhi

Dong W, Shah HC (1987) Vertex methods for computing functions of fuzzy variables. Fuzzy Sets Syst 24:65–78

Dou C, Woldt W, Bogardi I, Dahab M (1995) Steady state groundwater flow simulation with imprecise parameters. Water Resour Res 31(11):2709–2719

Dou C, Woldt W, Dahab M, Bogardi I (1997a) Transient groundwater flow simulation using fuzzy set approach. Ground Water 35(2):205–215

Dou C, Woldt W, Bogardi I, Dahab M (1997b) Numerical solute transport simulation using fuzzy sets approach. J Contam Hydrol 27:107–126

Freissinet C, Vauclin M, Erlich M (1999) Comparison of first order analysis and fuzzy set approach for the evaluation of imprecision in a pesticide groundwater pollution screening model. J Contam Hydrol 37:21–43

Guan J, Aral MM (2004) Optimal design of groundwater remediation systems using fuzzy set theory. Water Resour Res 40(1):1–20

Gupta R, Bhave PR (2007) Fuzzy parameters in pipe network analysis. Civil Eng Environ Syst 24(1):33–54

Kaufmann A, Gupta MM (1991) Introduction to fuzzy arithmetic: theory and applications. Van Nostrand Reinhold, New York

Moore R (1979) Methods and applications of interval analysis. SIAM Studies in Applied Mathematics

Prasad RK, Mathur S (2007) Groundwater flow and contaminant transport simulation with imprecise parametrers. J Irrig Drainage Eng 133(1):61–70

Revelli R, Ridolfi L (2002) Fuzzy approach for analysis of pipe networks. J Hydraul Eng 128(1):93–101

Ross TJ (1995) Fuzzy logic with engineering applications. McGraw-Hill, New York
Rossman LA (2000) EPANET 2, Users Manual. National Risk Management Research Laboratory, U.S. Environmental Protection Agency, Cincinnati, Ohio
Todini E, Plilati S (1987) A gradient method for the analysis of pipe networks. In: International conference on computer application for water supply and distribution. Leicester Polytechnic, UK
Xu C, Goulder IC (1999) Optimal design of water distribution networks using fuzzy optimization. Civil Eng Environ Syst 16(4):243–266
Yang HQ, Yao R, Jones D (1993) Calculating functions of fuzzy numbers. Fuzzy Sets Syst 55:273–283
Zadeh LA (1965) Fuzzy sets. Inf Control 8:338–353

Applications of Wavelet Transform Technique in Hydrology—A Brief Review

Khandekar Sachin Dadu and Paresh Chandra Deka

Abstract Recently, wavelet transform analysis has become a popular analysis tool due to its ability to elucidate simultaneously both spectral and temporal information within the signal. This overcomes the basic shortcoming of Fourier analysis, which is that the Fourier spectrum contains only globally averaged information. Therefore, a data preprocessing can be performed by time series decomposition into its subcomponents using wavelet transform analysis. Wavelet transforms provide useful decompositions of the main time series, so that wavelet-transformed data improve the ability of a forecasting model by capturing useful information on various resolution levels. The wavelet decomposition of a nonstationary time series into different scales provides an interpretation of the series structure and extracts significant information about its history, using few coefficients. For these reasons, this technique is largely applied to time series analysis of nonstationary signals. In terms of hydrologic applications, this modeling tool is still in its nascent stages. The practicing hydrologic community is just becoming aware of the potential of wavelet transform as an analyzing tool. This paper is intended to serve as an introduction to wavelet transformation for hydrologists. Apart from descriptions of various aspects of wavelet transform and some guidelines on their usage, this paper offers brief comparisons of the nature of wavelet transformations and other modeling philosophies in hydrology. The merits of wavelet transform applications have been discussed.

Keywords Wavelet transform · Artificial neural network · Hydrology · Streamflow · Time series · Forecasting

---

K.S. Dadu (✉) · P.C. Deka
Department of Applied Mechanics and Hydraulics, National Institute of Technology Karnataka, Surathkal, Mangalore 575025, India
e-mail: khandekarsd@yahoo.com

P.C. Deka
e-mail: paresh_deka@sify.com

A.K. Sarma et al. (eds.), *Urban Hydrology, Watershed Management and Socio-Economic Aspects*, Water Science and Technology Library 73,
DOI 10.1007/978-3-319-40195-9_19

1 Introduction

The hydrology system is a highly complex nonlinear system under the influence of rain-bearing system and underlying surface system. It is influenced by many factors, such as weather, land with vegetal cover, infiltration, evapotranspiration, so it includes the good deal of stochastic dependent component, multi-time scale, and highly nonlinear characteristics. Forecasting of hydrological time series can be done using stochastic models like Auto regressive (AR), Auto regressive moving average (ARMA), Auto regressive integrated moving average (ARIMA), etc. These models are basically time series models and have a limited ability to capture nonstationarities and nonlinearities.

A nonstationary time series can be decomposed into certain number of stationary time series by wavelet transform (WT). Then different single prediction methods are combined with WT to improve the prediction accuracy. In most of the hybrid models, WT is used as preprocessing technique. The wavelet-transformed data aid in improving the model performance by capturing helpful information on various resolution levels. Due the above-mentioned advantages of WT, it has been found that the hybridization of wavelet transformation with other models like ANN, FL, ANFIS, linear models, etc., improved the results significantly than the single regular model (Prahlada and Deka 2011).

Wavelet theory (Mallat 1989) is first developed in the end of 1980s of last century. Nowadays, it has been applied in many fields, such as signal process, image compression, voice code, pattern recognition, hydrology, earthquake investigation, and many other nonlinear science fields. The objective of this paper is to examine how successfully WT has been used in hydrologic problem. The researches and applications of wavelet analysis have already begun in hydrology and water resources. The document (Li et al. 1997) points out the potential applications of wavelet analysis to hydrology and water resources. Li et al. (1999) probed longtime interval forecast of hydrological time series with combining neural network models based on WT. Wang et al. (2000) have proposed a wavelet transform stochastic simulation model, which generates synthetic streamflow sequences that are statistically similar to observed streamflow sequences. The multi-time scale characteristics of hydrological variable have been studied by Wang et al. (2002). Wavelet analysis has been a hot research point in prediction of time series analysis due to its multiresolution function (Zhou et al. 2008). In this study, general applications of WT are discussed briefly. However, this study did not present any detail on hydrologic applications. Rather, it complements earlier studies.

2 Wavelet Transformation Basics

2.1 General

In the last decade, WT has become a useful technique for analyzing variations, periodicities, and trends in time series. A wavelet transformation is a strong mathematical signal processing tool like Fourier transformation with the ability of analyzing both stationary as well as nonstationary data, and to produce both time and frequency information with a higher resolution, which is not available from the traditional transformation. WT provides multiresolution analysis, i.e., at low scales (high frequency) it gives better time resolution and poor frequency resolution and at high scales (low frequency) it gives better frequency resolution and poor time resolution and in actual practice for all the time series signals such information is important. The lower scales (i.e., compressed wavelet) trace the abrupt change or high frequency of a signal and the higher scales (i.e., stretched wavelet) trace slowly progressing occurrences or low-frequency component of the signal.

Signals whose frequency content does not change with time are called stationary signals. In other words, the frequency content of stationary signals does not change in time. In stationary signals it is not necessary to know at what times frequency components exist, since all frequency components exist at all times.

Mathematical transformations (viz., Fourier transform (FT), Short Time Fourier transform (STFT), WT, etc.) are applied to time domain signals (raw signals) to obtain further information from that signal that is not readily available in the raw signals. The above-mentioned mathematical transformation techniques are briefly described in the following sections.

2.2 Fourier Transform (FT)

If the FT of a signal in time domain is taken, the frequency–amplitude representation of that signal is obtained. That is, we have a plot with one axis being the frequency and the other being the amplitude. This plot tells us how much of each frequency exists in the raw signal. But it does not tell about what spectral component exist at any given time instant, i.e., the time information is lost. So FT is not suitable for nonstationary data. The FT is defined by the following two equations:

$$F(\omega) = \int_{-\infty}^{\infty} x(t) \cdot \mathrm{e}^{-2j\pi\omega t} \mathrm{d}t \tag{1}$$

$$x(t) = \int_{-\infty}^{\infty} F(\omega) \cdot \mathrm{e}^{2j\pi\omega t} \mathrm{d}\omega \tag{2}$$

In the above equation ω stands for frequency, t stands for time, and $x(t)$ denotes time domain signal. Equation (1) is FT of $x(t)$ and Eq. (2) is inverse FT of $F(\omega)$. In Eq. (1), the signal $x(t)$, is multiplied with an exponential term, at some certain frequency "ω", and then integrated over all the times. This integral is calculated for every value of "ω". If the value of this integration is large, then this means that the signal have a major component of "ω" in it.

2.3 *Short Time Fourier Transform (STFT)*

The STFT is an improvement on the FT (frequency) because it provides a measure of time and frequency resolutions. The difference between STFT and FT is that in STFT, the signal is divided into small enough segments, where these segments (portions) of the signal can be assumed to be stationary. For this purpose, a window function "***w***" is chosen. The width of this window must be equal to the segment of the signal where its stationarity is valid. This window is first located to very beginning of signal. The window function and signal are then multiplied. This product is assumed to be another signal, whose FT is to be taken. In other words, FT of this product is taken, just like taking FT of any signal. The next step is shifting this window to new location, multiplying with the signal, and taking FT of the product. This procedure is followed until the end of the signal is reached.

STFT is defined as

$$\mathrm{STFT}(t,\omega) = \int_t [x(t) \cdot w^*(t)] \cdot \mathrm{e}^{-2j\pi\omega t} \cdot \mathrm{d}t \quad (3)$$

In the above equation $x(t)$ denotes raw signal, $w(t)$ denotes window function, and * is complex conjugate.

Wide window gives good frequency resolution, but poor time resolution. Narrow window gives good time resolution, but poor frequency resolution. The use of a fixed window size at all times and for all frequencies is a limitation of this method.

2.4 *Wavelet Transformation*

The wavelet representation addresses the above limitation, by *adaptively* partitioning the time–frequency plane, using a range of window sizes. At high frequencies, the WT gives up some frequency resolution compared to the FT. Figure 1 shows representation of the effect of using FT and WT.

The WT breaks the signal into its wavelets (small wave) which are scaled and shifted versions of the original wavelet so-called mother wavelet.

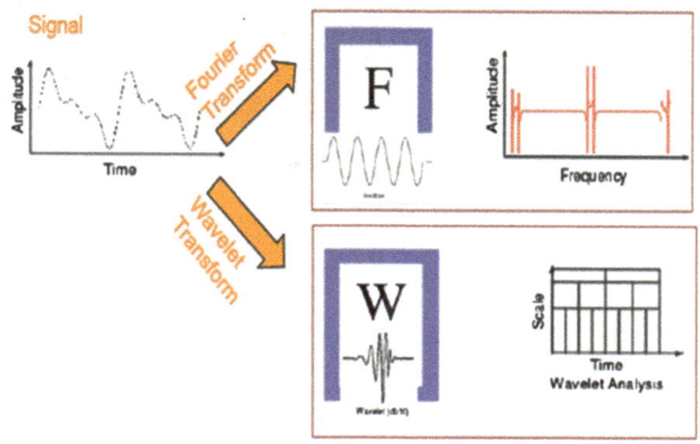

Fig. 1 Fourier Transform and wavelet transformation

The generation of wavelet coefficients for a time series involves five steps (The Mathworks 2010):

(i) Given a signal X_t and a wavelet function $\Psi j,k$ compares the wavelet to a section at the start of the signal (Fig. 2a).
(ii) Compute the coefficient, $c_{j,k}$, which is an indication of the correlation of the wavelet function with the selected section of the signal.
(iii) Shift the wavelet to the right and repeat steps (i) and (ii) until the entire signal is covered (Fig. 2b).
(iv) Dilate (scale) the wavelet and repeat steps (i) through (iii) (Fig. 2c).
(v) Repeat steps (i) through (iv) for all scales to obtain coefficients at all scales and at different sections of the original signal.

The wavelet transformation is divided into two types:

1. Continuous wavelet transform (CWT)
2. Discrete wavelet transform (DWT).

2.4.1 Continuous Wavelet Transform (CWT)

The Continuous Wavelet Transform (CWT) of a signal $x(t)$ is given by Eq. 4.

$$\mathrm{CWT}(a,b) = \frac{1}{\sqrt{a}} \int_{-\infty}^{\infty} x(t) \cdot \psi^{*}\left(\frac{t-b}{a}\right) \cdot \mathrm{d}t \tag{4}$$

In the above equation, the transformed signal is a function of two variables, a and b, the scale and translation factor, respectively, of the function $\psi(t)$. * corresponds to complex conjugate. $\psi(t)$ is the transforming function, and is called the mother wavelet, which is defined mathematically as

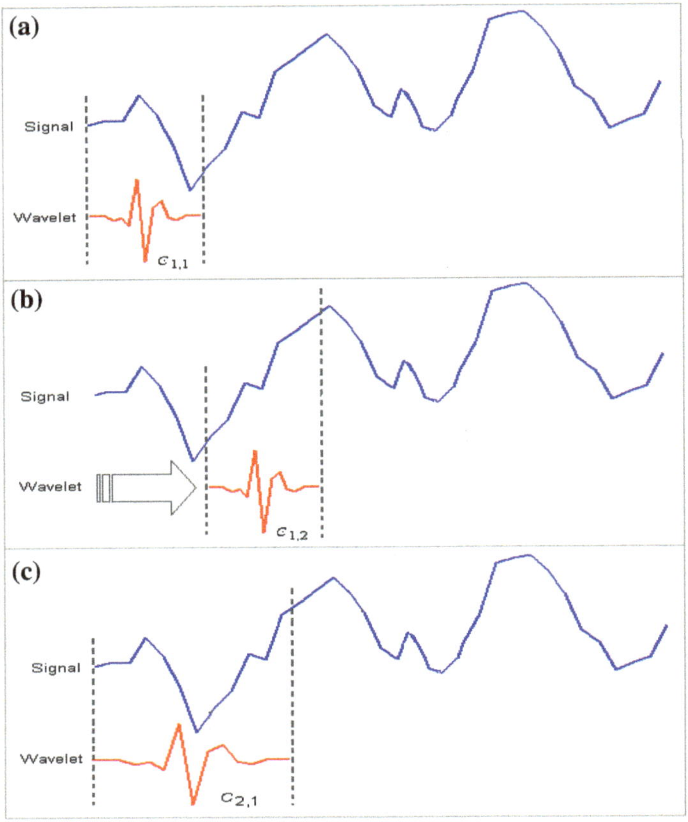

Fig. 2 Generating wavelet coefficients from a time series

$$\int_{-\infty}^{\infty} \psi(t)\mathrm{d}t = 0 \tag{5}$$

The term translation is related to the location of the window, as the window is shifted through the signal. This term, obviously, corresponds to time information in the transform domain. The scale parameter is defined as 1/frequency. Low frequencies (high scales) correspond to a global information of a signal (that is usually spans the entire signals), whereas high frequencies (low scales) correspond to a detailed information of a hidden pattern in the signal (that usually lasts a relatively short time).

The CWT is computed by changing the scale of the analysis window, shifting the window in time, multiplying by the signal, and integrating over all times.

The original signal is reconstructed using the inverse wavelet transform as

$$x(t) = \frac{1}{C_\psi} \int_{-\infty}^{\infty} \int_{0}^{\infty} \frac{1}{\sqrt{a}} \psi\left(\frac{t-b}{a}\right) \cdot \mathrm{CWT}(a,b) \frac{\mathrm{d}a \cdot \mathrm{d}b}{a^2} \tag{6}$$

where C_ψ is admissibility constant.

2.4.2 Discrete Wavelet Transform (DWT)

Calculating the wavelet coefficients at every possible scale is a fair amount of work, and it generates a lot of data. If one chooses scales and positions based on the powers of two (dyadic scales and positions) then the analysis will be much more efficient as well as accurate. This transform is called discrete wavelet, and has the form as

$$\psi_{m,n}\left(\frac{t-b}{a}\right) = \frac{1}{\sqrt{a_o^m}} \psi\left(\frac{t - nb_o a_o^m}{a_o^m}\right) \tag{7}$$

where m and n are integers that control the wavelet dilation and translation, respectively; b_o is the location parameter and must be greater than zero; a_o is a specified fixed dilation step greater than 1. From this equation, it can be seen that the translation step $nb_o a_o^m$ depends upon the dilation, a_o^m. The most common and simplest choice for parameters a_o and b_o are 2 and 1 (time steps), respectively. This power of two logarithmic scaling of the translations and dilations is known as the dyadic grid arrangement. The dyadic wavelet can be written in more compact notation as

$$\psi_{m,n}(t) = 2^{-m/2} \psi(2^{-m} t - n) \tag{8}$$

Discrete dyadic wavelets of this form are usually chosen to be orthonormal. This allows for the complete regeneration of the original signal as an expansion of a linear combination of translate and dilate orthonormal wavelets. For discrete time series x_i, where x_i occurs at discrete time i, the dyadic wavelet transform becomes

$$T_{m,n} = 2^{-m/2} \sum_{i=0}^{N-1} \psi(2^{-m} i - n) x_i \tag{9}$$

where $T_{m,n}$ = wavelet coefficient for the discrete wavelet of scale $a = 2^m$ and location $b = 2^m n$. Equation (9) considers a finite time series, x_i, $i = 0, 1, 2, \ldots, N-1$, and N is an integer power of 2: $N = 2^M$. This gives the range of m and n as, respectively, $0 < n < 2^{M-m} - 1$ and $1 < m < M$. At the largest wavelet scale (i.e., 2^m, where $m = M$), just one wavelet is needed to cover the time interval and only one coefficient is produced. At the next scale (2^{M-1}), two wavelets cover the time interval, therefore two coefficients are produced, and so on down to $m = 1$. At $m =$

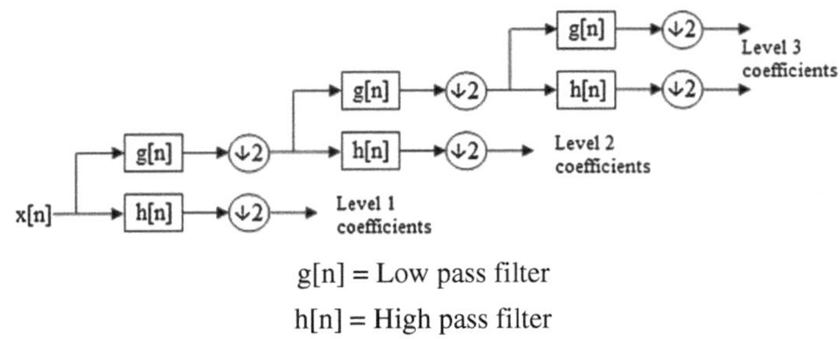

Fig. 3 Wavelet decomposition tree

1, the a scale is 2^1, i.e., $2^M/2^1$, i.e., 2^{M-1} or $N/2$ coefficients are needed to describe the signal at this scale. The total number of wavelet coefficients for a discrete time series of length $N = 2^M$ is then $1 + 2 + 4 + 8 + \cdots + 2^{M-1} = N-1$ (Addison et al. 2001).

DWT operates two sets of function viewed as high-pass and low-pass filters (see Fig. 3). The original time series are passed through high-pass and low-pass filters and separated at different scales. The time series is decomposed into one comprising its trend (the approximation) and one comprising the high frequencies and the fast events (the detail).

3 Applications of Wavelet Transform in Hydrology

Recently, WT analysis has become a popular analysis tool due to its ability to elucidate simultaneously both spectral and temporal information within the signal. Some of the recent works carried out in the Hydrology are discussed below.

Addison et al. (2001) used WT analysis to a variety of open channel wake flows. Feature location was undertaken using a continuous WT, and both turbulent statistical analysis and thresholding of the turbulent signal components are undertaken using a discrete WT. It was found that the CWT is the preferred method for feature detection within fluid velocity time signals.

Wensheng and Ding (2003) carried out a multi-time scale prediction of ground water level at Beijing and daily discharge of Yangte River Basin at China using Hybrid Model of Wavelet-Neural Network. Through a Trous algorithm and three-layer neural network forecasting results were carried out. Twelve years of shallow monthly ground water level data were used, 9 years for calibration, and 3 years for validation. Daily discharge data of 8 years were used for training and 2 years for testing. The comparisons revealed that the model increase the forecasted accuracy and prolong the length time of prediction. The proposed WLNN model focused on improving the precision and prolonging the forecasting time period.

Kim and Valdes (2003) developed nonlinear model for drought forecasting based on a conjunction of wavelet transforms and neural networks in the Conchos river basin in Maxico. The results indicate that the conjunction model using dyadic wavelet transform significantly improves the ability of neural network in forecasting.

Cannas et al. (2005) studied the river flow forecasting 1 month ahead with Neural Networks and Wavelet Analysis using monthly runoff data for the Tirso Basin, Italy. The dataset was split into three parts, first 40 years was used for training, next 9 years for cross validation, and last 20 years for testing. The reconstruction of the data was done by traditional feed forward, MLP networks. For the nonstationary and seasonal irregularity of runoff time series, the best results were obtained using data clustering and DWT combination. Tests showed that neural networks trained with preprocessed data showed better performance.

Zhou et al. (2008) developed monthly discharge predictor–corrector model based on wavelet decomposition using 52 years records of monthly discharge at Yichang station of Yangtse river. The decomposed times series data were used as input to ARMA model for prediction which improves the prediction accuracy.

Rao and Krishna (2009) carried out modeling using Hydrological Time Series data adopting Wavelet-Neural Network for four west flowing rivers in India namely Kollur, (22 years data from 1981 to 2002), Seethanadi (26 years data from 1973 to 1998), Varahi (26 years 1978–2003), and Gowrihole (25 years data from 1979 to 2003). The results of daily Streamflow and monthly Groundwater level series modeling indicated that the performances of WNN Models are more effective than ANN Models.

Nourani et al. (2009a, b) studied the rainfall–runoff modeling using Wavelet–ANN approach for predictions of runoff discharge 1 day ahead of the Ligvanchai watershed at Tabriz, Iran. The daily rainfall and runoff time series for 21 years were used. The time series were decomposed up to four levels using Haar, Daubechies (db2), Symlet (sym3), and Coiflet (coif1). The Study showed that both short- and long-term runoff discharges could be predicted considerably. The model results show the high merit of Haar wavelet in comparison with the others. Authors also recommended that WT could be used for trend analysis in watersheds.

Kisi (2009) developed neuro-wavelet (NW) model by combining two methods DWT and artificial neural network (ANN), for 1 day ahead intermittent streamflow forecasting and results were compared with those of the single ANN model. Intermittent streamflow data from two stations in the Thrace Region, the European part of Turkey, in the northwest part of the country were used in the study. In NW model, the original time series were decomposed into a five number of subtime series components by Mallat DWT algorithm. The correlation coefficients between each subtime series and original intermittent streamflow time series were found. These correlation values provide information for the determination of effective wavelet components on streamflow. The new subtime series having high correlation coefficient was used as input to the ANN model. The NW model was found to be much better than the ANN in high flow estimation. The test results showed that the DWT could significantly increase the accuracy of the ANN model in modeling intermittent streamflows.

Rajaee et al. (2010) investigated the Prediction of daily suspended sediment load 1 day ahead with wavelet and neuro-fuzzy combination model using time series data of discharge and suspended sediment load as input in a gauging station from the Pecos River in USA. Results showed that the wavelet analysis and neuro-fuzzy model performed better predictions rather than neuro-fuzzy and sediment rating curve (SRC). The cumulative suspended sediment load estimated by this technique was closer to the actual data. The WNF model considers periodic and stochastic characteristics of suspended sediment phenomenon and may provide suitable constructions not clearly seen in the suspended SRC. The model also could be employed to stimulate hysteresis phenomenon, while the SRC method is incapable in this event.

Shiri and Kisi (2010) studied short-term and long term streamflow forecasting using a wavelet and neuro-fuzzy conjunction model to investigate the daily, monthly, and yearly streamflow of Derecikviran station on Filyos River in the Western Black Sea region of Turkey using 31 years of streamflow data. The results obtained showed that the neuro-fuzzy (NF) and wavelet–neuro-fuzzy (WNF) models increased the accuracy of the single NF models especially in forecasting yearly streamflow. Also the single NF and WNF models were compared with each other by adding periodicity components into the inputs. The comparison results indicated that adding periodicity component generally increased the models accuracy.

Kisi (2010) developed neuro-wavelet models for daily suspended sediment estimation for two stations on tongue river in montana using daily streamflow and suspended sediment data. The comparison results reveal that the developed model could increase the estimation accuracy.

Adamowski and Sun (2010) investigated a method based on coupling discrete wavelet transform (WA) and ANN for flow forecasting applications in nonperennial rivers in semiarid watersheds at lead times of 1 and 3 days for two different rivers in Cyprus. The discrete trous wavelet transform was used to decompose flow time series data into eight levels of wavelet coefficients which are used as inputs to Levenberg Marquardt artificial neural network models to forecast flow. WA–ANN model provided more accurate results than regular ANN.

Nourani et al. (2011) studied two hybrids for two watersheds located in Azerbaijan, Iran. Artificial Intelligence approaches for modeling rainfall–runoff process. Two hybrid AI-based models which are reliable in capturing the periodicity features of the process are introduced for modeling. In the first model, the SARIMAX (Seasonal Auto Regressive Integrated Moving Average with exogenous input)–ANN model, an ANN is used to find the nonlinear relationship among the residuals of the fitted linear SARIMAX model. In the second model, the wavelet–ANFIS model, WT is linked to the ANFIS concept and the main time series of two variables (rainfall and runoff) are decomposed into some multifrequency time series by WT. Afterward, these time series are imposed as input data to the ANFIS to predict the runoff discharge one time step ahead. The obtained results showed that, although the proposed models can predict both short and long terms runoff discharges by considering seasonality effects, the second model is relatively more appropriate because it uses the multiscale time series of rainfall and runoff data in the ANFIS input layer.

Kisi and Shiri (2011) developed precipitation forecasting model using wavelet–genetic programming and WNF conjunction. They found that hybrid wavelet–genetic programming model was of better performance than hybrid wavelet–neuro-fuzzy model.

Rajaee et al. (2011) developed ANN, wavelet analysis and ANN combination (WANN), multilinear regression (MLR), and SRC models for daily suspended sediment load (*S*) modeling in the Iowa gauging station in the US. In the WANN model, DWT was linked to the ANN method. For this purpose, the observed time series of river discharge (*Q*) and *S* were decomposed into five levels by DWT which were imposed as input to ANN to predict 1 day ahead *S*. A complex Morlet wavelet technique was applied to analyze wavelet construction of daily *Q* and *S*. The number of nodes in the input in WANN model was determined by $(i + 1) \times 2$, because this model uses two variables (*Q* and *S*) and each time series is decomposed into *i*, $i = (1,2,\ldots,5)$ detailed time series and approximation time series.

This study was aimed at examining the effects of employed mother wavelet type on the proposed WANN model efficiency. Seven different mother wavelets were used [viz., Daubechies-2 (db2) (the most popular wavelet), the Haar wavelet (a simple wavelet), and some irregular wavelet such as Biorl.1, Rboil.1, Coifl, Syml, and Mayer wavelets].

It was found that, increasing the decomposition level, in levels over Level 1, decreases the model's performance, because high decomposition levels lead to a large number of parameters with complex nonlinear relationships in the ANN technique. The WANN model was more accurate in predicting the *S* and its performance was better than the ANN, MLR, and SRC models.

Wang et al. (2011) utilized wavelet transform method for synthetic generation of daily streamflow in Jinsha river of China. Daily streamflow sequences with different frequency components are decomposed into the series of wavelet coefficients at various resolution levels using wavelet decomposition algorithm. Based on these sampled subseries, a large number of synthetic daily streamflow sequences are obtained using wavelet reconstruction algorithm. They concluded that this newly developed method is able to generate streamflow sequences based on probability distributions and type of dependence structure.

4 Conclusion

This paper serves as an introduction to WT with emphasis on their application to hydrologic problems. It presents brief description of WT, the underlying concept, and mathematical aspects, and the role of WT relative to other approaches in hydrology. Guidelines for application of WT to hydrological problems are presented. The role of WT in various branches of hydrology has been examined here and found that WT is robust tool in analysis of many nonlinear and nonstationary hydrologic processes such as rainfall–runoff, streamflow, groundwater modeling, precipitation, evaporations. However, WT tends to be data (signal) intensive and prudent on statistical properties of dataset. For this emerging technique, still more questions arises which must be further studied.

References

Adamowski J, Sun K (2010) Development of a coupled wavelet transform and neural network method for flow forecasting of non-perennial rivers in semi-arid watersheds. J Hydrol 390: 85–91

Addison PS, Murrary KB, Watson JN (2001) Wavelet transform analysis of open channel wake flows. J Eng Mech 127(1):58–70

Cannas B, Fanni A, Sias G, Tronchi S, Zedda MK (2005) River flow forecasting using neural networks and wavelet analysis. Eur Geosci Union, Vienna, Ausrtia 7:24–29

Kim T, Valdes JB (2003) Nonlinear model for drought forecasting based on a conjunction of wavelet transform and neural networks. J Hydrol Eng 8(6):319–328

Kisi O (2009) Neural networks and wavelet conjunction model for intermittent stream flow forecasting. J Hydrol Eng 14(8):773–782

Kisi O (2010) Daily suspended sediment estimation using neuro-wavelet models. Int J Earth Sci 99:1471–1482

Kisi O, Shiri J (2011) Precipitation forecasting using wavelet-genetic programming and wavelet-neuro-fuzzy conjunction models. Water Resour Manage 25:3135–3152

Li X, Ding J, Li H (1997) Wavelet analysis and its potential application to hydrology and water resources. J Sichuan Union Univ (Eng Sci) 1(4):49–52

Li X, Ding J, Li H (1999) Combing neural network models based on wavelet transform. J Hydraul 2:1–4

Mallat SG (1989) A theory for multiresolution signal decomposition: the wavelet representation. IEEE Trans Pattern Anal Mach Intell 11(7):674–693

Nourani V, Alami TM, Aminfar MH (2009a) A combined neural-wavelet model for prediction of Ligvanchai watershed precipitation. Eng Appl Artif Intell 22:466–472

Nourani V, Komasi M, Mano A (2009b) A multivariate ANN—Wavelet approach for rainfall-runoff modeling. Water Resour Manage 23:2877–2894

Nourani V, Kisi O, Komasi M (2011) Two hybrid artificial intelligence approaches for modeling rainfall—runoff process. J Hydrol 402:41–49

Prahlada R, Deka PC (2011) Hybrid wavelet neural network model for improving forecasting accuracy of time series significant wave height. Int J Earth Sci Eng 4(5):857–866 (ISSN0974-5904)

Rajaee T, Mirbagheri SA, Nourani V, Alikhani A (2010) Prediction of daily suspended sediment load using wavelet and neuro-fuzzy combined model. Int J Environ Sci Tech 7(1):93–110

Rajaee T et al (2011) River suspended sediment load prediction: application of ANN and wavelet conjunction model. ASCE J Hydrol Eng 16(8):613–627

Rao YRS, Krishna B (2009) Modelling hydrological time series data using wavelet neural network analysis. IAHS Publication 333:101–110

Shiri J, Kisi O (2010) Short term and long term stream flow forecasting using a wavelet and neuro-fuzzy conjunction model. J Hydrol 394:486–493

The MathWorks (2010) Matlab and Simulink. www.mathworks.com

Wang W, Yuan P, Ding J (2000) Wavelet analysis and its applications to the stochastic simulation of daily discharge process. J Hydraulics 11:43–48

Wang W, Ding J, Xiang H (2002) The multi-time scale analysis of hydrological time series with wavelet transform. J Sichuan Univ 35(4):14–17

Wang W, Hu S, Li Y (2011) Wavelet transform method for synthetic generation of daily streamflow. Water Resour Manage 25:41–57
Wensheng W, Jing D (2003) Wavelet network model and its application to the prediction of hydrology. Nature Sci 1(1):67–71
Zhou HC, Peng Y, Liang G-H (2008) The research of monthly discharge predictor corrector model based on wavelet decomposition. Water Resour Manage 22:217–227

ANN and ANFIS Modeling of Failure Trend Analysis in Urban Water Distribution Network

Libi P. Markose and Paresh Chandra Deka

Abstract Pipeline leakage is one of the crucial problems affecting urban water distribution system from both environmental and economical point of view. Unfortunately, necessary large databases are not maintained in India for proper replacement of pipes. In this situation, this research purports at using two artificial intelligence techniques such as artificial neural network (ANN) and adaptive neuro-fuzzy inference systems (ANFIS) to access the present condition and to predict the future trend of pipeline network of Peroorkada zone in Trivandrum city, Kerala, where a huge amount is spent every year for leakage rectification. Using different influential input variables, four models (all diameter pipes) have been developed. Also, the effect of each parameter (length, age, and diameter) along with previous year failures and previous year failures alone to current year failures are analyzed. Another two models of selective pipe diameter for considering the influence of prefailures up to last year alone and prefailures up to last year along with length to current year failures are constructed. Prioritizing the pipeline replacement is done for mains having 400 mm diameter and above since network details pertaining to those diameters are available. The performance of the models is evaluated using coefficient of correlation and mean absolute error and is compared to multiple linear regression (MLR) models. Three of them perform well and almost in kind, even though ANN is slightly having an upper hand. The applicability and usefulness of ANN and ANFIS will surely become beneficial for the authorities to take decisions regarding the replacement of pipes and this can in turn increase the efficiency of pipes.

Keywords Pipeline leakage · Failure prediction · ANN · Urban water distribution system · MLR · ANFIS

L.P. Markose (✉) · P.C. Deka
Department of Applied Mechanics and Hydraulics, National Institute of Technology, Srinivasanagar, Surathkal 575075, Karnataka, India
e-mail: libimarkose@gmail.com

P.C. Deka
e-mail: paresh_deka@sify.com

A.K. Sarma et al. (eds.), *Urban Hydrology, Watershed Management and Socio-Economic Aspects*, Water Science and Technology Library 73,
DOI 10.1007/978-3-319-40195-9_20

1 Introduction

Even with the improvement in technologies related to urban water supply, it has become common for cities to have a legion of water pipeline breaks each year. To ensure that the authorities can manage their water distribution systems to provide an adequate supply of water in a cost effective, reliable, and sustainable manner, it is essential that they develop a clear understanding of water pipelines deterioration process. Predictive modeling includes a collection of technologies that can be based on the massive data including pipe physical property, environmental factor, operational condition, and historical failure record.

Recently soft computing techniques like artificial neural networks (ANN) and adaptive neuro-fuzzy inference systems (ANFIS) have usurped many conventional methods in successful modeling of complex water resources systems. Unlike many hydrological applications, there are limited applications of ANN in modeling pipeline failure trend. Tabesh et al. (2009) used ANN and ANFIS to model pipe failure rate (number of accidents per year per unit length) by considering five input parameters and the results are compared with multivariate regression approach. RaedJafar and Juran (2010) presented an application of artificial neural networks (ANN) to model the failure rate of 4862 urban mains using a 14-year database collected in a city in the north of France and to estimate the optimal replacement time for the individual pipes in an urban water distribution system. Abdel Wahal M Budtiena et al. (2011) introduced the effectiveness of ANN to model pipe line breaks by considering ten input parameters.

Even though literatures have explored the usage of ANN to predict pipeline failure, the variation in performance of ANN by changing inputs and length of dataset are not considered and very few have used ANFIS for failure determination. Moreover no study regarding the analysis of failure trend in pipelines have been attempted in Peroorkada zone of Trivandrum city.

The main objectives of the present study are to analyze the trend in pipeline failure using ANN and ANFIS and to prioritize the renewal of network containing larger diameter pipelines with the best model.

2 Artificial Neural Network

Artificial neural networks are also referred to as "neural nets," "artificial neural systems," "parallel distributed processing systems," and "connectionist systems." A neural network paving its origin in nineteenth century is an information processing system which is biologically inspired, i.e., they are composed of elements that perform in a manner that is analogous to the most elementary function of the biological neuron. Feed-forward neural network (Fig. 1) with backpropagation algorithm is used for analysis. In this, the nodes are arranged in layers starting from the first input layer and ending at the final output layer. There can be several hidden layers with each layer having one or more nodes. Information passes from the input to the outside side. The nodes in one layer are connected to those in the next, but not to those in the same layer. Thus output of a node in a layer is only dependent on the inputs it receives from previous layers and the corresponding weights. Backpropagation algorithm does its work in two steps. In the first step, each input pattern of the training dataset is passed through the network from the input layer to the output layer. The network output is compared with the desired target output and in the second step error is computed and is propagated back toward the input layer with the weights being modified. Backpropagation uses delta rule to adjust the connectivity weights. During training, weights need continuous adjustment from iteration t to $t + 1$. The adjustment $\Delta W(t+1)$, which is required in iteration $(t + 1)$, is assumed linearly related to the negative gradient of E with w in iteration t. The constant of proportionality in this linear relation is known as the learning rate (η). Mathematically this relation can be expressed as follows:

$$\Delta W(t+1) = \eta\left(-\frac{\partial E}{\partial W}\right)_{W=W(t)} \tag{1}$$

$$W(t+1) = W(t) - \eta\left(\frac{\partial E}{\partial W}\right)_{W=W(t)} \tag{2}$$

For improving the convergence the following modification is made in Eq. (2).

Fig. 1 Feed-forward backpropagation network

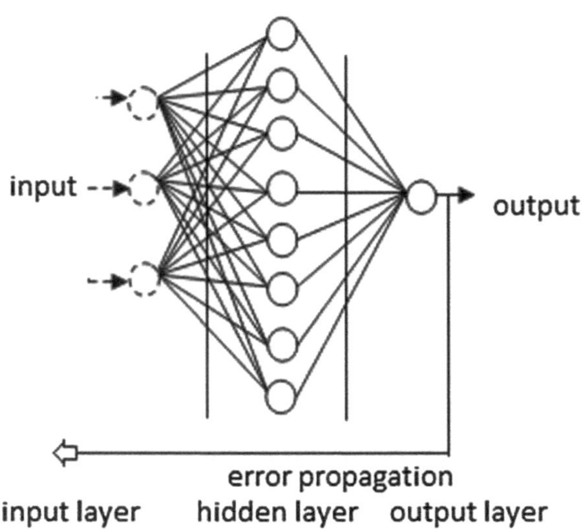

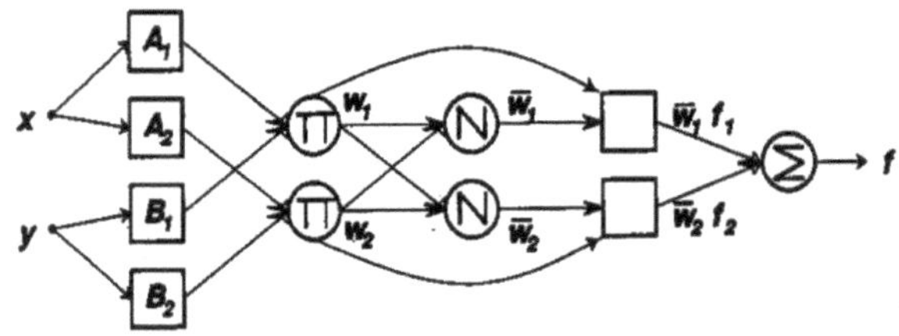

Fig. 2 Schematic of ANFIS architecture

$$W(t+1) = W(t) - \eta\left(\frac{\partial E}{\partial W}\right)_{W=W(t)} + \mu\Delta W(t) \tag{3}$$

where (μ) is the momentum factor as it imparts the momentum to the rate of convergence.

3 Adaptive Neuro-Fuzzy Inference System

Adaptive neuro-fuzzy inference system proposed by J.S.R Sang is a methodology to simulate complex nonlinear mappings utilizing neural network learning and fuzzy inference methodologies. It brings into service the learning capability of ANN for rule generation and parameter optimization. ANFIS architecture (Fig. 2) has five nodes out of which first and fourth are adaptive and others are fixed nodes. In the first layer each input (*x*/*y*) is converted to a membership values ranging from 0 to 1. Here A_1, A2, B1, B2 are linguistic labels. In the second layer all the incoming signals are multiplied together to get (w_1/w_2) and a normalized firing strength ($\bar{w}_1/\bar{w}_2$) is obtained in the third layer. In the subsequent layer firing strength will be multiplied with the function (f_1/f_2) and the last layer computes the summation (*f*) of all incoming signals.

4 Study Area

The study concerns the water distribution system of Peroorkada zone in Trivandrum city which comprises an area of 15.46 km^2. Total population as per 1991 census is 137714. Total water demand is 25.82 m^3/min. There is only one reservoir at Peroorkada which has a capacity of 8 million litres. The total domestic connection is 13700 and the number of household connections is approximately 18470. The Peroorkada network consists of 99 nodes and 114 pipes and 16 loops. The water distribution network map of Peroorkada zone is shown in Fig. 3.

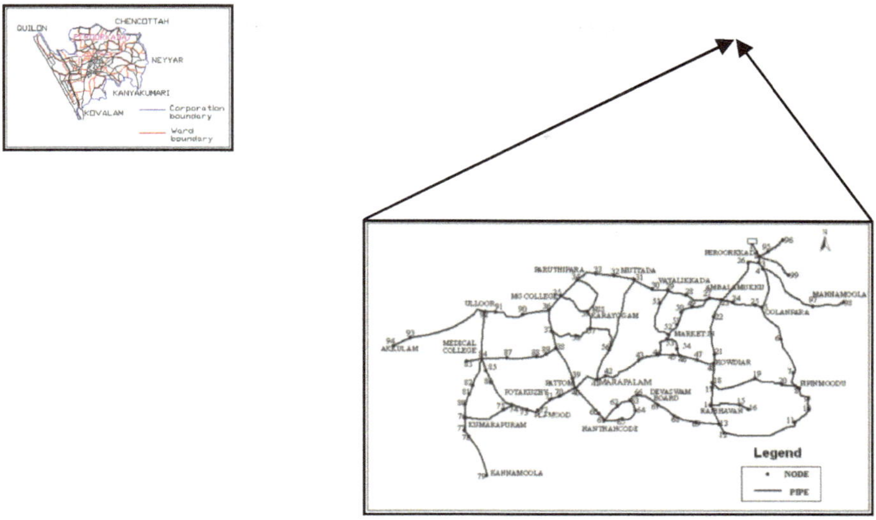

Fig. 3 Location and water distribution network of Peroorkada zone

5 Available Data and Model Structures

5.1 Available Data

The complete network containing diameters of pipe other than 400 mm prestressed concrete pipe is not available. The details (length, age, material, diameter, location, number of failures for the last ten years from 2000 to 2010) are available for 400 mm and for the main three pipes supplying water from Aruvikkara to Trivandrum city. These details are collected from the running contract leakage rectification available in Kerala Water Authority, Kowdiar, Aruvikkara, and Vellayambalam. For smaller diameter pipes, both the network details and the location of failure are not available even though failures happened in the past ten years for each of these pipes are there.

5.2 Model Structures

The dataset containing failures and prefailures are divided into two sets, i.e., testing and training data. As a total, 140 datasets are available for diameters lesser than 400 mm and 117 datasets are available for 400 mm diameter. Different lengths of testing data are considered to know the improvement in performance of the model.

In this study, for ANN a three-layered feed-forward backpropagation network (newff) has been selected based on the outcome of past literatures. To train the network the fastest training algorithm Levenberg Marquardt algorithm has been selected to achieve the training speed. To know the optimal number of neurons in the hidden layer different trials are conducted. After training the networks are simulated for unknown values using testing sets. The predicted values obtained from the network are compared with the observed values using various performance evaluators and the best network is ranked.

The developed models are also trained using ANFIS. Different membership functions and epochs are considered to get the best possible results.

For the performance evaluation of the models, two statistical criteria namely mean absolute error and coefficient of correlation are used.

Mean absolute error

$$\mathrm{MAE} = \frac{\sum |X - Y|}{N} \quad (4)$$

Coefficient of correlation

$$\mathrm{CC} = \frac{\sum xy}{\sqrt{\sum x^2} * \sqrt{y^2}} \quad (5)$$

where

N number of datasets

X target output pattern value

Y output from neural net work model

x $X - X_{\mathrm{mean}}$, $y = Y - Y_{\mathrm{mean}}$

6 Results and Discussion

The efficacy of ANN and ANFIS to analyze the trend of failure is verified by means of six models and is shown in Table 1. The six models are constructed by varying the factors responsible for failure. The fifth and sixth models are constructed for 400 mm pipes. The modeling of ANN and ANFIS are coded using MATLAB 2010a. The feed-forward backpropagation network with single hidden layer is

Table 1 The structure of forecasting models

Model type	Inputs	Outputs
M1	Failures in the previous year	Failures in the present
M2	Diameter, failures in the previous year	Failures in the present
M3	Age, failures in the previous year	Failures in the present
M4	Diameter, age, failures in the previous year	Failures in the present
M5	Length, number of failures up to 2009	Failures up to 2010
M6	Number of failures up to 2009	Failures up to 2010

Table 2 Comparison of performance of ANN models in training with the length of dataset

Dataset (D)	10 % D		30 % D		25 % D	
Model	CC	MAE	CC	MAE	CC	MAE
M1 (1-7-1)	0.91	42.16	0.95	27.91	0.9	25.74
M2 (1-4-1)	0.91	35.56	0.95	29.73	0.94	26.63
M3 (1-2-1)	0.92	41.34	0.98	21.57	0.94	23.05
M4 (1-3-1)	0.94	32.84	0.95	31.42	0.93	25.28
M5 (1-3-1)	0.98	0.09	0.99	0.17	0.99	0.11
M6 (1-3-1)	0.98	0.15	0.99	0.14	0.98	0.18

Table 3 Comparison of performance of ANN models in testing with the length of dataset

Dataset (D)	10 % D		30 % D		25 % D	
Model	CC	MAE	CC	MAE	CC	MAE
M1 (1-7-1)	0.95	13.33	0.7	47.75	0.93	33.66
M2 (1-4-1)	0.95	15.39	0.81	36.94	0.93	36.24
M3 (1-2-1)	0.91	22.97	0.79	45	0.93	36.24
M4 (1-3-1)	0.93	23.72	0.81	35.2	0.93	33.1
M5 (1-3-1)	0.96	0.41	0.97	0.34	0.98	0.25
M6 (1-3-1)	0.95	0.51	0.95	0.28	0.96	0.26

performing best in majority of applications in the field of hydrology for ANN modeling, so it is been used here. The other parameters used are tangent sigmoid and linear activation function. The results obtained are shown by means of graphs and tables. Tables 2 and 3 show the variation in performance indices for testing and training while considering different lengths of testing dataset. The 10 and 30 % dataset for testing are taken after leaving first 90 and 70 % dataset for training, but the 25 % of dataset for testing is considered by starting with the second point and taking every fourth point. In majority of cases with 25 % dataset the models are giving best performance due to better training obtained by the model. The length of dataset which gives the best results for each of the models is considered. ANN is performing well in case of normalized data when compared with raw data for the first model (Fig. 4) so all the dataset has been normalized before giving it to the remaining models. When analyzed by ANN among the first four models, the fourth model is performing well both in case of training and testing and from the last two models model 5 excels others. Thus in the case of ANN, the models which took majority of parameters is performing well (Tables 4 and 5). In the case of ANFIS, model 2 does well in training and testing among the first four and model 6 among the last two (Tables 4 and 5). From this it can be underpinned that diameter, length, and prefailures are important parameters responsible for failure. Table 5 gives the values of various performance indices of different models evaluated by ANFIS and ANN and MLR. In majority of models ANN bettered both ANFIS and MLR, even though in case of model 6 an exception occurred. ANFIS performs well in training

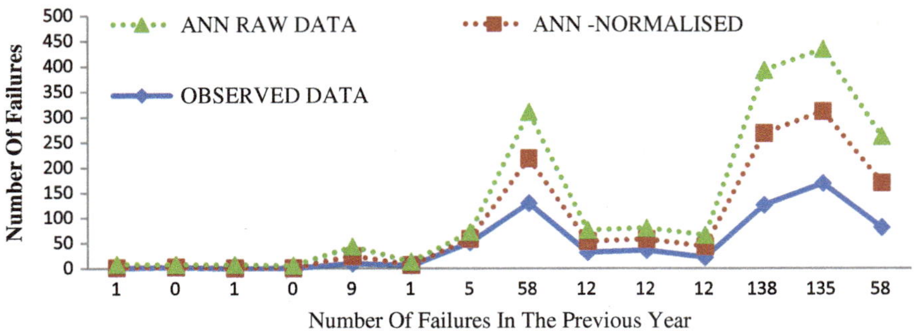

Fig. 4 Comparison of using normalized and raw data in ANN

Table 4 Performance of models analyzed by ANFIS and ANN

Dataset (D)	ANFIS		ANN	
Model	CC	MAE	CC	MAE
M1	0.89	37.43	0.91	42.16
M2	0.93	32.54	0.91	35.56
M3	0.98	18.31	0.94	23.05
M4	0.96	22.8	0.93	25.28
M5	0.99	0.05	0.99	0.11
M6	0.99	0.12	0.99	0.14

Table 5 Performance of models analyzed by ANN, ANFIS, and MLR

Dataset (D)	ANFIS		ANN		MLR	
Model	CC	MAE	CC	MAE	CC	MAE
M1	0.93	16.93	0.95	13.33	0.92	17.7
M2	0.92	13.21	0.95	15.39	0.93	14.86
M3	0.91	37.46	0.93	36.24	0.9	40
M4	0.86	49.49	0.93	33.1	0.9	42.45
M5	0.88	0.62	0.98	0.25	0.91	0.36
M6	0.95	0.17	0.95	0.28	0.94	0.88

phase for all models compared to ANN. But in testing phase its capability is not up to the mark. This may be due to over training. To prioritize the pipe network considering 400 mm diameter, the model M5 (ANN) is used since it outfits model M4 by giving good correlation (Table 6). From the first four models model 1 predicts well the failures in the testing for both ANN and ANFIS and among the last two, model 5 performs better. So these two models are taken for the comparing ANN, ANFIS and MLR (Figs. 5 and 6). From Fig. 5, it can be observed that ANN and ANFIS are predicting lower and medium number of failures in an accurate manner than higher. Compared to ANN and MLR models, ANFIS predicts higher

Table 6 Prioritization order for renewal based on the predicted number of failures by model M5 (ANN)

Link	Length (m)	Order of prioritization	Diameter (mm)
0-a	2422	1	1200
a-b	4905	2	1000
4_5	687	2	400
3_4	117	3	400
22_23	307	3	400
30_31	322	3	400
71-72	252	3	400
6_7	634	3	400

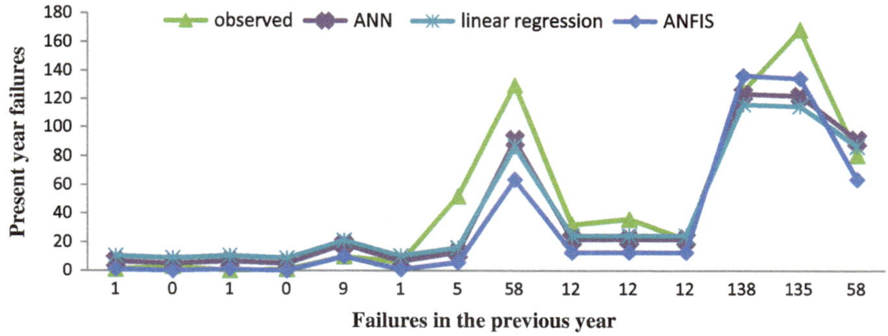

Fig. 5 Comparison between ANN, ANFIS, and MLR in predicting failures (model 1)

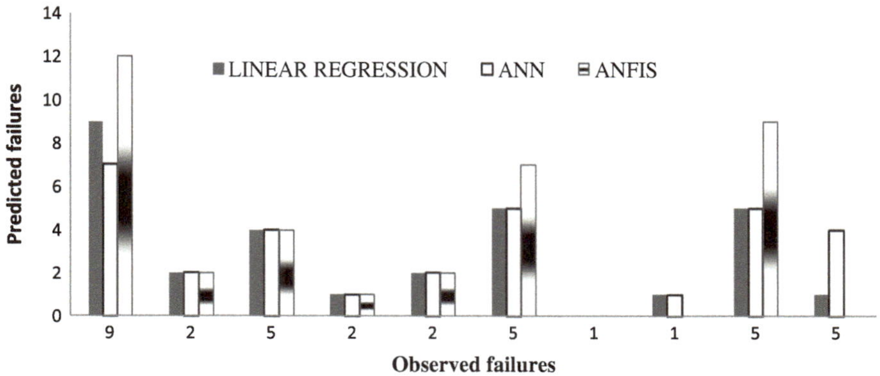

Fig. 6 Comparison between ANN, ANFIS, and MLR in predicting failures (model 5)

number of failures accurately in case of all diameter pipes. Figure 6 shows the betterment of ANN in prediction of failures than ANFIS and MLR for 400 mm and above prestressed concrete pipes.

7 Conclusion

This preliminary study looks into the capability of ANN and ANFIS for predicting failure in urban distribution network. Based on the leak detection study conducted by JBIC assisted Kerala Water Supply Project, the total physical loss of treated water is found to be 35.5 %. In this time of budget cuts and limited resources, the ability to optimize the use of maintenance money by employing predictive models in the planning stages is rapidly becoming a reality for urban underground infrastructure management. Planned maintenance facility in need of repair can yield significant savings over unscheduled or emergency repairs. The inputs to the model are selected based on relevant literature and data availability. The work can be extended with more number of inputs.

Acknowledgments The authors appreciate the enthusiasm taken by the officials of Kerala Water Authority, Kowdiar, Vellayambalam, and Aruvikkara.

References

Achim D, Ghotb F, McManus KJ (2007) Prediction of water pipe asset life using neural networks. J Infrastruct Syst 13(1):26–30

Ahmad Asnaashari IS (2007) Geostatistical analysis of water mains failure: a case study from Iran. Water Asset Manage Int 3(3):1–6

Bubtiena AM et al (2011) Application of artificial neural networks in modeling water networks. 7th international colloquium on signal processing and its applications

Ho CI, Lin MD, Lo SL (2010) Use of a GIS-based hybrid artificial neural network to prioritize the order of pipe replacement in a water distribution network. Environ Monit Assess 166:177–189

Jafar R, Shahrour I, Juran I (2010) Application of artificial neural networks to model the failure of urban water mains. Math Comput Model 51:1170–1180

Kleiner Y, Rajani B (2002) Forecasting variations and trends in water-main breaks. J Infrastruct Syst 8(4):4–122

Le Gat Y, Eisenbeis P (2000) Using maintenance records to forecast failures in water networks. Urban Water 2:173–181

Pelletier G, Mailhot A, Villeneuve JP (2003) Modeling water pipe breaks—Three case studies. J Water Res Plan Manage 129(2):2–115

Rogers PD et al (2009) Failure assessment modeling to prioritize water pipe renewal: two case studies. J Infrastruct Syst 15:162–171

Rogers PD (2011) Prioritizing water main renewals: case study of the denver water system. J Pipeline Syst Eng Pract 2:73–81

Sivanandam SN, Deepa SN (2011) Principles of soft computing. Wiley India Pvt. Ltd, Daryaganj, New Delhi-110002

Suja R, Deepthi N, Letha J (2011) Multi objective reliability based design of water distribution system. J Civil Eng (IEB) 39(1):19–31

Tabesh M et al (2009) Assessing pipe failure rate and mechanical reliability of water distribution networks using data driven modelling. J Hydroinformatics 11(1):1–17

Tian CH et al (2011) Pipe failure prediction. IBM Research, China Beijing 100193, 121–125

Effects of Data Pre-processing on the Prediction Accuracy of Artificial Neural Network Model in Hydrological Time Series

Aniruddha Gopal Banhatti and Paresh Chandra Deka

Abstract The accurate prediction of hydrological behaviour in both urban and rural watershed can provide valuable information for the urban planning, land use, design of civil project and water resources management. Hydrology system is influenced by many factors such as weather, land cover, infiltration, evapotranspiration, so it includes the good deal of stochastic dependent component, multi-time scale and highly nonlinear characteristics. Hydrologic time series are often nonlinear. In spite of high flexibility of Artificial Neural Network (ANN) in modelling hydrologic time series, sometimes signals exhibit seasonal irregularity. In such situation, ANN may not be able to cope with such data if pre-processing of input and/or output data is not performed. Pre-processing data refers to analysing and transforming input and output variables in order to detect trends, minimise noise, underline important relationship and flatten the variables distribution in a time series. These analysis and transformations help the model learn relevant patterns. Pre-processing techniques, which facilitates stabilisation of the mean and variance, and seasonality removal, are often applied to remove irregularities in data used to build soft computing models. In this study, different data pre-processing techniques are presented to deal with irregularity components existing in a hydrologic time series data of the Brahmaputra basin within India at the Pandu gauging station near Guwahati city using daily time unit and their properties are evaluated by performing one step ahead flow forecasting using ANN. The model results were evaluated using root mean square error (RMSE) and mean absolute percentage error (MAPE) and found that logarithmic-based pre-processing techniques provide better forecasting performance among various pre-processing techniques.

A.G. Banhatti (✉) · P.C. Deka
Department of Applied Mechanics, National Institute of Technology, Surathkal, Karnataka, India
e-mail: anibanister@gmail.com

P.C. Deka
e-mail: pareshdeka@yahoo.com

A.G. Banhatti
G.H.Raisoni College of Engineering and Management, Pune 412207, India

A.K. Sarma et al. (eds.), *Urban Hydrology, Watershed Management and Socio-Economic Aspects*, Water Science and Technology Library 73,
DOI 10.1007/978-3-319-40195-9_21

The results indicate that detecting irregularities and selecting an appropriate pre-processing technique is highly beneficial in improving the prediction performance of ANN model.

Keywords Daily water demand · Fuzzy logic · ANFIS · MLR · Time series

1 Introduction

A common assumption in the time series analysis is that time series data have constant mean and variance. That is, they are stationary. This is normally true except when shocks are administered to the system generating the series, resulting in non-stationary values in variance, or there is a trend in the series, resulting in irregular and non-stationary nature in the mean. Pre-processing techniques, which facilitates stabilisation of the mean and variance, and seasonality removal, are often applied to remove irregularities in data used to build soft computing models.

Data pre-processing is an important step in developing ANN application, which could affect model accuracy and results. Pre-processing data refers to analysing and transforming input and output variables in order to detect trends, minimise noise, underline important relationship and flatten the variables' distribution. These analysis and transformations help the model learn relevant patterns.

Before data is used by an algorithm, it must go through several transformations in order to prepare the raw data. The success of an algorithm greatly depends on the quality of input data. As different methods can handle only different samples, it is recommended to exploit certain data features with the purpose of finding out which pre-processing transformation works best.

There are number of reasons for identifying irregular data and trends, using appropriate pre-processing techniques (Virli and Freisleben 2000) such as data representation can help to prewire in the forecasting model information about the underlying structure (like trends and seasonal patterns) and to obtain better forecast.

Recently, ANN has shown great ability in modelling and forecasting nonlinear hydrologic time series. Although classic time series models like autoregressive moving average (ARMA) are widely used for hydrological time series forecasting, they are based on linear models assuming the data are stationary and have limited ability to capture nonlinearities in hydrologic data. ANNs are found suitable for handling huge amounts of dynamic, nonlinear and noisy data when underlying physical relationships are not fully understood (Nourani et al. 2009). In spite of high flexibility of ANN in modelling hydrologic time series, sometimes signals exhibit seasonal irregularity. In such situation, ANN may not be able to cope with irregular data if pre-processing of input and/or output data is not performed (Cannas et al. 2005). Based on study done by Nguyen and Chan (2004), data pre-processing is one of the most important steps for developing an ANN model for prediction. They have presented three data pre-processing strategies and gave the advantages, disadvantages and compare the results of each approach. Virli and Freisleben (2000), have demonstrated the effects of non-stationarity on ANN prediction on economic time series.

Simple ANN systems as well as complicated hybrid have been used to analyse real-world time series, which are usually characterised by mean and variance changes, seasonality and other local behaviour. Such real-world time series are not only invariably nonlinear and non-stationary, but also incorporate significant distortions due to both 'knowing and unknowing misreporting' and "dramatic changes in variance" (Granger'94). The presence of these characteristics in time series stress desirability of data pre-processing. These studies have focussed on investigating the ability of NN to model hydrologic time series and effect of data pre-processing on forecast performance of NN.

The forecasting methods vary greatly and depend on the data availability, the quality of models available, and the kinds of assumptions made, amongst other things. ANN has found increasing considerations in forecasting theory, leading to successful applications in various forecasting domains. ANN can learn from examples (past data), recognise a hidden pattern in historical observations and use them to forecast future values. In addition to that, they are able to deal with incomplete information or noisy data and can be very effective especially in situations where it is not possible to define the rules/relationships or steps that lead to the solution of a problem.

The attractive features of ANN to various forecasting domains are many. Being a data-driven learning machine as opposed to conventional model-based approaches, permitting universal approximation of arbitrary linear or nonlinear functions, and therefore offering great flexibility in learning the generator of noisy data from examples and generalising structure from it without priori assumptions (Zhang et al. 1998). Due to their flexibility, neural network lacks systematic procedure for model development. Therefore, obtaining a reliable neural network model involves selecting a large number of parameters experimentally through trial and error.

Despite many satisfactory characteristics (Zhang et al. 1998) of ANNs, developing an ANN model for a particular forecasting problem is a non-trivial task. Several authors such as Plummer (2000), Xu and Chen (2001) and Lam (2004) have provided an insight on issues in developing ANN model for forecasting. These modelling issues must be considered carefully because it may affect the performance of ANNs. Based on their studies, some of the discussed modelling issues in constructing ANN forecasting model are the selection of network architecture, learning parameters, and data pre-processing techniques applied to the time series data.

The performance of ANN in forecasting is influenced by ANN modelling that is the selection of the most relevant network architecture and network design. Poor selection of parameter settings can lead to slow convergence and incorrect output (Kong and Martin 1995). One critical decision is to determine the appropriate network architecture, that is, number of layers, the number of nodes in this layer and the number of arcs, which interconnect with the nodes. The network design decision include the selection of activation function in the hidden and output neurons, the training algorithm, data transformation or normalization method, training and test set and performance measures. Kong and Martin (1995) had focussed their studies on parametric effects on building a BP (backpropagation) network for a particular forecasting problem. The issues on modelling fully connected feedforward networks for forecasting had been discussed by Zhang et al. (1998). Maier and Dandy (2000) have reviewed the modelling issues and outlined the steps that should be followed in developing ANN model for predicting and forecasting water resources variables.

When BP algorithm was introduced in 1986, there has been much development in the use of ANN for forecasting by a number of researchers such as Kong and Martin (1995), Lopes et al. (2000), and Deka and Chandramouli (2005).

Thus, this study uses a backpropagation network to predict the hydrologic behaviour of the Brahmaputra basin within India at the Pandu gauging station using daily flow. In particular, due to high seasonal irregularity, typical of a Himalayan weather regime, the role of data pre-processing through logarithmic transformation and detrending transformation has been investigated. This study examines the effect of network parameters through trial and error by varying network structures based on the number of input nodes, activation functions and data pre-processing in designing BP network forecasting model.

The novelty in this work is that, unlike other schemes reported in the literature, our method explicitly takes the statistical properties of the time series into consideration, and only recommends Log-based pre-processing when the properties of the data indicate that such pre-processing is appropriate. If a sophisticated method is used without understanding the underlying properties of the time series, then ironically for certain classes of time series, the forecast are worse than by simpler methods.

2 Study Area

Brahmaputra is the fourth largest river in the world in terms of average discharge at mouth, with a flow of 19,830 cumecs. The hydrologic regime of the river responds to the seasonal rhythm of the monsoons and to the freeze-thaw cycle of the Himalayan snow. The rainy season (May–October) accounts for 82 % of the mean annual flow at Pandu. The discharge is highly fluctuating in nature. Discharge per unit drainage area in the Brahmaputra River Basin is among the highest of major rivers of the world. At Pandu, the Brahmaputra yields 0.0306 cumecs/km^2 and the mean annual flood discharge is 51,156 cumecs. When classifying the data into training and testing subsets, it is essential to check that the data represents the same statistical population. Table 1 indicates the large variation in data in the station Pandu. The maximum discharge being 61,000 cumecs and the minimum being 2432 cumecs for Pandu Station.

There are daily data of streamflow in cumecs starting from 1/1/1980 to 30/12/1998, a total of 6939 data points. When the same is arranged for one-day,

Table 1 Statistical analysis of daily discharge at Pandu station

Station Pandu	Statistical parameters	Training set Q (m^3/s)	Testing set Q (m^3/s)	All dataset Q (m^3/s)
Daily data	Min	2432	5539	2432
	Max	61,015	54,100	61,015
	Mean	17,520	18,897	17,904
	S_d	11,011	10,306	10,836
	C_{xx}	0.59	0.64	0.59

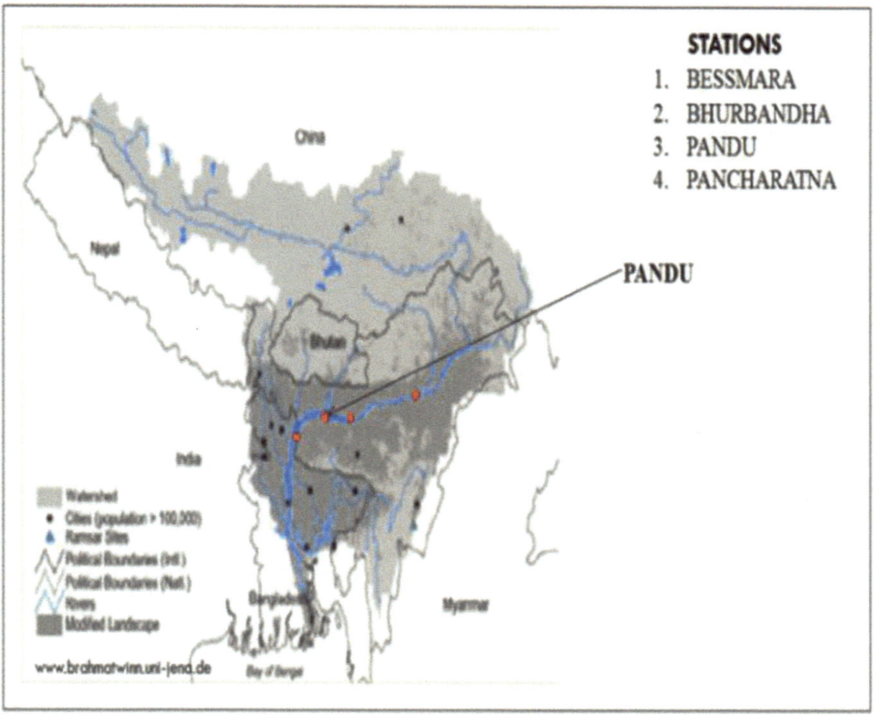

Fig. 1 Study area (Pandu gauging station)

two-day and three-day lagged data, we have to remove the last 3 data points which give 6936 data points. These are divided in the ratio of beginning two-third for training and validation and remaining one-third for testing. Thus, we have used 4624 data points for training and validation and 2312 data points for testing. The study area (station Pandu) is shown in Fig. 1.

The statistical analysis of the data in hydrology provides information about various parameters in general and its distribution in particular. The data statistics for both training and testing at Pandu gauging station are given in Table 1 which contains minimum, maximum, mean, standard deviation (S_d) and skewness coefficient. It is observed that high skewness coefficient has a considerable negative effect on ANN performance. It is obvious from the Table 1 that the extreme values of the available data fall in the training set.

The error analysis is carried out using the RMSE and MAPE in order to evaluate the prediction accuracy.

Root Mean Square Error:

$$\mathrm{RMSE} = \sqrt{\frac{\sum_{i=1}^{N} (X - Y)^2}{N}}$$

where X and Y are the observed and computed flow respectively with N number of observed data. The RMSE is used to measure forecast accuracy, which produces a positive value by squaring the errors. The RMSE increases from zero for perfect forecasts through large positive values as the discrepancies between forecasts and observations become increasingly large.

Mean Absolute Percentage Error (%):

$$\mathrm{MAPE} = \frac{1}{N}\sum_{i=1}^{N} \frac{|X - Y|}{X} \times 100$$

Mean absolute percentage error is defined as the absolute error divided by the true value. It is generally expressed as percentage and helps us to calculate the ratio between absolute error and the true value.

The objectives of the study are to design and develop ANN model, which combines sigmoid activation function in hidden layer and logarithmic activation function in output layer. This research attempts to understand the network parameters by varying them and observing their effect on the network. Specifically, the parametric effect of varying the data pre-processing technique was examined for developing an efficient forecasting model.

3 Methodology

An attempt has been made in this study to investigate the predictive power of neural network by adopting trial and error to tune the number of input nodes and activation functions. The layered feedforward (FFBP) with backpropagation (BP) training algorithm has been shown in the Fig. 2.

A function called activation/threshold/transfer function modifies the signal coming into the hidden layer nodes. The activation function is also referred to as a squashing function in that it squashes (limits) the permissible amplitude range of the output signal to some finite value. Typically, the normalised amplitude range of the output of a neuron is written as the closed unit interval [0, 1] or alternatively [−1, 1].

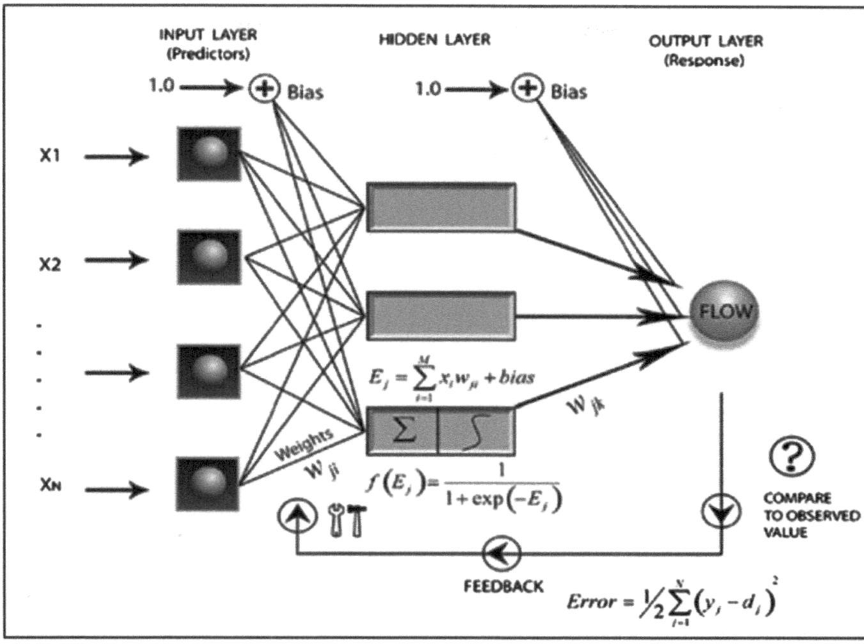

Fig. 2 Three-layered FFNN with BP training algorithm

The choice of activation functions may strongly influence complexity and performance of ANNs (Duch and Jankowski 1999). Most researchers like to use the sigmoid function because of its nonlinear nature which plays an important role in the performance of the ANN (Kong and Martin 1995). Following the convention, a number of authors such as Lopes et al. (2000) and Sreenivasulu and Deka (2011), etc., have simply used the sigmoid function for all hidden and output neurons. This study uses three ANN models with different activation functions. The various activation functions used in the study are shown in Fig. 3.

A total of 10 different network architectures on the basis of various input nodes (1–4) using lagged data are examined with different type of ANN model and pre-processed datasets in this study. A constant value of learning rate and momentum coefficient are used throughout the experiment. Various datasets such as raw data, log-transformed data and log transform + first difference data were used in the analysis.

4 Results and Discussion

The final task is to make a selection of an appropriate dataset or data pre-processing for the flow time series. The dataset with smallest RMSE and MAPE in test dataset is desirable. Before we search for an appropriate dataset, we evaluate the

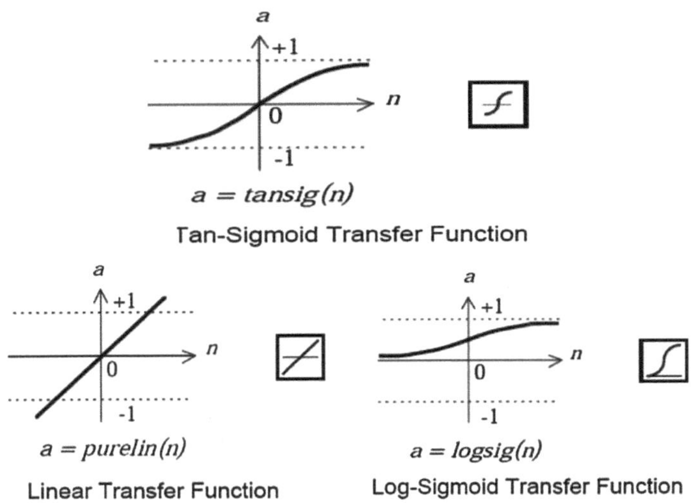

Fig. 3 Types of activation function

forecasting performances and generalisation abilities of various models. Out of four input combinations, three lagged flow inputs generates better performance, which are presented in Tables 2 and 3. Forecast results show that sigmoid model of both tansig and logsig outperformed other purelin model in every dataset. Also, proposed logsig model that comprised of 3 input and 7 hidden neurons provide better results.

Table 2 Training results of various dataset with different activation functions

Dataset	Best network structure	Tansig		Logsig		Purelin	
		RMSE (m^3/s)	MAPE (%)	RMSE (m^3/s)	MAPE (%)	RMSE (m^3/s)	MAPE (%)
Raw dataset	3-2-1	1205.86	4.51	1188.46	3.50	2121.11	19.22
Log transform	**3-7-1**	**1183.61**	**3.36**	**1184.73**	**3.30**	**1578.78**	**5.13**
Log + first difference	3-9-1	2052.69	5.44	2051.39	5.47	2164.69	6.48

Bold values denotes the chosen network configuration

Table 3 Testing results of various dataset with different activation functions

Dataset	Best network structure	Tansig		Logsig		Purelin	
		RMSE (m^3/s)	MAPE (%)	RMSE (m^3/s)	MAPE (%)	RMSE (m^3/s)	MAPE (%)
Raw dataset	3-2-1	1035.09	3.16	1025.48	2.80	1694.43	8.63
Log transform	**3-7-1**	**1019.19**	**2.68**	**1019.34**	**2.78**	**1481.86**	**4.44**
Log + first difference	3-9-1	1814.72	4.36	1808.04	4.30	1960.33	5.37

Bold values denotes the chosen network configuration

Tansig and Logsig models are performing better and almost similarly for log-transformed dataset compared to other type of dataset. Again, it was observed from both the result tables that the forecasting performances of all the models reveals similar trend in training and testing.

Overall, the forecasting results show that log-transformed pre-processed dataset provides better forecasting performance than other pre-processed and raw datasets for all ANN models.

Further, the variation of RMSE with respect to number of hidden neurons are shown in Figs. 4 and 5 to obtain the effects on model performance for best pre-processed dataset. The error variation trends are almost same for both training and testing revealing optimal training. Increasing the number of hidden neurons are not too much significant in improving the performance for best input structure. Small network decreases the complexity of the network in order to process the input from each network node. The drawback of small size network is it slows convergence. The small network seems to take more time and numbers of iterations to learn and understand the data patterns due to the less information provided to the network.

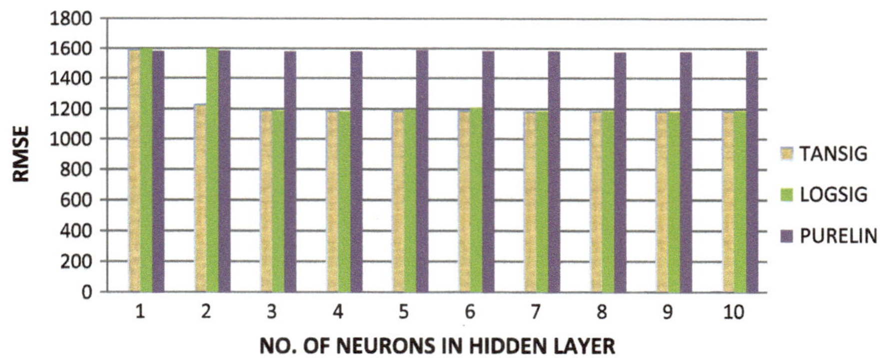

Fig. 4 Error comparison for neurons in hidden layer of log-transformed data (training results)

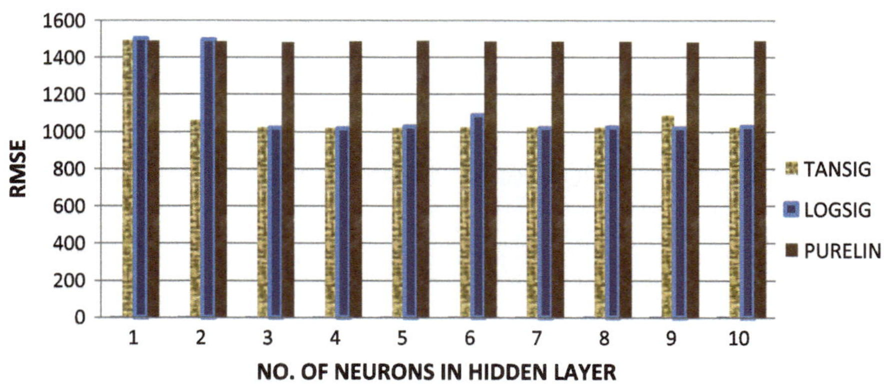

Fig. 5 Error comparison for neurons in hidden layer of log-transformed data (testing results)

Large input nodes mean more information is provided to the network and less time is taken by the network to learn the data patterns. Here, it is difficult to confirm what type of network may be optimal to generate best performance, as more analyses are necessary to make distinctive conclusion.

5 Summary and Conclusion

Thus, it is observed that pre-processing of the data by log transformation greatly improves the performance of networks. It is also observed that PURELIN or PL type of networks fare poorly giving high values of MAPE and RMSE compared to LOGSIG (LS) and TANSIG (TS) types. Using data with more number of lagged terms also seems to improve network performance.

The pre-processing technique of log transformation is highly beneficial to forecast an existing trend and if seasonal variations exist in the time series. Log transform + first difference datasets that are believed to improve an ANN forecast quality do not provide a promising result in this study. This problem may be due to the unsuitable network architecture and learning parameters used during the experiments. Further experiment could be done with different architecture and learning parameters to identify an appropriate ANN model. These results demonstrate that it is possible to improve the ANN performance with much effective activation function. According to our study that the both TANSIG and LOGSIG activation function can be used in the vast majority of ANN applications as a good choice to obtain high accuracy.

References

Cannas B, Fanni A, Sias G, Tronchi S, Zedda MK (2005) River flow forecasting using Neural Networks and Wavelet Analysis, vol 7. EUG, European Geosciences Union, Vienna, Austria, pp 24–29

Deka P, Chandramoulli V (2005) Fuzzy neural network modelling for hydrologic flow routing. ASCE J Hydrol Eng 10(4):302–314

Duch W, Jankowski N (1999) Survey of neural transfer functions. Neural Comput Surv 2:163–212

Kong JHL, Martin GPMD (1995) A back propagation neural network for sales forecasting. IEEE 2121–2124

Lam M (2004) Neural network techniques for fincial performance prediction:integrating fundamental and technical analysis. Decis Support Syst 37:567–581

Lopes MLM, Minussi CR, Lotufo ADP (2000) A fast electric load forecasting using neural networks. In: Proceedings 43rd IEEE Midwest symposium on circuits and systems, learning MI. 8–11 Aug 2000, IEEE, pp 1–4

Maier HR, Dandy GC (2000) Neural networks for the prediction and forecasting of water resources variables: a review of modelling issues and applications. Environ Model Softw 15:101–124

Nguyen HH, Chan CW (2004) A comparison of data preprocessing strategies for neural network modelling of oil production prediction. In: Proceedings of the third IEEE International conference and cognitive informatics (ICCI'04). IEEE Computer Science
Nourani V, Alami MT, Aminfar MH (2009) A combined neural-wavelet model for prediction of Ligvanchai watershed precipitation. Elsevier, Eng Appl Artif Intell 22:466–472
Plummer EA (2000) Time series forecasting with feedforward neural networks: guidelines and limitations. Master thesis, University of Wyoming
Sreenivasulu D, Deka PC (2011) A comparative study on RBF and NARX based methods for forecasting of groundwater level. Int J Earth Sci Eng 04(4):743–756
Virili F, Freisleben B (2000) Nonstationary and data preprocessing for neural network predictions of an economic time series. IEEE 129–134
Xu L, Chen WJ (2001) Short term load forecasting techniques using ANN. In: Proceedings of the 2001 IEEE International Conference of control Applications, pp 157–160
Zhang G, Patuwo BE, Michael YH (1998) Forecasting with artificial neural networks. IEEE 909–912

Urban Water Consumption Estimation Using Artificial Intelligence Techniques

H.J. Surendra and Paresh Chandra Deka

Abstract The general objective of an urban water supply authority is to match supply and demand at a level of service acceptable to the consumers. In this context, predicting water consumption in urban area is of key importance for water supply management. Unfortunately necessary large database are not properly maintained in a country like India. In this situation, this research work propose at using ANFIS (Adaptive Neuro Fuzzy Inference System) to assess the present condition of daily water consumption and predict the future trend in the urban environment. In the field of water resources and hydrology, Artificial Intelligence system approaches such as fuzzy logic (FL), Artificial Neural Network (ANN), etc. have been used successfully for nonlinear and non-stationary time series modeling. For complex system, both FL and ANN techniques are combined together so as the individual strength can be exploited in a better manner for the construction of powerful hybrid system. The purpose of this study was to develop hybrid (ANFIS) model for the estimation of daily water consumption. The performance and potential of the various developed ANFIS models are examined using a four-year's time series water consumption data collected from New Mangalore Port Trust in Mangalore city, India. The influence of lagged variables with respect to daily water consumption and their combination are analyzed. The performances of the ANFIS models are evaluated by using three performance indices namely CC, MSE and MRE and best model was selected. The best ANFIS model reveals higher correlation with recorded data when compared to single fuzzy logic model and statistical Multi-Linear Regression model (MLR). Based on this analysis, the concern authority can take an effective decision regarding the short term as well as long term efficient water management strategy in urban environment of this coastal city.

Keywords Daily water demand · Fuzzy logic · ANFIS · MLR · Time series

H.J. Surendra (✉) · P.C. Deka
Department of Applied Mechanics and Hydraulics,
National Institute of Technology, Surathkal, Karnataka, India
e-mail: careof.indra@gmail.com

P.C. Deka
e-mail: pareshdeka@yahoo.com

A.K. Sarma et al. (eds.), *Urban Hydrology, Watershed Management and Socio-Economic Aspects*, Water Science and Technology Library 73,
DOI 10.1007/978-3-319-40195-9_22

1 Introduction

Water is known as the most important resources in any urban development program. Proper planning, design and management of water resources in urban area depends on water demand forecasting. Urban water demand forecasting is the process of making predictions about future water use based on knowledge of historic water use pattern. Water demands are highly dynamic and are affected by many factors such as population, economic cycles, technology, weather and climate condition, price and conservation program, industrial establishment, cost of supply etc. Careful analysis and forecast of urban water demand can improve the timing of expensive urban water utility system operation, which is many times the cost of the forecast. Short-term water demand forecast support water system operations as well as budgeting and financial management, also it provides valuable trigger in determining the time and capacity for new water resources development. It also helps in providing a simulated view of future and identifies suitable management alternatives to balance water supply and demand (Mohamed and Mualla 2010). There are different approaches to water demand forecasting including statistical or mathematical techniques. Aijun et al. (1996) used a rough set approach for water demand prediction to analyze a set of training data and generate decision rules and it was found to be useful for incomplete and deterministic information. DurgaRao (2005) used Multicriteria spatial decision explanatory variables for water demand forecasting. Herrera et al. (2010) developed predictive models for forecasting hourly water demand using ANN, projection pursuit regression (PPR), multivariate adaptive regression splines (MARS), random forest and support vector regression (SVR). They also used Monte Carlo simulation designed to estimate predictive performance of model obtained on data set and found that support vector regression model is most accurate one followed by MARS, PPR. Hongwei et al. (2009) used system dynamic approach for water demand forecasting based on sustainable utilization strategy of the water resources.

Although conventional time series modeling methods have served the scientific community for a long time and they provide reasonable accuracy, but suffer from the assumption of stationery and linearity (Kermani and Teshnehlab 2008). Many new methodologies are developed for modeling the data but current trend seems to be model the data rather than physical process. For modeling the data, artificial intelligence techniques (AI) such as fuzzy logic (FL), artificial neural network (ANN) and adaptive neuro fuzzy inference system(ANFIS) are probably the attractive techniques used among the researchers because of their capability to handle imprecise, fuzzy, noise and probabilistic information to solve complex problem in an efficient manner.

Altunkaynaket al. (2005) used fuzzy logic approach for water consumption prediction of the Istanbul city, using Takagi Sugeno method for time series data by considering only one lag as input for the analysis. Kermani and Teshnehlab (2008) used normalized data for water consumption prediction using ANFIS method and also further, auto regressive model is employed for the analysis and they found that ANFIS model is better than autoregressive model. Yurdusev and Firat (2009) used ANFIS method to forecast monthly water consumption modeling and they have adopted cross correlation method for selection of the input variables. Sen and Altunkaynak (2009) used Mamdani inference system for modeling of drinking water prediction using different fuzzy sets and rules in the analysis. In addition, there were many reports of using ANN in forecasting water demand (Babel and Shinde 2011; Jain et al. 2001; Firat et al. 2009, 2010).

Most of the literatures were related to ANN and ANFIS using various input variables. Here, using time series data, various fuzzy logic and ANFIS models has been developed and their performances were evaluated for the selection of better model. Further, the analysis has been extended to develop multilinear regression model for comparison. The aim of this research work is to demonstrate the advantage of artificial intelligence technique such as fuzzy logic and ANFIS method in modeling and prediction of daily water demand.

1.1 Study Area

The performance of artificial intelligence models along with multiple linear regression model was tested for the case of water consumption in New Mangalore Port Trust (NMPT). It has an average elevation of 22 m above mean sea level. NMPT experiences moderate to gusty winds during daytime and gentle winds at night. NMPT receives about 95 % of its total annual rainfall within a period of about six months from May to October, while remaining extremely dry from December to March. Humidity is approximately 75 % on average and peaks during May, June and July. The maximum average humidity is 93 % in July and average minimum humidity is 56 % in January. Temperature during the day stays below 30 °C and drop to about 19 °C at night. The current water supply system of the NMPT includes several ground water developments and also from the surface sources. They are utilizing most of the water from their own sources in addition to municipal supply. It has a population of approximately 6000 consumers with an average water demand of 3.5MLD. Due to irregularity in supply from the municipal, it is necessary to search for a new source. Therefore, water planners should give more attention to demand management, as the new resources are getting more expensive. So this study area is employed for forecasting the future water consumption in our research work. The variation of monthly water demand for a time series data is shown in Fig. 1.

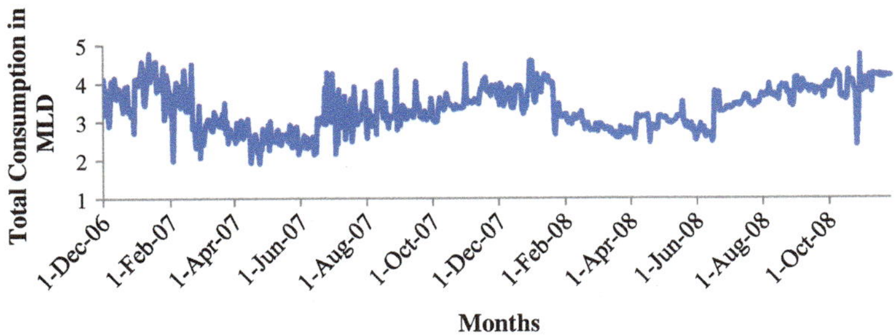

Fig. 1 Time series monthly water consumption

2 Fuzzy Logic

Fuzzy logic is capable of modeling vagueness, handling uncertainty, and supporting human type reasoning. They estimate a function without any mathematical model and learn from experience with sample data. Fuzzy logic starts with the concept of a fuzzy set. A fuzzy set is a set without a crisp; clearly defined boundary. Fuzzy set theory provides a systematic calculus to deal with such information linguistically and it performs numerical computations by using linguistic labels stipulated by membership functions. Moreover, a selection of fuzzy if then rules forms the key components of a fuzzy inference system that can effectively model human expertise in a specific application. Although the fuzzy inference system has a structured knowledge representation in the form of fuzzy if-then rules. A fuzzy inference system (FIS) is an inference mechanism establishing a relationship between a series of input and output sets. The inference system uses fuzzy sets theory, fuzzy logic principles when establishing such a relationship. Fuzzy inference system (FIS) is a rule based system consisting of three conceptual components. These are: (1) a rule base containing fuzzy if–then rules, (2) defining the membership functions (MF) and (3) an inference system, combining the fuzzy rules and producing the system results. The rule combination used for the fuzzy logic analysis is shown in Table 1. The General structure of the Fuzzy Inference System is shown in Fig. 2.

Table 1 Structure of rules criteria used for the analysis

R1	IF inp1 = mf1 AND inp2 = mf1 AND inp3 = mf1 THEN out1 = mf1
R2	IF inp1 = mf2 AND inp2 = mf2 AND inp3 = mf2 THEN out1 = mf2
R3	IF inp1 = mf3 AND inp2 = mf3 AND inp3 = mf3 THEN out1 = mf3

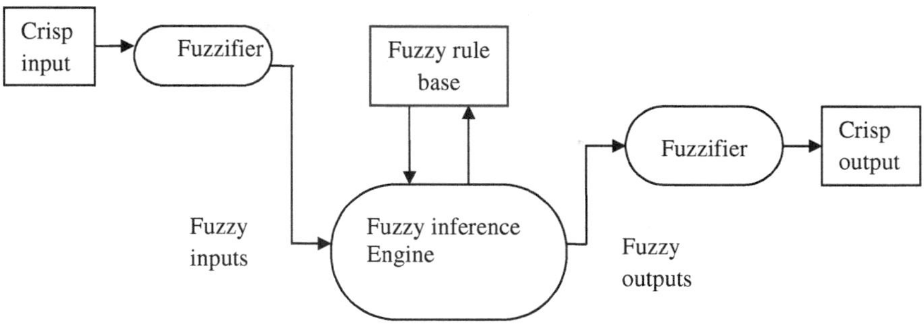

Fig. 2 General structure of fuzzy inference system

2.1 Adaptive Neuro Fuzzy Inference System (ANFIS)

In recent years, the integration of neural network and fuzzy logic has given birth to new research into neuro-fuzzy system. In fuzzy logic, there is no systematic procedure to define the membership function parameters. The construction of fuzzy rule necessitates the definition of premises and consequences as fuzzy set. On the other hand, ANN has the ability to learn from input and output pairs and adapt to it in an interactive manner. ANFIS eliminates the basic problem in fuzzy system design, defining the membership function parameters and design of fuzzy if-then rules, by effectively using the learning capability of ANN for automatic fuzzy rule generation and parameter optimization (Yurdusev and Firat 2009). Neuro fuzzy system has a potential to capture the benefits of both neural network and fuzzy logic in a single frame work. For this reason in this study, the ANFIS method is adopted for daily water consumption prediction. It has the advantage of allowing the extraction of fuzzy rules from numerical data, for the first order Takagi-Sugenofuzzy inference system. The general structure of ANFIS used in the analysis is shown in Fig. 3.

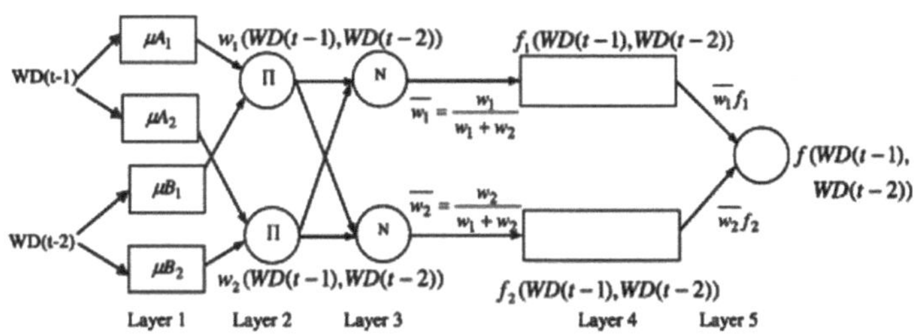

Fig. 3 Structure of adaptive neuro fuzzy inference system

Table 2 Represent the input and output structure of the models

Model	Inputs	Output
M1	WD(t)	WD(t+1)
M2	WD(t)WD(t-1)	WD(t+1)
M3	WD(t)WD(t-1)WD(t-2)	WD(t+1)
M4	WD(t)WD(t-1)WD(t-2)WD(t-3)	WD(t+1)

2.2 Model Development

One of the most important steps in developing a satisfactory prediction model is the selection of appropriate input variables, as these variables determine the structure of the model and affect the results of the model. Conventionally, the choice of appropriate input variables can be made using the cross-correlations between the variables. The structures of the model are given in Table 2. The performances of all models are evaluated according to criteria such as, Mean Relative error (MRE), Mean Square Error (MSE) and Correlation Coefficient (CC). The definition of these performance indices are given below:

$$\mathrm{CC} = 1 - \frac{\sum (X - Y)^2}{\sum x^2} \quad \mathrm{MRE} = \frac{1}{N}\sum_{i=1}^{N} \frac{|X - Y|}{X} \times 100$$

$$\mathrm{MSE} = (1/\mathrm{N}) \sum_{i=1}^{n} [X - Y]^2$$

where, X = observed values, Y = predicted values, N = total number of values, and $x = X - X_{\mathrm{mean}}$.

2.3 Data Set

The data employed in this study consist of Daily water demand in million liter (ML) at NMPT campus. The water consumption data were obtained from New Mangalore Port Trust (NMPT). The water consumption data were available for a period from December 2006 to August 2011. As a result, 1727 day wise data were used in this work. Out of these data, 1127 is used for training and 600 are used for testing. Statistical parameters of training and testing dataset are represented in Table 3. The CC ratio of all input and output variable are represented in Table 4.

Table 3 Represent the statistical parameters of training and testing data set

WD (MLD)	X_{max}	X_{min}	X_{mean}	S_x	C.V	Skewness
Training	4.79	1.29	3.29	0.544	0.165	−0.149
Testing	5.53	2.7	3.66	0.463	0.126	0.814

Table 4 Represent the CC value of all inputs and output variables

CC ratio	WDt-3	WDt-2	WDt-1	WD	WDt+1
WDt-3	1	0.85	0.8	0.74	0.7
WDt-2	0.85	1	0.85	0.8	0.74
WDt-1	0.8	0.85	1	0.85	0.8
WD	0.74	0.8	0.85	1	0.85
WDt+1	0.7	0.74	0.8	0.85	1

3 Result and Discussion

The results of testing part for the four different models of ANFIS methods are compared with the fuzzy logic and multiple linear regressions. The comparisons of different models are represented in Table 5. From the table, it is found that ANFIS method shows better performance compared to fuzzy logic and multiple linear regressions considering various performance indices such as CC, MRE and MSE. Comparison among all models shows that water demand of three days lag and current day produces better performance than other inputs parameters in forecasting future demand for time series data. In fuzzy logic model, trapezoidal membership function is used along with three rules criteria and centroid defuzzification methods was adopted to obtain crisp results. Here for M1 model, CC value is less and MRE value is high compare to other models. So for all developed fuzzy models, M4 model is selected as the best one based on the performance criteria of CC and MRE. Among ANFIS model, performance of M4 model is selected as best one based on the performance criteria of MRE. In addition to this, Regression model performance was also checked and it is found that performance of M4 model is better compare to other MLR models and also it is better compared to fuzzy model. For all the models in MLR and ANFIS method, CC value is almost equal, during this type of situation the best model is selected based on the performance criteria of MRE. Compared to fuzzy logic and multiple linear regression method, the performance of ANFIS method is better. So, totally 12 models are used, among all models M4 model performance is treated as the best one. The time series plot of observed and predicted value of M4 Fuzzy Model and M4 ANFIS model are shown in Fig. 4a, b. The higher deviations of fuzzy model outputs were appeared in Fig. 4a compared to

Table 5 Represent the comparative performance of all the models

Methods	Fuzzy logic method			Multiple linear regression			ANFIS method		
Models	CC	MSE (MLD)	MRE (%)	CC	MSE (MLD)	MRE (%)	CC	MSE (MLD)	MRE (%)
M1	0.93	0.18	7.98	0.99	0.06	0.43	1.00	0.06	0.38
M2	0.97	0.15	4.07	0.99	0.06	0.42	1.00	0.06	0.19
M3	0.98	0.16	3.39	0.99	0.06	0.42	1.00	0.06	0.16
M4	0.98	0.18	2.78	0.99	0.06	0.42	1.00	0.06	0.10

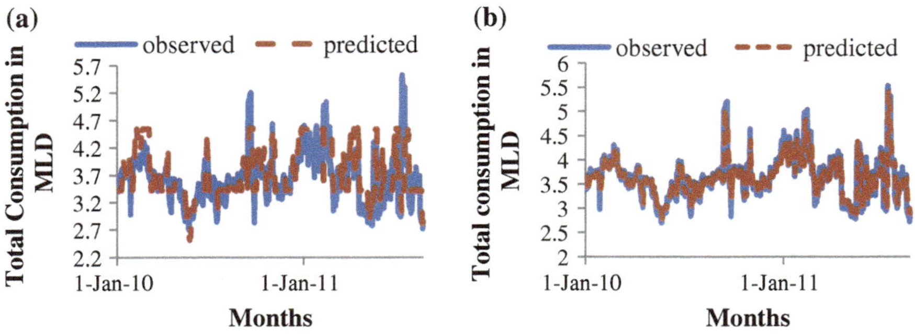

Fig. 4 Time series plot of observed and predicted value of **a** M4 fuzzy model and **b** M4 ANFIS model

observed data revealing poor performance. The close agreement between the ANFIS and observed value was obtained throughout the entire study periods both dry and wet season (Fig. 4b). Therefore, from the results we can conclude that M4 ANFIS model performance is better than the other models and methods. The better performance of MLR over the fuzzy model may be due to the lesser non-linearity characteristics of the used data. Also, less number of fuzzy sets along with limited rules could be the reason of lower forecasting performance of fuzzy models. Again, the forecasting performance of ANFIS model was found better than other methods due to the advantages of hybrid modeling which eliminates the limitations of both fuzzy system and artificial neural network and utilizing their strengths effectively.

4 Conclusion

In this study, applicability of the artificial intelligence techniques such as fuzzy logic and ANFIS method are adopted for water consumption modeling and prediction. Further, the model results were also compared with MLR method for various input scenarios. Comparing the performances of all models in fuzzy logic, M4 model performance is reasonably good. In multiple linear regression based on performance criteria of MRE M4 model is selected as the best one. In ANFIS method, based on MRE performance criteria, M4 model is selected as bestone. Totally, among artificial intelligence techniques ANFIS method with three days previous water consumption and current day water consumption (M4) model performed better and also found better than multiple linear regression method. The results of M4 ANFIS model shows that it can be successfully applied to establish a daily water consumption prediction. It was found that hybrid model performed better than single model in reliable forecast. However, it should be kept in mind that such forecasting method is mainly dependent upon the length and quality of data.

Acknowledgement The authors are grateful to Director, Executive Engineer (civil) and Assistant Engineers (civil) of New Mangalore Port Trust, for their valuable support and access to data for the research work.

References

Aijun AN, Shan N, Chan C, Cercone N, Ziarko W (1996) Discovering rules for water demand prediction: AN enhanced rough-set approach. PII:SO952-1976(96):000590
Altunkaynak A, Ozger M, Cakmakci M (2005) Water consumption prediction of Istanbul city by using Fuzzy logic approach. Water Resour Manag 19:641–654
Babel M, Shinde R (2011) Identifying prominent explanatory variables for water demand prediction using artificial neural network: a case of Bangkok. Water Resour Manag 25: 1653–1676
DurgaRao KHV (2005) Multicriteria spatial decision analysis for forecasting urban water requirement: a case study of Dehradun city, India. Landsc Urban Plan 71:163–174
Firat M, Yurdusev M, Turan M (2009) Evaluation of artificial neural network techniques for municipal water consumption modeling. Water Resour Manag (2009) 23:617–632
Firat M, Tarun M, Yurdusev M (2010) Comparative analysis of neural network technique for predicting water consumption time series. J Hydrol 384(2010):46–51
Herrera M, Torgo L, Izquierdo J, Garcia R (2010) Predictive models for forecasting hourly urban water demand. J Hydrol 387(2010):141–150
Hongwei Z, Xuehua Z, Bao Z (2009) System dynamic approach to urban water demand forecasting: a case study of Tianjin. Tianjin Univ 15:070–074 (Springer)
Jain A, Varshney K, Joshi U (2001) Short term water demand forecasting modeling at IIT Kanpur using artificial neural networks. Water Resour Manag 15(2001):299–321
Kermani Z, Teshnehlab M (2008) Using adaptive Neuro fuzzy inference system for hydrological time series prediction. Appl Soft Comput 8(2008):928–936
Mohamed M, Mualla A (2010) Water demand forecasting in Ummal-Quwain using the constant rate model. Desalination 259(2010):161–168
Sen Z, Altunkaynak A (2009) Fuzzy system modeling of drinking water consumption prediction. Expert Syst Appl 36(2009):11745–11752
Yurdusev A, Firat M (2009) Adaptive neuro fuzzy inference system approach for municipal water consumption modeling. J Hydrol 265(2009):225–234

Part VI
Socio-economic Aspects and Role of Society in Urbanization Impacts

Estimating Sustainable Carrying Capacity of Flood Prone Hilly Urban Areas

Arup K. Sarma, Banasri Sarma and Subhasish Das

Abstract The carrying capacity of an area is the maximum number of people that can be supported by the environment in an eco-friendly manner utilizing the available resources. Cities of the developing countries are expanding at a rapid rate without bearing in mind its carrying capacity limit. Due to lack of emphasis on this aspect of development planning, urban areas are suffering from severe environmental problems like flood and erosion. The paper gives an overall insight coupled with scientific overviews providing a method for the assessment of carrying capacity of an urban watershed. This method finally accommodates a sustainable population iteratively through trial allocation and feedback analysis and therefore it is named as "Sustainable Accommodation through Feedback Evaluation (SAFE model)". The model also incorporates technological intervention to improve the carrying capacity of the area through application of Ecological Management Practices (EMPs) to obtain a sustainable development regime.

Keywords Urban carrying capacity · Sustainable development · Ecological management practices · Urbanization · Optimization

1 Introduction

The earth can only hold a definite amount of human growth for a definite time (Malthus 1798). Various methods are available for carrying capacity estimation. The famous ecologist Odum proposed a concept of computing carrying capacity based on quantitative measure of the resources used by the supporting population

A.K. Sarma (✉) · B. Sarma · S. Das
Department of Civil Engineering, Indian Institute of Technology, Guwahati, Assam, India
e-mail: aks@iitg.ernet.in

B. Sarma
e-mail: banasris@gamil.com

S. Das
e-mail: dassubhasish02@gmail.com

A.K. Sarma et al. (eds.), *Urban Hydrology, Watershed Management and Socio-Economic Aspects*, Water Science and Technology Library 73,
DOI 10.1007/978-3-319-40195-9_23

(Odum 1988). The "Pressure–State–Response (PSR)" model developed by Organisation for Economic Co-operation and Development (OECD 2003) is another approach of computing carrying capacity. The basis of this method is that human exert "pressure" on the environment by their activities, which results in the radical change of the "state" of the environment. Graphical model, Uni-constraint model, IPAT equation are some of the existing models, which give idea about the carrying capacity. The graphical model uses demographic data and determines carrying capacity from the population growth curves considering the environmental resistance. This is a popular and simple method; however, this method cannot incorporate the factors that limit population growth while estimating carrying capacity. Therefore, methods focusing on carrying capacity calculation considering some constraints that limits population growth were being proposed. The Uni-constraint model is one such method, where a constraint is considered and the entire estimation is based on the assumptions circling the considered constraint. The population growth constraints can be availability of food, shelter, clothing, etc. However, population may put pressure on the environment in several ways and in such scenarios; several constraints need to be considered while calculating the carrying capacity. The IPAT equation is such a multi-constraint model that uses different factors in calculating the carrying capacity (Chertow 2001). The IPAT equation gives an idea about the cumulative or associated impacts of the population, its resource usage patterns and technological interventions on the environment. It does not give optimum carrying capacity in an explicit manner. It provides an indirect approach of estimation of carrying capacity by understanding the level of environmental degradation.

One of the most accepted models to calculate carrying capacity was developed by Mathis Wackernagel as "appropriated carrying capacity" model and was later renamed as "Ecological footprint" model (Ree 1992). Ecological footprint is a measure of the human demands on the biosphere. This model gives the amount of biologically productive land and water required to produce all the resources needed by the population for its consumption and developmental activities as well as to absorb the waste generated. The model thus estimates carrying capacity based on the data generated from resource accounting. Ecological footprint is always a function of socio-economic status, technology and resource availability. It keeps on changing with time and space. Xie et al. (2011) in their study for Yellow River Delta found increase in ecological footprint per capita from 0.1885 hm^2 in 2001 to 0.4639 hm^2 in 2008.

Carrying capacity of course is not a fixed entity. It can increase or decrease phenomenally (Arrow et al. 1995). Many factors are there that can influence the carrying capacity of a region. In addition, potential carrying capacity can be improved from comprehensive elements such as the level of economic development, science and technology and environmental protection (Xu et al. 2010). The pattern and extent of resource usage serves to be the primary factor that affects the carrying capacity significantly. This indeed depends highly on the socio-economic status of the people. The use of technology also influences the carrying capacity, i.e. technological interventions can enhance the carrying capacity by optimizing resource use and reducing waste generation. Also, environmental quality problems may become more severe with urban size; however factors such as land use, transportation system and spatial layout of a city are also critical factors for determining the "urban environmental carrying capacity" (Munda 2001). Oh et al. (2005) assessed urban carrying capacity by integrating urban management goal to maintain sustainable air and water quality standards. The study developed relationships among air and water quality with energy consumption, land uses and infrastructure to obtain indexes for urban management in terms of population, population density, development type, development density and land use.

Thus, concept of urban carrying capacity can help sustainable development and can very well be used to characterize urban ecosystem health, urban ecological security, and urban ecological risk (Xu et al. 2012). Unscientific haphazard development in hilly terrain is causing ecological disturbances leading to multiple hazards like flood, soil erosion, slope instability and pollution. This calls for a more structured approach of estimating carrying capacity of the hilly areas to have a sustainable population that would not promote/trigger human-induced hazards. Here, we formulated a framework to assess the carrying capacity of the hilly urban areas that comply with sustainable development and planning in such areas. This method finally accommodates a sustainable population iteratively through trial allocation and feedback analysis and therefore it is named as "Sustainable Accommodation through Feedback Evaluation (SAFE)". In this method, sustainable accommodation of carrying capacity is determined keeping the target to mitigate environmental hazards specific to the particular area under study.

Though, there are different methods for computing carrying capacity, no method specifically stands for determining sustainable urban population for hilly urban watershed giving emphasis on mitigation of hydrological implications of urban development. For many urban areas, particularly in India, the plains are already saturated with settlements and industries. The hills located within/near the urban area are therefore now becoming the target of development. The anthropogenic pressures are thus increasing day by day in the fragile hilly ecosystems; the result is the upsurge in the urban hubs in the hilly areas. People from low-income group generally encroach into such less favourable hilly area, as they cannot afford costly land of plains. Therefore, in this study, water-related hazard induced by unplanned urbanization are considered for determining the urban carrying capacity in a hilly urban watershed with the application of SAFE model.

Guwahati, the capital city of Assam, being situated in a region of heavy rainfall, suffers from the problem of flood almost every year during monsoon. Increasing pace of urbanization, witnessed in this city during the last two decades, has aggravated the problem to a great extent. Denudation of surrounding hills, various constructional activities and filling-up of low-lying areas are some of the prime factors responsible for the present flood scenario of the city. The alarming rate of soil erosion on the hills and its subsequent deposition in the drains has given a bigger dimension to this urban flood problem. This urban flood is disrupting the normal life of the urban community. Therefore, a degraded urban hilly watershed located in Guwahati city is considered for applying the SAFE model to mitigate the soil erosion and flash flood hazard. The sustainable population and infrastructural requirements for different socio economic scenarios along with ecological measures were determined for the study area.

2 The Safe Model

The SAFE model quantifies the maximum population that a hilly area can accommodate targeting mitigation of some environmental hazard. Here, the basic concept of ecological footprint is first used to decide a trial sustainable carrying capacity of the area under consideration. This trial carrying capacity is determined by allocating population and infrastructures iteratively, so that the infrastructures provided remain sufficient to cover the virtual footprint of the allocated population. Feedback of the urban watershed in terms of hydrologic alteration is then analyzed through application of hydrological model with virtual accommodation of the trial carrying capacity (Fig. 1). Feedback is assessed in terms of several case-specific performance criteria such as sediment yield and peak discharge at watershed outlet so that the area remains hazard free.

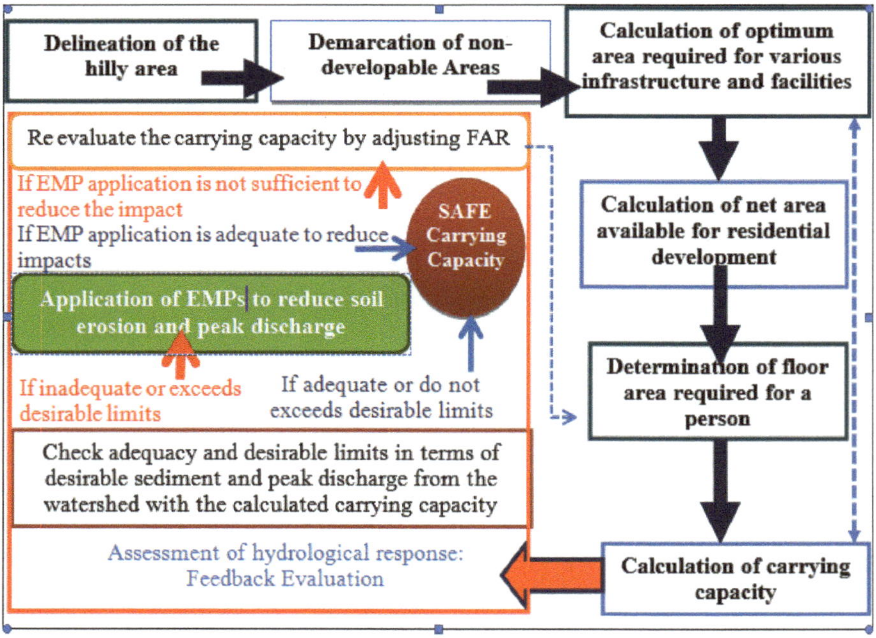

Fig. 1 The SAFE model for determination of carrying capacity with emphasis to mitigate flood and soil erosion hazard

The steps of SAFE are as follows:

Step I: Delineation of the watershed: In this step the hilly watersheds covering the potential urban area are delineated from digital elevation model (DEM) or marked from the city master plan following natural drainage network. In absence of high resolution DEM for an area, detail topographic survey should be carried out to improve the accuracy of watershed delineation.

Step II: Demarcation of the developable and non-developable area: The hills consist of both developable areas and areas having less scope for development, i.e. non-developable areas. In this step, the non-developable areas of the delineated hilly region are demarcated by field survey or by using geospatial tools from satellite imagery. The non-developable areas mainly consist of land with high slope, reserved forest areas, water bodies, stream lines, drainage channels, springs, depressions, etc. Thus, the developable areas for various developmental activities can be computed as:

$$A_D = A_T - A_{\mathrm{NDA}} \tag{1}$$

Here, A_D is the net developable area, A_T is the total hilly urban area, and A_{NDA} is the net non-developable area.

Step III: Determination of area required for different infrastructure and facilities: Now within the developable area, several sub areas are allotted for various urban infrastructure and facilities development like water treatment plants, sewage treatment plants, drainage, commercial hubs, health centres, educational institutions, recreational areas, transport facilities, etc. For calculating these areas the regional planning approach is adopted. For example, an urban centre with a population of 1000 will not require a solid waste dumping site; rather a provision of solid waste dumping can be kept by tying with the regional dumping site. Space required for different infrastructure can be determined through site specific requirement. The standard space requirement index prepared by concerned authority can be used for this purpose. So,

$$A_D = A_{\mathrm{IF}} + A_R \tag{2}$$

Here, A_{IF} is the maximum area earmarked for infrastructure development and A_R is the area for residential requirements.

Area required for infrastructure depends on the population, in other words, infrastructural requirements can be expressed as a function of carrying capacity.

$$A_{\mathrm{IF}}(\mathrm{CC}) = \sum_{i=1}^{n} A_{\mathrm{IFi}}(\mathrm{CC}) \tag{3}$$

where n is the number of infrastructure considered.

Step IV: Calculation of the available residential area: The net residential area available for settlement development can be calculated using the following equation:
From (1) and (2)

$$A_R = A_T - (A_{\text{NDA}} + A_{\text{IF}}) \tag{4}$$

Step V: Socio economic survey of the urban region and calculation of the floor area requirement of the people: A thorough demographic and socio-economic survey of the hilly urban area should be done to estimate an average floor area requirement per head of the people dwelling there. The floor area requirement of the people will greatly vary with respect to economy and lifestyle of the people living there.

Step VI: Determination of the Floor Area Ratio: The floor area ratio (FAR) is defined as:

$$\text{FAR} = \frac{A_F}{A_P} \tag{5}$$

where, FAR is the floor area ratio, $\mathbf{A_F}$ is the total floor area and $\mathbf{A_P}$ is the area of the plot.

Step VII: Calculation of carrying capacity: Based on the overall study, the carrying capacity of the area with respect to urban development can be calculated using the following equation:

$$\text{CC} = A_T - (A_{\text{NDA}} + A_{\text{IF}}(\text{CC})) \times \frac{\text{FAR}}{S} \tag{6}$$

Here, S is the floor area requirement per head.

Step VIII: Assessment of hydrological response and feedback evaluation: Check for adequacy of drainage system, sewerage system, water quality, etc. were not explicitly considered during carrying capacity calculation in view of the safe disposal of generated sediment and water yield. To consider this aspect, the sediment and water yield for different scenarios are calculated and then the values are compared with their allowable limits. If the sediment and peak discharge values go beyond the allowable limit, following two options need to be tried in sequence:

(i) Apply possible Ecological Management Practice to bring sediment yield and peak discharge within permissible limit. Optimal selection of EMPs is required in this step.
(ii) Re-evaluate the carrying capacity by reducing FAR.

3 Case Study: Garbhanga Watershed of Guwahati City of Assam, India

We selected a micro watershed "Garbhanga" for estimating its carrying capacity using SAFE model. The study watershed is located near the Games village area of Guwahati, a premier city of Northeast India. The watershed comprises of both habitat and undisturbed area. The site is very close to the main national highway and therefore is a suitable site for urban development from administrative and logistic point of view. The watershed has also a mixed terrain of both hills and plains. Degradation of this watershed has led to problems of soil erosion, drainage congestion, water logging and flooding. Site surveys were conducted from 16th to 24th July 2013 and it revealed that the watershed is highly vulnerable to mass degradation from all the aspects of conveyance, drainage and ecological balance. The situation of settlement pattern of the area has worsened over the last few years and even during the period of survey new settlers started cutting the earth and clearing of vegetation for construction of their houses. Constant clearing of trees and vegetation has worsened the situation of soil erosion. Most of the land lay barren and are highly vulnerable. The site is in a degraded condition and affects the adjacent highway adversely even at lightest of rainfall. During the survey, discussions regarding continued degradation of the area were conducted with the stakeholders and the people staying around. The survey of area from the socio-economic point of view revealed the existence of undeveloped open spaces, an urban Primary Health Centre, a dairy farm, and electrification of a section of the area, a school and a marketplace in its adjustment area.

Topographic survey of the area was done in this area and the triangulated irregular network (TIN) model was developed in ArcGIS 9.3 by using the elevation contours. The TIN was converted to raster DEM and then the watershed delineation for Grabhanga was carried out in ArcSWAT. The area is located within an elevation range of 61–168 m. The slope ranges from 0 to 60 %. The major soil type of this watershed is fine and coarse loamy type as obtained from soil map of Assam Remote Sensing Application Centre (ARSAC). The Garbhanga watershed has a total area of 18.5 ha with an approximate population of 2000–3000. Figure 2 shows the study area map.

3.1 Consideration for Developable and Non Developable Areas

After extensive field survey, the non-developable areas of the watershed were identified. There is 8.21 ha reserve forest in the watershed; out of which unauthorized residential developments have taken place 4.5 ha. There is also a low lying area near to the outlet of the watershed which is presently in water logged

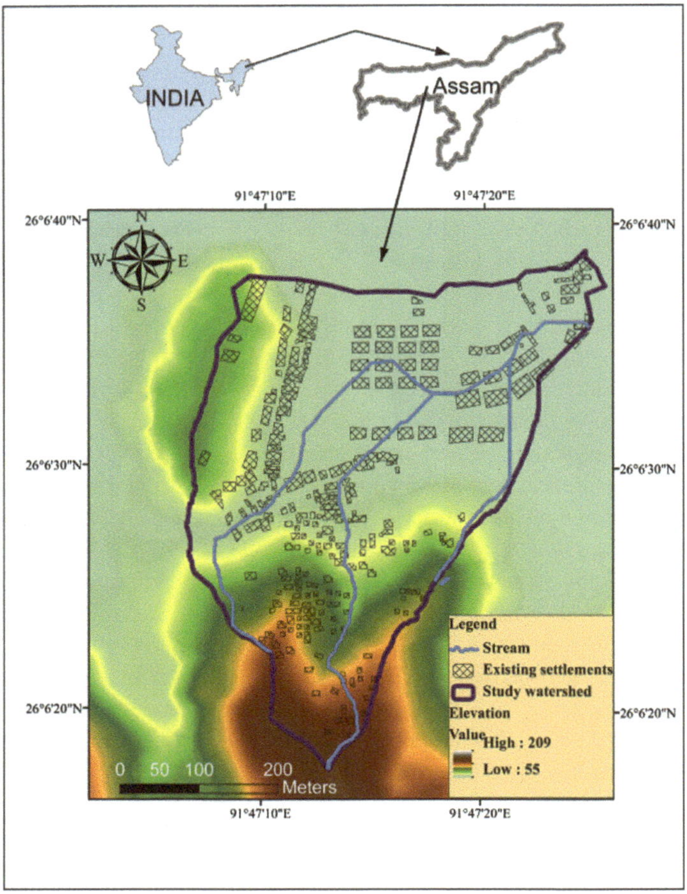

Fig. 2 The study area

Table 1 Existing facilities of Garbhanga watershed

Milk booth	½ km away from the watershed
Postal service	1½ km away from the watershed
Health care	1½ km away from the watershed
School	½ km away from the watershed

condition. There is 1.8 km of water flow path in the study area. A buffer of 5 m on both sides to this flow path is considered as non-developable, which gives an area of 1.7 ha.

The details of the existing facilities and non-developable areas are presented in the Tables 1 and 2.

Table 2 Area demarcated as non developable area of the watershed

Forest	8.21 ha
Buffer area to the existing water flow path	1.7 ha
Local water harvesting pond	0.0012 ha
Total non-developable area	9.9112 ha
Area available for development activities	8.6 ha

3.2 Consideration for FAR

Floor area ratio (FAR) indicates the density of housing in an area and indirectly allowance of a particular FAR can regulate no of storey of a building covering permissible coverage area of a plot. The ratio is generated by dividing the building area by the plot area, using the same units. Higher FARs tends to indicate more urban (dense) construction.

FAR need to be determined by considering various aspects like provision of intended free space, safe bearing capacity of soil, economy of people for affording earthquake resilient structures, drainage and transportation requirement and so on. While the proposed "SAFE" method itself will determine an acceptable FAR, one need to provide an initial value of FAR. This value can be given from guidelines provided by different organization including Urban Local Bodies. In absence of any such guidelines, a value of 1.5 can be used for initial trial value. This value is suggested based on the general trend observed so far in Indian condition.

This indicates that if one decides to construct house by covering 50 % of the land area, then he can go maximum up to three storey building. However, acceptability or possible scope of enhancement of FAR may be reviewed for hilly area from consideration of maximum permissible surcharge load from slope stability point of view.

The socio-economic survey has shown that the people residing in the area are of lower middle class. Based on the survey, the value of the floor area per person is considered as 20 and 10 m^2. The higher the value of floor area per person better is the economic condition. Considering the possibilities different combination of FAR and floor area requirements per person, we applied the SAFE model for three scenarios:

Scenerio1: FAR = 1, Floor area per person = 0.002 ha
Scenerio2: FAR = 1.5, Floor area per person = 0.002 ha
Scenerio3: FAR = 1, Floor area per person = 0.001 ha

3.3 Consideration for Infrastructural Facilities

Based on primary and secondary data collected on required area for different infrastructure and also considering some minimum value for some of the required

Table 3 Relationship between population and infrastructure requirements

Infrastructure	Relationship
Water treatment plant	$y = (x \times 0.2) + 11060$
Sewage treatment plant	$y = (x \times 0.2) + 11060$
Park	$y = (x^2 \times 0.0000004) + (x \times 0.3) + 4550$
Socio cultural facility	$y = \log(x) \times 0.372 - 3.349$
Road and drainage	$y = a \times 0.1$
Business and commerce	$y = a \times 0.15$
Open space	$y = a \times 0.1$

[x population, y area of infrastructure (in ha), a total area (in ha)]

infrastructures, logical statistical relationship between infrastructure and population were developed (Table 3) and used for calculation of required infrastructural areas. While using the relations care were also taken to assign a minimum area for essential facilities even if the population is quite low. Again, as some of the facilities like school, milk booth, health care and postal services are existing near to Garbhanga, area requirement for these facilities were not considered while calculating the carrying capacity.

With availability of adequate data, an improved statistical relationship between population and infrastructural area may be developed.

4 Results and Discussion

4.1 Calculation of Carrying Capacity for Different Socio Economic Scenarios

The SAFE model was applied for the three scenarios of development and in all these three scenarios, it is seen that the existing population has already crossed the carrying capacity limit and as such further enhancement of population is not possible within the area. This calls for evaluating the feedback of present and future developmental state and also needs some technological interventions to reduce the undesirable impacts of development. The results of the SAFE model are described in Table 4.

4.2 Hydrological Response

The development of infrastructure and other facilities in hilly area generally cause land levelling, cutting of vegetation and thereby rendering open or barren land. Considering the allowable coverage as per New Revised Guwahati Building

Table 4 Infrastructural areas and carrying capacities for different scenarios

Results of the SAFE model (area in ha)	Scenerio1	Scenerio2	Scenerio3
Road area + drainage area	0.85	0.85	0.85
Park	0.50	0.51	0.54
Open space	0.85	0.85	0.85
Business and commerce	1.29	1.29	1.29
Socio cultural activity	0.01	0.01	0.01
Water treatment plant	1.09	1.10	1.12
Sewage treatment plant	1.09	1.10	1.12
Total infrastructural area	5.69	5.70	5.78
Total residential area	2.89	2.84	2.80
Calculated carrying capacity (number of persons)	1448	2136	2801

Byelaws (GMDA 2006), for the present study the total concrete/paved area is projected to be 5.02 ha resulting into an additional barren land output of 3.584 ha. Now the biggest concern arises with the remediation of higher overland flow resulting into massive discharge at the downstream as well as high sediment yield.

4.3 Sediment and Water Yield Scenarios with the Calculated Carrying Capacity

We estimated the sediment and peak discharge of the watershed with Revised Universal Soil Loss Equation (RUSLE) method and Rational method with the parameter values listed in Table 5. Also, RUSLE cover factor and rational method runoff coefficient of different land covers were obtained from the available literatures (Renard et al. 1997; Toy et al. 1998; Sarma et al. 2005; Wall et al. 2002; Iowa Storm Water Management Manual 2008). For USLE model, the cover factor for the built-up area was taken as 0, and the cover factor of the barren land was taken as 1. The Practice factor (P factor) was considered equal to 1, as impact of different land uses is introduced indirectly by modifying the C factor of USLE. For applying the rational method, the runoff coefficients for various land covers were considered based on the available literature (Sarma et al. 2005; Iowa Storm Water Management Manual 2008). The runoff coefficient for the built-up area was taken as 1 and for that of the barren land as 0.5.

Table 5 Model parameters of rational method and RUSLE for the study watershed

[a]Designed rainfall intensity of rational method (I)	[b]Rainfall erodibility factor (R) of RUSLE	Calculated average soil erodibility factor of RUSLE	Calculated average slope—length factor of RUSLE
50 mm/hr	544	0.17	5.31

[a]Sarma et al. (2013)
[b]Sarma and Goswami (2006)

Table 6 Sediment and water yield at different developmental state

Development state	Considering undisturbed condition of the watershed	Considering present development pattern	Considering haphazard development in the whole watershed	Considering development with calculated carrying capacity
Annual sediment yield	2244 tonnes	6507 tonnes	10,322 tonnes	6059 tonnes
Peak discharge	0.51 cumec	1.26 cumec	2.08 cumec	1.41 cumec

The estimated the sediment and water yield scenario of the watershed are given in Table 6.

The projections derived after analyzing the models considering different scenarios (existing development, unplanned development and planned development with calculated carrying capacity) direct towards a higher runoff and sediment yield when compared to natural or undisturbed areas. Even we consider the planned development with designed carrying capacity; annual sediment yield is still very high compared to natural condition and does not have a significant difference with the present sediment yield value. Peak discharge scenarios do have a similar trend.

Taking into consideration the ever increasing development regime of the city, it becomes really difficult to control the runoff and sediment yield without management practice. Moreover, it is not possible to lower the population than the existing. In such situation, sediment and water yield from the area can be retained within sustainable limit by implementing Ecological Management Practices (EMPs) like vegetation, pond, forest, geogrid with vetiver grass, rainwater harvesting system, etc. (Sarma et al. 2013). Considering the economic status of the people, location of the plot and factor of cost effectiveness, the EMPs like grass cover, garden (with vegetables, flowers, fruit plants), forest (with grass cover) and geogrid with vetivers are being considered for this study. This will provide the following benefits:

- The vegetative cover over the land will lower erosion, sediment yield and run off. Thus minimizing the chances of occurrence of hazards.
- The people can get monetary benefits from the fruit gardens and fishery thus enhancing their economic status.
- Vegetative EMPs also provide carbon sequestration potential
- The aesthetic beauty of the plot is increased which will enhance value of the property

However, rational selection of EMPs for an urban watershed often become challenging when ecological, economic and social issues of that area need to be satisfied considering their survivability in long run. Therefore, we used the OPTEMP-CSMO (Sarma et al. 2013) model for selecting the optimal EMP combinations for Garbhanga watershed. The model selects optimal EMP combination at minimum possible cost that also maintains maximum possible carbon sequestration

Table 7 OPTEMP-CSMO model parameters

Name of EMPs	Unit cost per m^2 (Rs)	USLE cover factor	Rational method cover factor value	Carbon sequestration potential (tonnes/m^2/year)
Grass	41	0.01	0.20	0.000087
Garden	50	0.015	0.3	0.00429
Geogrid + vetivers	297	0.01	0.2	0.001524
Forest	2	0.2	0.40	0.00092

potential from the EMPs. The model also consider the constraints of restricting sediment yield and peak discharge within desirable limit along with EMP allocation feasibility criteria like area suitability constraints; maximum available area constraints and owners choice constraints.

Based on the value of sediment and water yield in natural condition and also considering the drainage capacity of downstream drain, the sustainable sediment yield and peak discharge from the area is considered to be within the range of 0–2000 tonnes/year and 0.5–1.5 cumec, respectively. Considering the barren land, there is 3.584 ha of land within the watershed available for EMP application. Again, we considered that the 4 ha of forest land where sparse residential development has just started should be renovated as forest and further development should not be allowed. As such, this 4 ha of land can be considered as compulsory area for EMP application with forest as the EMP. Besides, from the site survey 0.9 ha of land were found suitable only for application of grass. There is also 0.3 ha of steep slope land with vertical cutting where application of geogrid along with vetivers is must to retain the slope stability. Again as per the dwellers choice, a minimum of 0.4 ha of land have to be kept for garden. All these conditions were considered while applying the OPTEMP-CSMO model as suitability, area availability and owners choice constraints (Table 7).

4.4 Results of the OPTEMP-CSMO Model

The OPTEMP-CSMO model suggests a total of 5.04 ha of area for EMP applications which comprises of 0.9 ha of grass, 0.84 ha of garden and 0.3 ha of geogrid with grass/vetivers and 3 ha of forest. The results shows that, application of EMPs lowers sediment yield to natural level, but do not have a significant peak discharge reduction (Table 8). However, application of rainwater harvesting system in the residential areas lowers these values significantly to a desirable limit which can be disposed off safely with the designed drainage system.

Table 8 Optimal combinations of selected EMPS as obtained from the OPTEMP-CSMO model

Name of the EMP and their installation cost (construction, maintenance and material)	Results
Grass	0.9 ha
Garden—vegetables and flowers, bushy vegetations	0.84 ha
Geogrid + grass/vetivers	0.3 ha
Forest	3.0 ha
Total cost	Rs. 17.37 lakhs
Carbon sequestration from EMPs	68.72 tonnes/year
Annual sediment yield	1361 tonnes/year
Peak discharge for designed rainfall intensity (without rainwater harvesting system)	1.38 cumec
Peak discharge for designed rainfall intensity (with rainwater harvesting system)	1.17 cumec

5 Conclusions

Development yield benefits in true sense if and only if is done with a holistic approach without comprising the ecology. The assessment of carrying capacity is thus an imperative towards planning a sustainable development project. Also, one of the most distinct points which are being dramatically rendered unseen by the policy makers and developers is the ecological disturbance induced by the growing population leading to multiple hazards. Again based on the trend of population growth, the demands of the people with regards to infrastructure and other facilities will also increase. Hence, the carrying capacity should be periodically calculated so as to check haphazard, unplanned or illegal development which will harm the ecosystem in the long run by calling hazards or natural calamities.

The available carrying capacity models are comprehensive and specific towards determination of sustainable population under some specific ecosystem sustainability conditions. However none of the methods address hydrological consequences of urban development in particular, which is a growing concern for developing countries. SAFE model is basically encompasses a watershed-based approach, and it enables to address two interrelate and independent hydrological issues-soil erosion and high runoff generation from hilly urban watershed. In view of the growing concern about urban swell development, this model can be used at micro watershed level to address unwanted hydrological implications of development. This paper also highlighted the hydrological impacts of urban development in the hilly urban areas and how to mitigate the problems with the SAFE model considering the application of Ecological Management Practices to nullify the dwindling ecological status. The optimal EMPs were selected by applying the OPTEMP-CSMO model (Sarma et al. 2013) and their applicability were tested with two hydrological models (RUSLE and rational Methods). The model selected the best suited EMPs at minimum possible cost subjected to carbon sequestration maximization of the considered EMPs.

As a case study, the model is applied to a partially developed hilly urban watershed of Guwahati city. While applying the SAFE model, existing infrastructural facilities were considered and additional facilities required for the designed population were also suggested. Though area consideration for different facilities was taken from limited field data, care was taken to obtain a logical area from all required infrastructural facilities. The results of the SAFE model implied that the existing population has already crossed the carrying capacity limit and even the planned development is considered with all adequate infrastructural facilities (like road, drainage, water harvesting structure, etc.), soil erosion and peak discharge from the area will exceed the desirable limit. As such the downstream of the watershed is suffering from severe flash flood and severe soil erosion from barren areas. Therefore to check these issues, SAFE model suggested application of four EMPs (namely grass, geogrid + vetivers, garden and forest) in a logical manner within the watershed area. The study also suggested application of rainwater harvesting system at household level to reduce the peak discharge to a lower level than earlier.

This paper basically conceptualized how sustainable urban watershed development can be achieved considering the carrying capacity and how one can mitigate undesirable consequences of unplanned developmental activities through application of EMPs. The case study presented here is based on limited field data on infrastructures. However more accurate picture can be obtained by incorporating detailed analysis on these aspects. Also, focusing on additional benefits of EMPs (like return income) can provide a better dimension for further study. Also, sensitivity analysis of various model parameters in estimation of sustainable carrying capacity can be carried out. Thus, the SAFE model describes in this paper paves ways as how to use some EMPs to lower the after effects of urban development with respect to run off, sediment yield and eventually the resulting hazards.

References

Arrow K, Bolin B, Costanza R, Dasgupta P, Folke C, Holling CS, Jansson BO, Levin S, Mäler KG, Perrings C, Pimental D (1995) Economic growth, carrying capacity and the environment. Science 268:1–2. Accessed at: http://www.precaution.org/lib/06/econ_growth_and_carrying_capacity.pdf

Chertow MR (2001) The IPAT equation and its variants; changing views of technology and environmental impact. J Ind Ecol 4(4):13–29. Accessed at: http://mitpress.mit.edu/journals/pdf/jiec_4_4_13_0.pdf

GMDA (2006) New revised building byelaws for Guwahati metropolitan area. Developed by Guwahati metropolitan development authority, Guwahati. Accessed at http://www.naredco.in/notification/pdfs/revised_building_byelaws_GMDA.pdf

Iowa Stormwater Management Manual (2008) Manual prepared by Iowa State University (2C-4). Version2." Accessed at http://www.intrans.iastate.edu/pubs/stormwater/Design/2C/2C-4%20Rational%20Method.pdf

Malthus T (1798) An essay on the principle of population. Printed for J. Johnson, in St. Paul's Church-Yard, 1998 London, Electronic Scholarly Publishing Project. Accessed at http://www.esp.org

Munda G (2001) Indicators and evaluation tools for the assessment of urban sustainability. 10/2001-UHE/UAB. 11 Dec 2001 at: http://www.h-economica.uab.es/wps/2001_10.pdf

Odum HT (1988) Energy environment and public policy: a guide to the analysis of system. UNEP regional seas reports and studies no 95. Accessed athttp://www.unep.org/regionalseas/publications/reports/RSRS/pdfs/rsrs095.pdf

OECD (2003) OECD environmental indicators: development, measurement and use. Published by organization for economic cooperation and development. Accessed at http://www.oecd.org/environment/indicators-modelling-outlooks/24993546.pdf

Oh K, Jeong Y, Lee D, Lee W, Choic J (2005) Determining development density using the urban carrying capacity assessment system. Landscape and urban planning 73:1–15

Rees WE (1992) Ecological footprints and appropriated carrying capacity: what urban economics leaves out. Environ Urbanization 4(2):121–130. doi:10.1177/095624789200400212

Renard KG, Foster GR, Weesies GA, McCool DK, Yoder DC (1997) Predicting soil erosion by water: a guide to conservation planning with the revised universal soil loss equation (RUSLE). USDA agricultural handbook. No 703. Accessed athttp://www.ars.usda.gov/SP2UserFiles/Place/64080530/RUSLE/AH_703.pdf

Sarma AK, Goswami P (2006) Developing intensity duration curve with limited rainfall data. Predictions in ungauged basins for sustainable water resource planning and management. Jain Brothers, New Delhi, Nov 2006, 187–194, ISBN: 81-8360-044-1, 187-194

Sarma AK, Chandramouli V, Singh B, Goswami P, Rajbongshi N (2005) Urban flood hazard mitigation of Guwahati city by silt monitoring and watershed modelling. Report submitted to ministry of human resources department (MHRD) by Department of Civil Engineering, IIT Guwahati

Sarma B, Sarma AK, Singh VP (2013) Optimal ecological management practices (EMPs) for minimizing the impact of climate change and watershed degradation due to urbanization. Water Resour Manage 27(11):4069–4082

Toy TJ, Foster GR, Galetovic JR (1998) Guidelines for the use of the revised universal soil loss equation RUSLE) Version 1.06. Accessed at http://www.greenfix.com/Channel%20Web/pdfs/RUSLE%20Guidelines.pdf

Wall GJ, Coote DR, Pringle EA, Shelton IJ (2002) RUSLEFAC—revised universal soil loss equation for application in Canada: a handbook for estimating soil loss from water erosion in Canada. Research Branch, Agriculture and Agri-Food Canada. Ottawa. Contribution No. AAFC/AAC2244E. 117

Xie F, Zheng M, Zhang H (2011) Research on ecological environmental carrying capacity in Yellow River delta. Energy Procedia 5:1784–1790

Xu L, Xie X (2012) Theoretic research on the relevant concepts of urban ecosystem carrying capacity. Proc Environ Sci 13:863–872

Xu L, Kang P, Wei J (2010) Evaluation of urban ecological carrying capacity: a case study of Beijing, China. Proc Environ Sci 2:1873–1880

A Perceptual Analysis of Living Environment and Academic Performance of Geography Students, Benue State University, Makurdi, Nigeria

Irene D. Mngutyo and Benjamin Mngutyo

Abstract Declining quality of university education is a real issue in Nigeria today. The quality of living quarters has effect on performance of students; therefore, this paper has tested the hypothesis that students' academic performance is affected by how they perceive their living environment. The living environment quality is measured using parameters of surrounding environment, building type, cross ventilation in rooms, noise level, water supply, and waste disposal environment. Student's perception was measured using a comparative scale of good to bad, while student performance was measured by the cumulative grade point average (CGPA). Structured questionnaire were administered on students of 2008–2011 set from the Geography Department of the Benue State University systematically sampled (n = 68). Environmental parameters were used as variables and tested against the CGPA using the student's t test-to-test hypothesis and Pearson moment correlation coefficient to determine relationships. Results indicated that student's perception of their living quarters has an inverse influence on their academic performance. Improving students living quarters should be taken into consideration in policy decisions for improving the quality of education in Nigeria.

Keywords Student's quarters · Environmental quality · Perception · Academic performance

I.D. Mngutyo
Urban and Regional Planning Department, Benue State University, Makurdi, Nigeria
e-mail: datachi4dan@yahoo.com

B. Mngutyo (✉)
Department of Mass Communication, Benue State University, Makurdi, Nigeria
e-mail: bemben_2006@yahoo.co.uk

A.K. Sarma et al. (eds.), *Urban Hydrology, Watershed Management and Socio-Economic Aspects*, Water Science and Technology Library 73,
DOI 10.1007/978-3-319-40195-9_24

1 Introduction

Education is a powerful driver of the development of individuals and societies and key to achieving the Millennium Development Goals (MDGs) (World Bank 2011) the quality and level of educational attainment of a nation determines the sociopolitical and economic development level in the nation. Education is defined as the act or process of imparting or acquiring general knowledge, developing the powers of reasoning and judgment, and generally of preparing oneself or others intellectually for mature life.

Nigeria has the largest and most complex higher education system in the continent. It is made up of three levels of primary, secondary and tertiary education. Tertiary education is obtained from Universities, Polytechnics and Colleges. There are over 35 federal universities, 37 state universities, 67 private universities and over 200 other Tertiary institutions in Nigeria (Fapohunda 2014). The Federal Government has majority control of universities. The Nigerian Universities Commission (NUC) monitors quality of university education in Nigeria.

A World Bank report by Dike (2002) on education in Nigeria has attributed the growth in the education sector as being mainly in size and not in quality. The report says "employers complain that the quality of university graduates [and secondary school graduates], especially their communication skills, has fallen continually for two decades" (Bollag 2002 in Dike 2002). He cites problems in the implementation of policies for the expansion of the education system, lack of capacity for planning and management, limited financial resources, inadequate information systems and monitoring systems, etc., as some of the problems that have led to rapid and unplanned growth in Universities in Nigeria. He concludes by saying that the infrastructure and facilities remain inadequate for coping with a system that is growing at a rapid pace.

Other factors contributing to the decline in quality of education in Nigeria are the unstable environment due to frequent strikes by students or staff, the quality of students admitted to programs, and the quality of the academics recruited. Admission of more students than the university infrastructure can cope with causes overcrowding. This is also a factor in declining quality of university education in Nigeria. The university system has far outgrown the resources available to it to continue offering high-level quality education. Inadequate funding has resulted in problems such as the breakdown and deterioration of facilities, shortages of new books and current journals in the libraries, supplies for the laboratories and limited funding for research.

Overcrowding of facilities in Universities is being experienced in overcrowded hostel accommodations. This has resulted in most of the students having to live in residential neighbourhoods immediately surrounding the universities.

The neighbourhoods that students spill over into are often times already in existence before the universities came up and have to adapt to the accommodation needs of students by providing or remodelling existing structures to suite the unique need of students. The demand for hostel accommodation for students also provides opportunities for private investors to develop properties in the vicinity of the universities for student accommodation. This development often leads to dense, haphazard, and uncontrolled developments around the universities. These accommodations are often furnished with the minimal housing amenities for students. Students' accommodations often make for very depressing and less conducive environments for study and according to this paper should be looked at as another potential consequence of the decline in the quality of university education in Nigeria.

Benue State University was established in 1992 as a response to a genuine need by the state government to provide essential impetus to economic, cultural, social and vocational development (Students' Information Handbook 2011). The university is made up of six faculties of Arts, Education, Science, Social Science, Law, Management Sciences and a College of Health Science. The student body is made up of 17,000 students spread among the undergraduate, postgraduate, sub degree and technical education programs. By the University's policy, only final year students, first year students, disabled students, outstanding sports men and women and foreign students are accommodated on the campus because of lack of hostel accommodation. This results in the excess of students spilling over into the neighbourhoods of Akpehe, Logo and Gyado villa in the vicinity of the university.

These neighbourhoods are low-lying, densely and haphazardly developed. Student accommodations are varied due to the diversities of housing options. Accommodation types range from mud huts to one room apartments in compounds to singlestand alone self-contained apartments. Access roads and surroundings are very poor. Water and electricity supply are also inconsistent and poor. Some of the housing environments students live in and how they perceive these houses have a negative impact on the performance of students. This perception is also affected by relationships with neighbours and roommates, students' background, the building condition and the environment itself. The extent to which this perception affects student's academic performance is the main thrust of this study.

The study aims, therefore, to test hypothesis concerning the perceived household environment on the academic performance of Benue State University students who live in these surrounding neighbourhoods as compared to their colleagues who stay on campus and who live with their parents in town and the nature of the relationship. The alternate Hypothesis H1 says "student's academic performance is affected by the perception of their living environment". The Null hypothesisHo = "student's academic performance is not affected by the perception of their living environment". The study goes further to test the kind of relationship by using the Pearson Moment Correlation Coefficient.

Specific objectives used to test this hypothesis are determining the demographic statistics of final year students of the department of Geography, measuring the living quarter's environment, obtaining from the registrar academic records of the students and determining their perception of their environment. This study is significant because it examines a salient often ignored cause for declining quality of university education in Nigeria. Information from this study can be relevant in policy decisions on improving quality of university education in Nigeria.

The study is limited in size and scope probably; a more accurate picture of the situation can be achieved with a bigger sample size spread among other departments and faculties in the school and also covering other tertiary institutions in the country.

1.1 The Study Area

Benue State University is located near the southern bridgehead of the Benue River. The university occupies 6 km^2 of land between Gboko road and river Benue; approximately 1.5 km wide and 4 km long. It is bounded to the west by the Benue links headquarters and to the east by Tilley Gyado house. To the north by the river Benue and to the south by the residential neighbourhoods of Akpehe and Logo. These neighbourhoods are situated within Wailomayo ward of Makurdi town. The population in these neighbourhoods has grown steadily, and in addition, more intense structural growth has occurred over the last few years because of the presence of the university.

The climate of Akpehe and Logo is the same with the rest of Makurdi town, which is hot and humid tropical rainforest climate (AF) according to Koppens classification. High temperatures are experienced especially in the months of March and April. Students are usually in session during this period. The heat brings a lot of discomfort to people especially if buildings are constructed with materials that absorb heat. Rainfall occurs between the months of April and October. With average rainfall amount ranging between 1000 and 1300 mm spread over 5–7 months. Akpehe and Logo are poorly drained and low-lying, and as a result they are susceptible to flooding during the rainy season. These floods have adverse consequences on accessibility, the surrounding environment and even on the buildings (see Figs. 1 and 2). Because of high demand for land for development, the residential neighbourhoods of Akpehe and Logo are heavily and haphazardly developed without regard to development control. The neighbourhoods are not connected to the Makurdi mains. This means that water sources are boreholes, surface well, water tankers from Makurdi water works and water vendors called *mai ruwa.* Electricity is also not consistent leading many who can afford it to rely on generators for power supply.

2 Theoretical Framework

2.1 Housing

Housing has been universally acknowledged as one of the most essential necessities of human life and is a major economic asset in every nation. Adequate housing

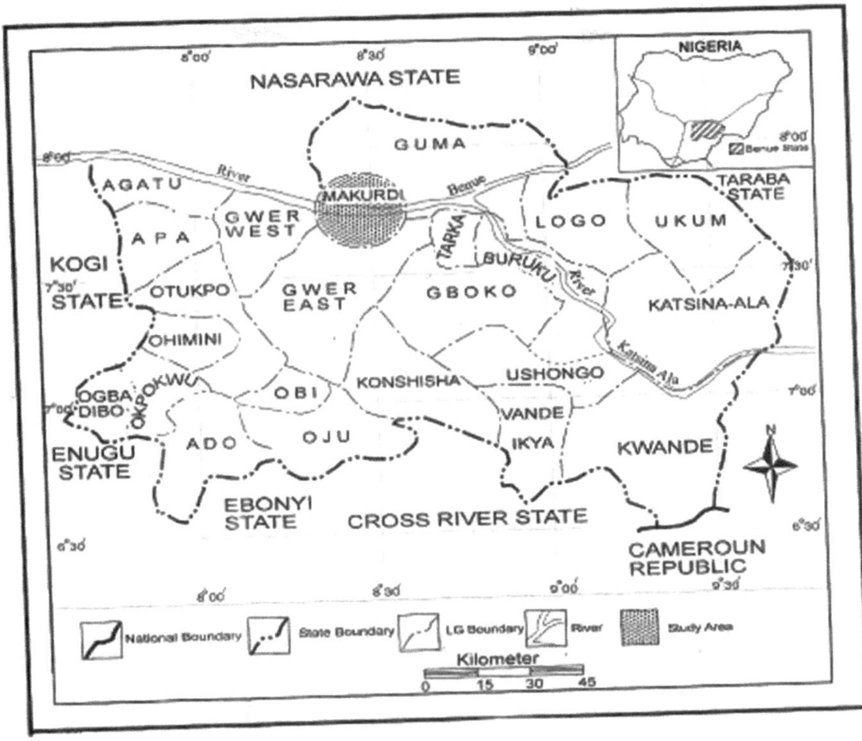

Fig. 1 Benue state showing Makurdi local government area

provides the foundation for stable communities and social inclusion (Oladapo 2006 in Jiboye 2010).

Housing is a broad term that is not easily defined because it encompasses many aspects of a building (Reynolds 2009). He makes two important conclusions; that a house is basically a compartment that shields from the elements and that existing housing is non-functional without systems. Systems such as water supply, power supply and waste disposal systems. He compares the vulnerability, limitations and dependency of occupants on housing systems to a patient on life support in a hospital. Ezenagu (2000) as quoted in Chigh (2008) also says a house is not only a unit in the built environment but incorporates its environment, facilities as well as connectivity to other houses and the wider community in which it is situated. Student housing in context therefore is seen as the physical shell of the building, the support systems like electricity, water supply and waste disposal, the environment surrounding the house and the relationship the house has with the community.

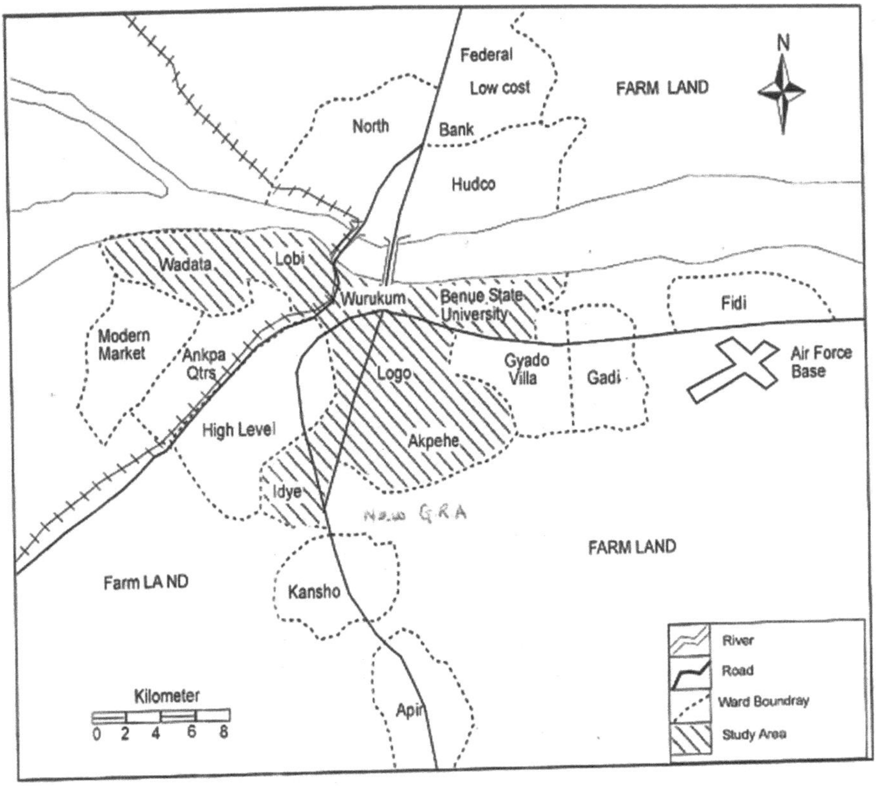

Fig. 2 Makurdi town showing the wards that make up Makurdi

2.2 *Housing in Nigeria*

Housing provision is one of the major challenges facing developing countries experiencing rapid urban growth like Nigeria (Olotuah 2005). The private sector is the major provider of housing in Nigeria. Public sector intervention is usually in the form of quarters for government staff (Salau 1992; Olotuah 2000 in Olotuah 2005) the provision of housing does not really match the growth of the population in most urban centres. This accounts for the monumental deficiency in urban housing quantitatively and qualitatively (Jagun and Olotuah 2000, 2003 in Olotuah 2005).

Quantitative housing needs means the extent to which the quantity of existing housing stock falls short of that required to provide each household in the population with accommodation of a specified minimum standard. While qualitative housing needs is the quality of housing that makes it satisfy the basic requirements of the users. Housing often does not meet needs of users for the following reasons—a wrong perception of housing needs of the low-income earners who make up the vast majority of urban dwellers, typical housing that is not rooted in Nigeria's climatic, cultural and socio-economic environment, improper planning, unrealistically high cost of building materials, insensitivity of government in regulating policy regarding private sector housing delivery and poor execution of housing policies.

Housing needs of different income groups within the society are dependent on basic needs and economic standing. For low-income earners, proximity to school or work places more than high-quality architectural delights are important. Turner (1977 in Olotuah 2005) suggested that basic needs of the different income groups, which he states as security, identity and opportunity, determine their preferences for houses, which meet their aspiration and needs. Student housing are an example of low-income group housing that hardly meets the requirements of users.

2.3 Housing Conditions in Nigeria

The development patterns in urban centres of Nigeria today are haphazard with many parts of cities being developed by the private sector without any form of regulation (Fadamiro Bobadoye and Adelowo 2005). This has led to high rates of uncontrolled and unplanned urban growth in Nigeria, resulting in poverty and poor quality of the environment. Consequently, urban dwellers in many Nigerian cities are living in distress because of over stretched urban systems leading to slums, congestion, unemployment, despair and diseases (Onibokun 1974 in Fadamiro et al. 2005). In the same vein, the quality of housing is affected by this lack of control. Developers who are mainly in the private sector build without regard for human comfort and standards. This problem is further compounded by the lack of cooperation between government agencies in development control. This leads to many houses that are poorly constructed. Many neighbourhoods are unplanned with inadequate and sometimes nonexistent systems sustaining them. These conditions characterize most neighbourhoods in Nigeria. They are also most pronounced in neighbourhoods housing students.

2.4 Housing Satisfaction

Conceptually, housing satisfaction according to Djebarni and Al-Abed (2000 in Jiboye 2010) refers to the degree of contentment experienced by an individual or family with regard to the current housing situation. It is an index for determining the level of contentment with housing and refers to an entire continuum of satisfaction.

Studies done by Konadu-Agyemanyg et al. (1994), Gilbertson et al. (2008) and So and Leung (2004 in Jiboye 2010) have established a strong correlation between housing, good health, productivity and socio-economic development. These studies have also observed that there is a significant association between housing conditions and physical and mental health of an individual. Thus establishing a significant correlation between the quality of life and the comfort, convenience and visual acceptability of the house.

Studies have also observed that a dwelling that is adequate from the physical or design point of view may not necessarily be adequate or satisfactory from the users' point of view. Thus, the concept of habitable and satisfactory housing is related to the physical, architectural and engineering components of the house, as well as to the social, behavioural, cultural and personal characteristics of the inhabitants, the components of the environment of which the house is a part; and the nature of the institutional arrangements under which the house is managed (Onibokun 1974 in Jiboye 2010).

According to Oladapo (2006 in Jiboye 2010), qualitative housing is the major challenge of urban housing in Nigeria. This is because of the failure of many public and private housing projects due to lack of adequate thought and consideration given to adequate housing, as relevant factors or parameters, which combine to determine tenants' housing satisfaction. He says the criteria guiding design and development have been based on developers' standard rather than users' preferences and needs. It is observed that student's preferences as end-users are hardly considered in housing.

2.5 Measurement of Housing Satisfaction

The works of Galster and Hesser (1981), Amerigo (2002) and Potter and Cantarero (2006) highlight two general approaches to empirical research in residential satisfaction. One is to view residential satisfaction as a criterion of quality of life, while the other is to view it as a predictor of a variety of behaviours. Other studies done by Onibokun (1974), Western et al. (1974), KelleKc and Bebkoz (2005), Oladapo (2006 in Kamaruddin Zainal and Aminnudin 2009) shows that tenants satisfaction could be measured by housing attributes such as the function and physical adequacy of the dwelling, quality and adequacy of social and community facilities, the nature and effectiveness of official policies and personnel attitudes, convenience for living, the condition and maintenance of the home environment, maintenance of the dwelling facilities, privacy, territoriality and neighbourhood security among many others.

This study places emphasis on the housing attributes, function and physical adequacy of the dwelling, quality and adequacy of community and social facilities and personal attitudes as means of evaluating the perception of satisfaction of students of Geography Department of Benue State University, Makurdi.

2.6 Perception of Housing Environment and Academic Performance

Academic achievement of a student is not determined by intelligence alone. Other contributory factors include: learning environment, student's motivation, family and living environment. A study by Niebuhr (1995 in Kamaruddin et al. 2009) examined relationships between several variables and student academic achievement. His findings suggest that the elements of both school climate and family environment have a stronger direct effect on academic performance. Furthermore, a study done by Hammer (2003) also shows that the home environment is as important as what goes on in the school.

Schrager (1986 in Kammaruddin et al. 2010) show a relationship between living situations and college academic performance. The study showed that the interpersonal environments that rise within different living groups have an impact on academic performance. For example, students living in campus tend to perform better academically than those living at home with parents. Several factors could be responsible for this. One of the factors could be the stress associated with living situation as living in campus is less stressful than living off campus. Other studies including those done by Blimling (1999 in Kammaruddin et al. 2010) have shown that place of residence is related to academic performance, but few of the studies have focused on specific factors that impact on the specific factors related to perception of living environment and its impact on academic performance.

This study, therefore, studies the perception of living environments of the Geography Students of Benue State University and how this perception is linked to their academic performance.

Research in this area should increase the awareness on provision of better accommodation facilities for students in Nigerian Universities, thereby enhancing their performance and improving the quality of university education in Nigeria.

3 Methodology

Questionnaires were the major instrument for data collection and were administered to final year students of Geography Department Benue State University, Makurdi. The questionnaire was divided into three parts. The first and second parts, deals with the background of the students and on the housing conditions—type of house, number of people in the room, water source, waste disposal methods, noise levels in dwelling and toilet types. The third part measured perception of household environment using a 5 point Likert scale that ranged from "strongly disagree" (1 point) to "strongly agree" (5 points). Academic achievement was measured using the cumulative grade point average of the students.

Student's t-test was used to test the hypothesis. The household environment was broken down into variables of surrounding environment, noise level; cross ventilation in rooms, water and electricity source, building type and waste disposal systems. Perception of students based on these variables was measured using a Likert scale. In analysis, the numerical values of this scale was reversed (ranging from 1 to five) of questions so that the higher numbers always indicated poor perception. For each participant, the six questions regarding perception were averaged together resulting in a score for that participant. The scores for all participants in a group were collapsed to get an average score for the different groups. The average perception score for the sample was 2.74 (between very slightly dissatisfied and neutral) for those off campus, 2.72 (between slightly satisfied and neutral) for those on campus, and 2.31 from those living at home with parents. This measurement scale was modified from a study by Ivie (2010) the relationships among stress of living situation, health and academic performance. The average perception score for each student is then tested against the students CGPA for a relationship using Pearson Product Moment Correlation Coefficients. This information is then grouped into the three groups: on campus, off campus and at home students.

4 Results and Discussion

4.1 Demographic Profiles

Demographic profiles of the students surveyed reflect their age, marital status, gender, ethnic groups and sponsorship. Hence, 68 students were surveyed and only 44 returned the questionnaires hence 44 final year students of the department of Geography were used for the study. The sample was made up of both male and female students and 59.1 % of the respondents fall within the age bracket of 15–25 years. This indicates that on the average the students are in the virile active stage of life when they have the energy, time and intelligence to focus on their studies and do well academically. A total of 36.36 % are above 25 years and 2 % are between the age range of 36–40. All the age ranges are in the active stage of life. Barring other issues like health, bad habits and peer pressure they should do well academically. Attitudes to learning determine academic performance and these attitudes are imbibed from socio- cultural environments one is exposed to Table 1 indicates that the most frequently occurring ethnic group are the indigenous Tiv (68.1 %) followed by the next largest ethnic group in Benue the Idoma (25 %) followed by other minorities like Igede (2.7), Eggon and Uffia (2.7 %) each. All these ethnic groups inculcate a learning culture that if followed should translate to good academic performance. Also because they are indigenous the students are in their familiar territory and so should perform well academically because the stress and financial implications of travelling far from home to school are reduced. Apart from ethnic groups ones gender determines responsibilities in the home and therefore the time devoted to study, from Table 1, 73 % are males while only 27 % are females. This indicates that more of the students should be on good academic standing because males devote more time to academics while females share their time with learning and other household chores. Marital status translates to additional responsibility especially for females who are home managers. Table 1 also shows data indicating that 25 % of the students are married while 75 % are single. Meaning that for those who are married their attention will be divided between their

Table 1 Shows demographic profiles of students of Geography, Benue State University

Items	Frequencies	Percentages
Age		
15–25	26	59.1
26–35	16	36.36
36–40	2	4.54
Total	44	100
Ethnic groups		
Tiv	30	68.1
Idoma	11	25
Igede	1	2.27
Eggon	1	2.27
Uffia	1	2.27
Total	44	100
Gender		
Male	32	72.73
Female	12	27.27
Total	44	100
Marital status		
Married	11	25
Single	33	75
Total	44	100
Sponsorship		
Self	16	36.36
Parents	24	54.54
Guardians	4	9.1
Total	44	100

homes and their academics. This negatively affects academic performance and also this group must have to stay off campus because the university does not provide accommodation for married students.

Self-sponsored students have to engage in other pursuits in addition to academic work in other to support themselves. This negatively affects academic performance and determines the type of accommodation the student can afford. Table 1 shows data indicating that most of the students are sponsored by parents and guardians (45.46 %). While 36.36 % of the students are self sponsored. This factor diminishes academic performance. On the other hand, it can also increase academic performance by increasing the determination in these self-sponsored students to quickly complete their education and reduce the financial burden they bear. Academic performance is enhanced when there is conducive learning and living environment. Information on the living environment of students of Geography Department Benue State University is shown below.

4.2 Household Environment

According to Table 2 59.10 % of the final year students of Geography are staying off campus, 18.18 % are living on campus and 22.72 % are living with parents/spouses in town. This indicates that more of the students are living off campus (81.82 %). This is an indication that more students are experiencing the stress of commuting to school, and living in environments unconducive for study. This has negative impact on academic performance. From Table 2, 40.90 % of the students live in huts which are locally constructed with mud and roofed with thatch.

Reasons include cost and non-availability of housing. This means that 41 % of the students are living in very poorly constructed buildings negatively impacting on their academic performance and even on their health because these huts are often damp and have only one window. 36.38 % are living in single rooms in compounds where they share facilities like toilet, bath and kitchen with other residents. This factor also decreases comfort and increase exposure to stress from neighbours. Hence, the conditions are most times not optimal for study. This negatively affects academic performance. 22.7 % living in self-contained apartments correspond with the percentage living at home with parents/spouses. This factor also reduces academic performance because living at home means the students have to contribute their quota to the smooth running of the home and are less motivated to study being in an atmosphere that is not the school atmosphere. Of the students surveyed, 75 % have cross ventilation in their rooms. This factor improves the living condition in the rooms because Makurdi is hot and humid. Cross ventilation in a room ensures the air is circulated in a room thereby improving air quality and reducing dampness and smell. The 25 % who do not have cross ventilation in their rooms are experiencing conditions that do not make for conducive study and healthy living. This reduces academic performance. Table 2 also shows that 56.82 % of the students rely on surface wells for water supply. These wells are often used by many people increasing the chances for contamination. The wells are also more often not properly located in the compounds and so easily contaminated by the human waste disposal systems of soak aways and septic tanks. Finally, the factor of hauling and carrying water takes up time that could be invested in study. All these factors contribute to decreasing academic pursuit. This is because when health is compromised by poor water; academics have to be interrupted until the student is well again. This has negative impact on academic performance. Also, 34.08 % rely on the more safe water from the boreholes and public water supply. Their health is not compromised. While 9.1 % rely on water from water vendors that can be sourced from anywhere. This increases the risk of students to water borne diseases. From Table 2, 68 % of students use the water closet type of toilet. This is safest but most difficult to maintain because it requires a lot of water. 27.7 % use the pit latrine, which increases their exposure to disease that in turn reduces academic excellence. Noise level has 59 % of the students of Geography department affirming that noise levels in their living quarters are ok for study and 40 % saying the noise level in their living quarters interfere with their study time.

Table 2 Information on household environment

Item	Frequencies	Percentages	Average
Accommodation			
On campus	8	18.18	
Off campus	26	59.10	Off campus
At home with parents	10	22.72	
Total	44	100	
Type of building			
Hut	18	40.90	
Single room in compound	16	36.38	Hut
Self contained apartment	10	22.72	
Total	44	100	
Cross ventilation in rooms			
Yes	33	75	
No	11	25	Yes
Total	44	100	
Water source			
Borehole	7	15.90	
Public water mains	8	18.18	
Surface wells	25	56.82	Surface wells
Water vendors	4	9.10	
Total	44	100	
Toilet type			
Wc	30	68.18	Wc
Pit	12	27.27	
Bush	2	4.54	
Total	44	100	
Surrounding environment			
Impervious	7	15.90	
Plain	26	59.10	Plain
Covered with grass	11	25	
Total	44	100	
Waste disposal			
Empty plots nearby	18	40.9	Empty plots nearby
Communal dumps	15	34.1	
Street	11	25	
Total	44	100	
Noise level			
Ok	26	59.1	Ok
Not ok	18	40.9	
Total	44	100	

4.3 Students' Perception of Their Household Environment

Table 3 shows information on how final year students of Geography department perceive their living environment. The living environment is broken down into variables for easy measurement. The variables are water supply, surrounding environment, noise level, waste disposal, and building type and cross ventilation of room. These variables relate with and enhance student comfort and conduciveness for study, which brings about good academic performance from the table, the students staying at home find noise levels very satisfactory. This is also true for students staying in campus while those staying off campus are only slightly satisfied with the noise level in their rooms. This could probably be as a result of interference from neighbours.

Students staying at home find their surrounding environment very satisfactory. Those staying off campus are only slightly satisfied with their surrounding environment while those staying on campus are very dissatisfied with their surrounding environment. This is so because hostel conditions are often cramped and poor.

Students staying at home perceive the water supply to their homes as neutral while those staying off campus and on campus are very dissatisfied with the water supply to their living quarters.This is most likely because even though many parts of Makurdi do not have water piped directly to the homes,water is usually available nearby in a surface well or borehole. On campus water is hardly available elsewhere if the university authorities do not supply water. Most of the times the supply from the University is not consistent resulting in very dirty environments especially in the toilets on campus.

Students staying at home perceive their waste disposal system as neutral. The students living off campus are slightly satisfied while those staying on campus are very dissatisfied with how waste generated by them is disposed. On campus waste disposal is most of the times left to the students who leave heaps of unsightly refuse around the hostel. This factor accounts for the level of dissatisfaction with waste disposal by students on campus.

Students staying at home are slightly dissatisfied with the type of building in which they live. While those living off and on campus are very dissatisfied with the type of buildings in which they live. For those staying off campus the hut is the most frequently occurring building type. This could be a reason for the dissatisfaction. Students staying on campus live in hostels, which are cramped with students in very poor and unsanitary spaces. This could be the reason for strongly perceived dissatisfaction with building type by students living on campus.

Students staying at home are very satisfied with the cross ventilation in rooms while those off campus are slightly dissatisfied with the cross ventilation in rooms as against those in campus who are neutral to the cross ventilation in their rooms. Cross ventilation is a requirement for registration of buildings with the development control board in Makurdi on account of the heat and high humidity in the town. Hence, most houses get cross ventilation.This could account for the high level of satisfaction and neutrality of students living at home and on campus respectively. For those off campus their level of dissatisfaction could be as a result of the

Table 3 Perception of satisfaction with housing environment of final year students of Geography Department, Benue State University

Level of satisfaction of students at home				Noise level	Explanation
1	2	3	4	5	Very satisfied
1		1	1	7	
Off campus					
1	2	3	4	5	Slightly satisfied
8	3	4	13		
On campus					
1	2	3	4	5	Very satisfied
1	1	1	2	3	
Surrounding environment of students at home					Very satisfied
1	2	3	4	5	
1		1	1	7	
Off campus					
1	2	3	4	5	Slightly dissatisfied
7	9	4	6	2	
On campus					
1	2	3	4	5	Very dissatisfied
4	2	1	1		
Water supply at home					
1	2	3	4	5	Neutral
		4	3	3	
Off campus					
1	2	3	4	5	Very dissatisfied
10	6	4	6	2	
On campus					
1	2	3	4	5	Very dissatisfied
4	2	1	1		
Waste disposal home					
1	2	3	4	5	Neutral
1	1	4	1	3	
Off campus					
1	2	3	4	5	Slightly dissatisfied
8	9	8	2		
On campus					
1	2	3	4	5	Very dissatisfied
4	2	1	1		
Building type of students at home					
1	2	3	4	5	Slightly satisfied
	1	1	4	4	

(continued)

Table 3 (continued)

Level of satisfaction of students at home				Noise level	Explanation
Off campus					
1	2	3	4	5	Very dissatisfied
10	3	5	7	3	
On campus					
1	2	3	4	5	Very dissatisfied
3		2	1	2	
Cross ventilation in rooms at home					
1	2	3	4	5	Very satisfied
	1	1	2	6	
Off campus					
1	2	3	4	5	Slightly dissatisfied
6	11	6	4	1	
On Campus					
1	2	3	4	5	Neutral
2		4		2	

haphazardly built structures which they occupy which in most cases are hurriedly built and so not registered by the approving bodies.

4.4 *Hypothesis Testing and Correlation Between Perception and Household Environment*

The hypothesis Ho stated that student's perception of their living environment "does not affect academic performance. The students' *T* test was used to test this hypothesis. The test accepted the H1 stating that "academic performance of final year students of Geography department is affected by the perception of their living environment" The test showed a significant relationship between academic performance and perception of household conditions for students living at home and on campus. Results also show a strong relationship between students living off campus and academic performance. These findings are at variance with other studies that show an inverse relationship between students staying off campus and academic performance. Students staying on campus should perform better, this study indicates otherwise. The situation may be so because the living conditions on campus are very poor and in some cases worse than what obtains off campus. Also, study results indicate that students staying off campus perform better academically than those on campus and at home probably because they are more motivated to finish their course and leave the constraining environments they are in. Findings that students living at home are not performing well are consistent with studies that show that home environments do not enhance learning.

5 Conclusions

The study tested hypothesis that student's academic performance is affected by the perception of their living environment. Three groups of living situations were identified. Students living on campus, those living off campus and those living at home with parents. The living environment was broken down into variables, viz, surrounding, water supply, waste disposal, building type, and noise level and cross ventilation in rooms. Perception of students of these variables were averaged and tested against cumulative grade point average as an indication of academic performance. The results show an acceptance of the hypotheses by the students' T test. Results also indicate that students staying off campus are performing better than those on campus and those living at home. These results are not consistent with other studies that show that students staying in campus perform better because of a conducive study environment. Yet they are very interesting because they highlight very salient issues that on campus accommodation are so poor that they cannot compare with off campus accommodation, which is also not optimal. Results indicating that students staying at home are not performing well academically agree with other research findings. This is because the home does not have conducive for environment for learning and is often the source of many distractions.

6 Recommendations

Based on the results of this study the following recommendations have been made. The University authorities should look into partnerships with private investors to build and operate accommodations on and off campus. Student accommodations should be designed with the student's preferences as end-users in mind. These preferences should be documented in a prototype design that should be strictly enforced by agents of development control. Development control agents should enforce strict compliance by developers to a uniform standard of buildings and environment for buildings housing students. Where basic and minimal standards of health, safety and sanitation cannot be achieved by the university in hostels, hostel accommodation should be totally eliminated on campus. The government should provide incentives like reducing the price of land for private developers wishing to develop student's villages. Finally, development control agents should monitor private developments for students and commend developers who maintain good standard accommodations for students as motivation.

References

Amerigo M (2002) A psychological approach to the study of residential satisfaction. Residential Environments, London

Bollag B (2002) Nigerian Universities Start to Recover From Years of Violence, Corruption, and Neglect. The Chronicle of Higher Education, A40–A42

Chigh NR (2008) Students perception of housing conditions in logo II. Unpublished students project. Departrment of Geography, Faculty of Environmental Sciences, Benue State University, Benue State, Nigeria

Djebarni R, Al-Abed A (2000) Satisfaction level with neighbourhood in low-income public housing in Yemen. J Constr Manage 18(4):230–242

Dike R (2002) World Bank Report on Education in Nigeria. Accessed from world wide web address http://www.afbis.com/analysis/education10204234737.htm

Ezenagu VC (2000) Fundamentals of housing. Awka: Fountain Publishers. ore. A Inst Plann J May 201–208

Fadamiro JA, Bobadoye SA, Adelowo W (2005) Urban poverty and environmental quality in Akure city: the challenges for architectural profession in Nigeria. Paper presented at the African Union of Architects Congress, Abuja, Nigeria

Fapohunda O (2014) List of Nigerian universities and websites. Internet document assessed at http://www.justnaira.com/2012/03/list-of-nigerian-universities-and.html

Galster G, Hesser G (1981) Residential Satisfaction: Compositional And Contextual Correlations. Environ Behav 13:735–758

Gilbertson J, Green G, Ormandy D, Thomson H (2008) Good Housing and Good Health? A Review and Recommendations for Housing and Health Practioners: A sector study housing cooperation. UK. Available at http://www.healthhousing20060816144328.pdf. Accessed March 2009

Hammer B (2003) ETS identifies affecting student achievement-Washington update

Ivie RM (2010) The relationship among stress of living situation, health and academic performance. Accessed from eyhammer@loyno.edu

Jiboye DA (2010) The correlates of public housing satisfaction in Lagos, Nigeria. J Geogr Reg Planning 3(2):017–028. http://www.academicjournals.org/JGRP

Kamaruddin R, Zainal NR, Aminuddin ZM (2009) The Quality of Learning Environment and Academic Performance from a Student's Perception. J Bus Manage

Kellekc DOH, Bebkoz L (2005) Determinants of Users' Satisfaction and Environment Quality: Sample of Istanbul Metropolitan. URP Department, Faculty of Architecture, Istanbul TechnicalUniversity. Turkey (www.Feweb.vu.ul/ersa (2005))

Niebuhr K (1995).The effect of motivation on the relationship of school climate, family environment, and student characteristics to academic achievement (Report No.EA 027467). East Lansing, MI: National center for Research on TeacherLearning. (ERIC Document Reproduction Service No. ED393202)

Olotuah AO (2005) Urbanization, urban poverty and housing inadequacy. Paper presented at the African Union of Architects Congress, Abuja, Nigeria

Onibokun P (1974) Evaluating consumers satisfaction with housing: An application of a systems approach. Am Inst Planners J 40(3):189–200

Potter J, Cantarero R (2006) How does increasing population affect residential satisfaction?: A small community case study. Environ Behav 38(5):605–625

Reynolds (2009) A look at the existing concept of housing. Available from http://knol.google.com/k/michael-reynolds/a-look-at-the-existing-concept-of/eg1dlyqec92d/7#

Students Information Handbook (20011) Published by Ashitech Nigeria Limited

So AP, Leung AYT (2004) Survey of a ttitudes towards buildings in three Chinese cities: HongKong, Shanghai and Taipai. Facilities 22(3)(4):100–108

Western JS,Weldone PD, Haung TT (1974) Housing and Satisfaction in Singapore. A Inst PlannJ May 201–208

World Bank Report (2011) Education in Nigeria. Internet document assessed from the internet at www.worldbank.org/education

Empowering Community for River Basin Management

Bhaskar Chandrakar, N.P. Dewangan, Suraj Verma and Amit Mishra

Abstract There is a need to involve community in both technical and administrative aspects for river basin management. The community participation is vital and can be enhanced by educating and increasing awareness. Various methods like training packages, community action plans, conducting dramas and movies and staging mass media campaigns may be adopted for educating people. The basin management can be well implemented by the involvement of the local communities, the beneficiaries. Studies verify that block watershed samitis and village groups work satisfactorily. In fact, community-based watershed management has become the guiding principal of natural resources specially land, forest, and water in rural areas. The importance of community participation cannot be overemphasized for sustainability of any development effort. As a matter of fact, the Ministry of Rural Development, Government of India, has been striving to bring this vital component into the Watershed Development Programs by periodically reviewing the guidelines and tuning it with rural socioeconomic conditions. This paper stresses the fact that the basin management is closely related with the environmental protection and sustainable development. Therefore, it should be incorporated in the development plans to ensure efficient follow up measures at the community, sub-regional, regional, national and international levels.

Keywords Watershed management · Sustainable development · Government policies · Community empowerment

---

B. Chandrakar (✉) · S. Verma
Government Engineering College, Raipur, India
e-mail: chandrakar_bhaskar@yahoo.com

N.P. Dewangan
Sarguja University, Ambikapur, India

A. Mishra
Raipur Institute of Technology, Raipur, India

A.K. Sarma et al. (eds.), *Urban Hydrology, Watershed Management and Socio-Economic Aspects*, Water Science and Technology Library 73,
DOI 10.1007/978-3-319-40195-9_25

1 Introduction

A watershed is the area of land where all of the water that is under it or drains off it, goes into the same natural drain. Each watershed has its own network of rivers, streams and channels that drain water from and through the basin. This characteristic of drainage network plays a great part in determining how water moves through the basin and consequently impact upon issues, such as water quality and quantity in the given basin. Watershed management, therefore, controls watershed relevant characteristics for sustainable distribution of its resources by implementing plans, programs, and projects to sustain and enhance watershed functions including plants, animals, and human communities within the watershed boundary. Features of a watershed that agencies seek to manage include water supply, water quality, drainage, storm water runoff, water rights, and overall planning utilization of watershed. To succeed, watershed management has to be participatory. This is one of the lessons coming out of decades of failures of centrally planned programs. Watershed management is not so much about managing natural resources as it is about managing human activities, as it affects these resources. The drainage area of river provides natural boundary for managing and mitigating the human and environmental interaction. Because human activities include actions by government, municipalities, industries and land-owners, watershed management must be a cooperative effort. Effective watershed management can prevent community water shortages, poor water quality, flooding, erosion, and other environmental improvements.

2 Significance of Community Participation

Rajim is a place in Raipur district approximately 170 km from the study area in the same Mahanadi basin. Rajim is situated in the bank of Mahanadi River and has set an ideal example where community participation proved to be successful as there is rise in ground water table, rejuvenation of old dried wells, hand pumps, etc. (Fig. 1).

3 Types of Watershed

The common mode of classification is the size, drainage, shape, and land pattern. The classification can also be based on the size of streams or rivers, points of interception of streams and rivers, drainage density, and its distribution. The usual classification based on the geographical area are Macro watershed (>50,000 ha), Sub-watershed (10,000–50,000 ha), Milli-watershed (1000–10,000 ha), and Mini watershed (1–100 ha).

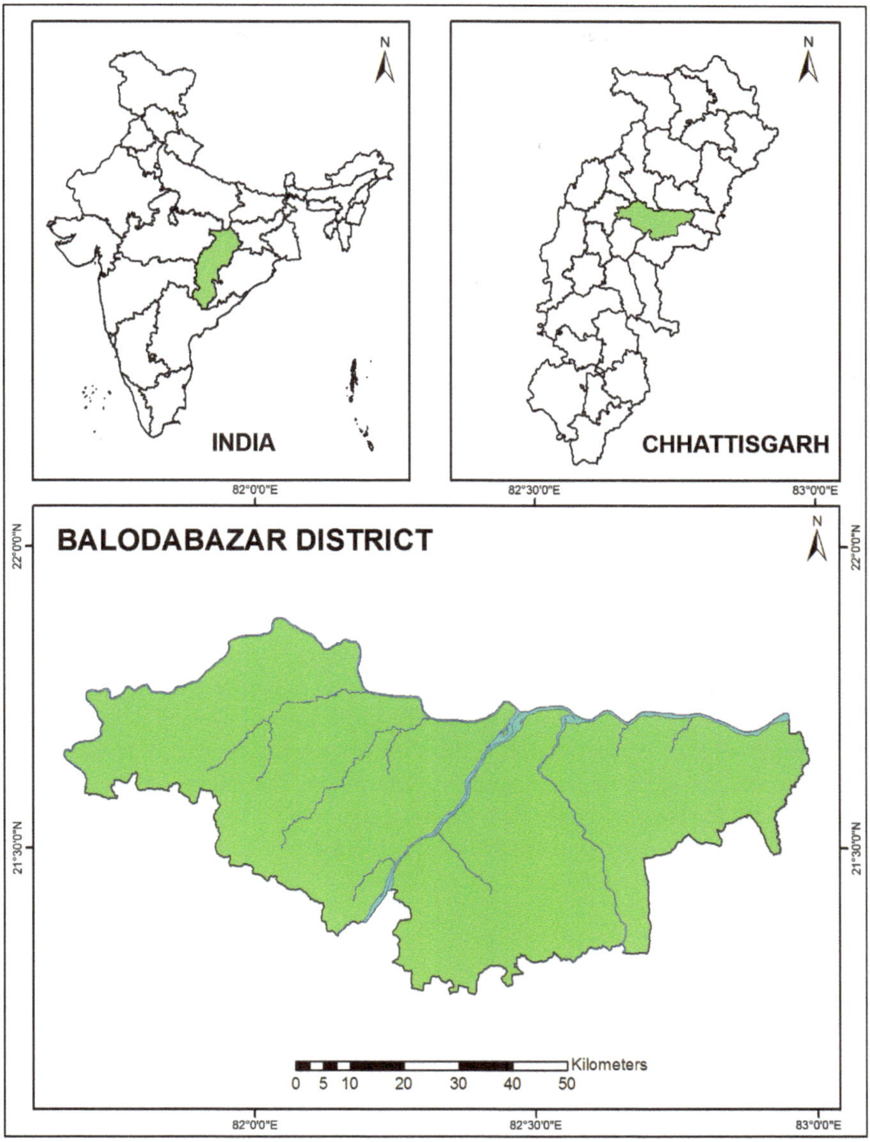

Fig. 1 Location of study area

4 Empowering Community

The areas at or near a watershed are prone to floods. To ensure the safety of the local people, it is essential that they should be given certain administrative powers. The local administrative bodies such as Gram Panchayat should be constituted to meet the needs of people and to make them aware of the consequences of the flood and its preventive measures. The different ways in which the people can be empowered are summarized below:

- The Gram Panchayat should have certain administrative powers so that they can take decisions in order to prevent flood hazards.
- Advertisements should be telecasted in the radios and televisions that provide information regarding the preventive measures for flood control.
- Door-to-door awareness programs should be conducted to ensure that every family in the area participates in the campaign.
- Developing forecasting and early warning systems with responsible safe communication support.
- Campaigns and programs should be organized to spread awareness amongst the people.

5 Baloda Bazar—A Case Study

Mahanadi River was selected for the case study because Rajim has been studied earlier which is situated on the bank of the same river. Rajim has proved to be an ideal example for the water basin management by community empowerment. This was known through direct interaction with the local farmers. During field visits, the local people who were benefitted from this type of program described us the method they adopted for the basin management. Learnings from Rajim made us select Baloda Bazar for the case study with similar geographical locations. As also, there was problem of fall in ground water table in the area, so it was necessary to take some action for its improvement.

6 Details of Baloda Bazar

The project area is 4911 m^2 and consists of 12 villages (8 micro watersheds) located in a 10 km radii. The project being implemented is well managed by adopting the norms for treating the land involving the beneficiaries, watershed institutions, etc. The record at the district, block, watersheds samitis, and village level were found to be satisfactory. The area is 100–120 km from Raipur. The project area is situated in between 82°24' latitude from 21°34' longitude. Chhattisgarh region falls in agro climate zone and semi humid climate. The average rainfall of project area is 1120 mm annually. The climate is hot and humid. The project area had red yellow soil to medium black soil. The availability of ground water is poor. The general slope of Raipur District is towards south-east direction and most of water flows towards the catchment of the Mahanadi River.

7 Examples of Community Participation

- Womens' Participation: In milli-watershed of Baloda Bazar, there are eight watershed committees converging eight micro-watersheds spread over 12 villages. There are 155 members out of total member 35 women which is 22 % of total watershed constitute committee membership.
- Constitution of **SHG (Self Help Groups)**: Under this Milli-watershed, there are 55 active SHG. A loan of rupees 5000–10,000 has been disturbed to 28 SHGs. Loan is being regularly returned by the respective SHG.
- Regular Meeting: There is provision to conduct one meeting in a month for the watershed committee members for the approval of works. Conflict resolution and issues related to watershed implementation can be discussed in the meeting. However, as explained by project officer, that the minimum of four meetings takes place for each Gram Sabha.

8 Implementation of Empowerment

Apart from framing the policies for efficiently managing the watershed basin, the most important part is to ensure that those policies are implemented effectively. The stages of implementation are:

- The State Government of Chhattisgarh has organized various awareness programs for the benefit of the people.
- Various NGO have come forward to spread awareness among the local population.
- Demonstrations of the preventive measures have been done using roadside plays or shows known as Nukkads.
- The Gram Panchayat should periodically discuss the importance of efficient watershed management.
- Representatives from the community are being trained to convey the significance of watershed management.
- Advertisements by the government are telecasted in the radio channels helping in proper and wide circulation of the information.

9 Action and Benefits

- Under actions like those that protecting natural areas are the measures for the municipal zoning, regulation can be used to alter or restrict land uses in wetlands, food plains, upland forests, etc. Such measures prevents contamination in overland flows from streams, may increase infiltration and preserve species and ecosystem. The benefits accrued from such measures are the reduced risk of water supply contamination, improved yield of ground water for water supply and the reduced risk of exposure to water borne pathogens, improved water quality for recreation.

- Under actions like Water Conservation are the various measures including optimization of urban and agricultural irrigation practices, replace water use appliances, restrict municipal water use, modify water use habits, reuse wastewater etc.

These reduce water withdrawals and can also improve the performance of wastewater treatment plants.

9.1 Benefits

Increased capacity to serve new customers and reduced the cost of water supply and wastewater system.

10 Result

The rise in ground water table is clearly visible in village area. The water level went upwards by 0.3 m only in 2 years. The reason was storage of surface water in Nalabandhan (stopping the flow of water in narrow stream) near the village. Two wells have been rejuvenated in the watershed area and digging of two wells has been sanctioned in 2007–2008. The area under cultivation also increased. Approximately 50 ha. of land has been brought under cultivation which was earlier left barren. This proved to be beneficial for the families of farmers as their cultivable area increased. New types of crops were cultivated and there were some changes in the cropping pattern adopted in the villages. Some farmers have started the cultivation of non-paddy crop including vegetables, gram, arhar (pigeon peas), mustard, wheat, potato, onion, etc. It was clear that the productivity also has been increased.

11 Discussion

The results of community empowerment come out to be fruitful as the ground water level increased, wells were rejuvenated, crops types, and production increased. However, initially the local people were not ready for lending their hands as they considered it to be Governments' responsibility. Later, they were explained benefits and the case of Rajim, which inspired the farmers for their involvement. As the farmers are the direct beneficiaries, the one who would avail its benefits, they cooperated in every possible manner. The local governing body permitted and some administrative powers were also made to be brought under public consensus for constructing bounds in streams, etc. Finally, the task was accomplished and favorable results were seen. Such type of community program can be conducted in large scale also.

12 Conclusion

Watershed prioritization is an important aspect of planning for implementation of the watershed management program. First, demonstration has been made about implementation of the hydrological model at the watershed scale to generate the estimates of water and sediment yield at the micro watershed level to be used in the planning process. Such information is crucial and invariably not available at such scale. Some peripheral interfaces have been designed to help the planners of the watershed program. Two such applications, one for finding the interaction between the administrative and watershed boundaries that shall help in allocation of financial resources with respect to the watershed boundaries and the other to locate the watershed harvesting structures, which is a common features in the watershed management program, have been formulated and demonstrated.

Effect on Migrants Due to Urbanization: A Study of Slum Area in New Delhi

Simona Sarma and Bhaswati Choudhury

Abstract Agrarian structure has been the backbone of India's economy since time immemorial. However, present trends have been toward greater levels of industrialization ultimately causing urbanization. Urbanization refers to increase in the physical growth of the urban sectors due to the global urge of different nations to link itself to the processes of modernization, westernization and the sociological process of rationalization. Migration seems to be the 'push' factor, which has accelerated the process of urbanization that has resulted in changes in physical and social organizations of the migrants as well as spatial organization of the cities. The purpose of the study is to analyze the living conditions of the migrants in cities and to figure out their problems. The study is based on the materials obtained through survey and observation. The survey composed of personal interviews which were conducted in a slum area near Mukherjee Nagar, New Delhi. The study showed that urban centers become heavily populated with substandard housing and poor living conditions giving rise to land insecurity, unemployment and increase in crime. This situation often leads to decrease in proximity making survival problematic for the vast majority. Moreover, as the agnates of the family (male members) leave for the cities for better economic gains leaving behind their kin group, it has often led to disintegration of joint families in the rural areas. Imparting education and providing the opportunities in the vicinity of the village and making them aware of the various governmental policies can help promote their well-being.

Keywords Migrants · Urbanization · Women · Income · Education

S. Sarma (✉) · B. Choudhury
Department of Sociology, Delhi University, New Delhi, India
e-mail: simona.sarma@gmail.com

B. Choudhury
e-mail: bchoudhurymh@gmail.com

A.K. Sarma et al. (eds.), *Urban Hydrology, Watershed Management and Socio-Economic Aspects*, Water Science and Technology Library 73,
DOI 10.1007/978-3-319-40195-9_26

1 Introduction

A majority of Indian population today resides in the rural areas. Due to "development" (however though this development may not necessarily lead to a greater good), the rural–urban gap in the country is widening. Its further consequence has been a significant clash between 'tradition' and 'modernity' where the "modern" seems inevitably better. Thus, a relatively higher standard of living, benefits of a "modern" city and the possibilities of finding better employment opportunities continue to attract thousands of poor migrants into Delhi every year, making it the second largest hub of in-migrants. This migrant population consists largely of Punjabis, Biharis, Jats, Bengalis, Bhojpuris, Malayalis, etc. This surge of in-migrants unfortunately, consists largely of unskilled workers as a result of which a majority of the migrants find employment only in the unorganized sector. Such high level of migration has therefore resulted in increased "informalization" of Delhi's labor force. The greater informalization of work force combined with the high in-migration into the city has further led to perpetuation of inequities in living standards. The slum-dwellers lack even the basic public amenities needed for living a decent life. For instance, they face a severe housing shortage, which manifests itself in several forms like squatter settlements and illegal subdivisions such as 'jhuggis' and 'jhopdis'. Apart from that, there are only a few migrant households with access to piped water supply, power supply, and sanitation facilities. Homeless women in particular face several insecurities. Although millions of Delhi's residents have benefited significantly from the available medical facilities, large number of poor still lacks the basic health services. Conscious of these poor physical conditions of the migrants and their not having access to basic amenities, many of the Governments in Asia have launched programs at national and local levels to improve the microenvironment in slums and squatter settlements. Unfortunately, however, resource allocation and their spatial coverage have gone down in recent years under the new system of governance and more recently due to global meltdown. Though there has been avowed concern for socioeconomic upliftment of the workers in unorganized sector (consisting mainly of migrants in most countries), not much have come up in terms of programmatic interventions to facilitate their absorption in urban centers. In census 2001, the reasons for migration have been classified into seven broad groups—work/employment, business, education, marriage, moved at birth, moved with family and others. It is observed that employment among males and marriage among females are the main reasons for migration. Thus, the main objective of this study is to identify some of the basic problems associated with such processes along with specific insights into their living conditions. We have mainly taken a "bottom-up approach", i.e., we have tried to bring the grievances of the migrants onto the bigger platform and hence understand the problem from their vantage point (Webster 1990). On the basis of ten parameters we have derived the living conditions of the migrants and tried to understand their state from their perspective.

2 Study Area

Delhi, the capital of the country, attracts thousands of population every year. As per census 2001, its area of 1438 km^2 is home to 13,850,507 persons with a density of 9340 persons/km^2. It is the second largest hub of in-migrants within the country with 5.6 million migrants. The area chosen for our study is a slum in Mukherjee Nagar which comes under the north–west district of Delhi. It is the main hub of civil service coaching center where thousands of aspirants flood in every year. Ironically, in that same place we also find migrants mainly blue-collar workers who engage in struggle to find a day's labor.

3 Materials and Methods

The paper is based on the method of qualitative analysis. It is primarily based on a survey involving the use of interview techniques with the aid of questionnaire and observation method. The first step involved reviewing of secondary sources of information related to the topic of our study in order to understand the existing discourses. The archives chosen for this purpose included a whole range of essays, articles, books, newspapers, etc. Thereafter, the draft of the questionnaire was prepared which covered 10 questions that focussed on the theme of the study. The questionnaire comprised of only open-ended questions (Given in Appendix).

The target population of our study was people in the age group of 20–35. These were migrants who had moved particularly from the rural areas of Delhi and its surrounding states in the hope for better opportunities. In the selection of samples, attempt was made to keep them uniform for both males and females to get reliable and valid data. For this, probability sampling techniques were employed which ensured that every element in the universe had an equal chance of getting included in the sample. A proportionate stratified sample of about 50 males and females was used.

This was then followed by collection of data from the area of study. The data was collected on the basis of interview method with a definite set of questions which enabled systematic and meaningful generalizations. In order to gain more information, interviews were followed up with systematic focus group discussions. The interview was kept flexible enough that enabled in-depth case-study analysis. Further, in the span of three months of conducting the study, nonparticipant observation was made which unfolded valuable insights.

The final step involved processing and analyzing the collected data.

4 Results and Discussions

4.1 Opportunities Attained

Many did come to the city to get some better facilities and opportunities in their life. However, coming here they have seen that the reality seems to be far away from their dreams.

The first man we interviewed replied:

> "*Jo socha tha woh toh kabhi nahi hua aur na hoga!*" (the thing that we thought have never or will never happen!).

The city has in fact moved them into misery and made their life more devastating. The attitude of the residents, i.e., the non-migrants further cause problems and tensions as they often consider the migrants as a burden on city life (who try to consume the scarce resources of the city and in turn disturb its "beauty"). In fact, several slums in the name of "development" were uprooted. This can be seen as part of the efforts to "aestheticise" and legally rationalize the city as a way of creating a world-class urban environment (Mehra 2012). This was in stark contrast to the egalitarian life in the villages. Thus, although there have been relatively better opportunities, they have not been able to utilize them to the fullest. For instance, many jobs refuse to hire them because of the background to which they belong. Most importantly they are pushed into a particular 'bracket' with stereotypes formed about them.

On the other hand, some people, mainly construction workers, paid on a daily basis, have nevertheless been able to better their conditions by being involved in the informal work that cities provide. Moreover, many small-scale jobs of plumbers, carpenters, drivers, etc., are available in the cities, which are not possible in the villages. Thus, as compared to the villages, city has at least enabled them to "partially improve" their life. Therefore, they want to continue in the cities with the hope that they would be able to make their both ends meet. Therefore, those who can make use of those opportunities are in fact happy to be there in the cities.

4.2 Work to Income Proportion

From the income point of view, majority of the migrants remain unsatisfied. Their pay is not proportional to the work done. They work for the whole day and some even for the night, but the money that they get is not even enough to feed the entire family. Moreover, these families often have large numbers (their fertility remains high firstly due to lack of knowledge regarding contraceptive measures and secondly because they feel more members of the family will bring more income). But because of this many parents cannot even afford clothing for all the members. When asked if they are aware of the various governmental income policies they say that those policies are not helping them in anyway.

Raju, a construction worker, clearly mentioned: "*Sarkar kaha kuch kar rahe hain hamare liye? Hum toh unke hisab se sirf gandagi failate hain seharo mein*" (What is the government even doing anything for us? According to them, we only just spread dirt in the cities.)

Even if they try to get advantage from those policies, it is not worth their problems and many of them have had to return empty handed.

4.3 Services Provided by Managers

There are rarely any services provided to the migrants by the managers. The managers neglect the workers and concentrate only on gaining profits. They ignore the cries of the workers and pay little or no attention to their awful condition. Some say that though certain services are provided but they are not according to what they require.

In case of housing facilities, single room is provided for a family of six to fit in. Although there is facility for running water the time specified for its flow is very limited. The public toilet too is left in a very precarious condition. There is no system of cleaning them. Such poor levels of sanitation and hygiene coax them into bad health and diseases. Consequently, however, medical facilities are not provided as such, leaving the people ill and sick.

There are very less options for special grants (bonus, gifts, leave, etc.) in times of celebrations. Moreover, the conditions in which they live, constrain them from performing any kind of recreational activities. Educational facilities for the children of these slum-dwellers are bleak. Such lack of education, forces the migrants to further carry on in their parents' steps. Hence, they are doomed to live in their appalling state.

4.4 Education of Children

Majority of the migrants are reluctant to let their children have basic educational qualification as they feel that education cannot bring them a better future. They are preoccupied with the thought that doing such menial work will at least make them earn money, rather than wasting time to study. They believe that their children too must indulge in their work to keep the service alive. Thus children are forced into these menial work from a very early age that results in 'child slavery'. In other words, the kids are compelled to carry out the work of their ancestors.

Nonetheless, a few lament that they themselves were deprived of what they can now give to their children. They believe that education *can* bring about a change in the society and they want their children to be those harbingers of change.

However, the scenario seems to be different in case of the girl child. Majority of the families want their daughters to engross themselves in household activities rather than going to school. This results in incidence of child marriage ultimately leading to premature child bearing. This further affects the health of both the mother and the baby.

4.5 Consequences of Agnates Leaving the Village

As per findings, most of the agnates (male members) leave their villages in search of work in the cities. As a result their families are left behind. Some young husbands and fathers seem to be satisfied with their decision, as their age allows them to travel once a week to their villages due to various traditional and social festivals which they cannot afford to miss. This is because high level of absenteeism from such community events in villages, tend to raise questions regarding their "commitment level". However, in reality, commitment of the management must also be taken into cognizance. Some aged men, however, seems to be in agony for not having the chance to keep their family near. Yet, there are still others who have walked in with the entire family leading to the increase of population and spatial problem in the area, ultimately causing "problematic urbanization".

In the villages, the scenario seems to be worse. The agnates in the cities do not get the sufficient income that would enable them to provide the necessary conditions of living to their families back home. As a result of which there has been reports of malnutrition, polio, goiter, etc.

4.6 Increasing Crime in the Area

The number of crime seems to have gone up in these areas. Because of weak monetary conditions and lack of satisfaction and happiness, many people have taken recourse to alcoholism, gambling, robbery, etc. Even in some cases, the police tend to be on the "wrong side". In fact, it has been noticed that even if the people report cases to the police without any evidence, it is always the migrants who would be exploited to the worst of their conditions. Hence, some say that the exploitativeness of the police authority along with unfriendliness of the local people is another impetus to the increase in crime rates. In some cases, people noted that there has been an increase in domestic violence as the males come in drunken late night and torture the females and the children. However, only a few of these incidents are reported with the fear that their "only service provider" will leave them unattended. So very few of such violent patriarchs end up in police custody.

4.7 Women Migrants

According to the data, 70 % of the migrant women in Delhi hail from Uttar Pradesh and Bihar (Source: Associated Chambers of Commerce and Industry). In Mukherjee Nagar majority of the women come in search of employment in the unorganized sector. This is more prominent when the male member is sick or dead because of which the whole responsibility of running the family falls on the women (It is very rarely thought by them that women can run the family even with her husband alive or working.). Women mostly are found engaged in construction work, small-scale business like weaving, etc., or as domestic helpers, tailors, beauty parlor assistants, vegetable vendors, scrap dealers and laundry assistants. The female work conditions are, however, very insecure with no proper training or guidance. Moreover, the amount paid to the women is less than what men get for the same kind of work. Apart from this, cases also has been recorded where a lot of women are sent back to the village by administrative authorities for the fear of involvement in trafficking and illegal activities despite the presence of their genuine identity cards. So, most of the females are hesitant to continue their work in the urban setting.

4.8 Policies for Migrants

According to the sources, the Indian government provides no such policies for migrants. The case is even worse in Delhi. In fact Delhi's Chief Minister Sheila Dikshit on 28th December, 2012 at a meeting of National Development Council, blamed the migrants for causing problems to the permanent residents by placing constraints on supply of water and electricity. Even the Indian home minister P. Chidambaram on 13 December, 2010, accused the migrants for increasing crime rates in Delhi. He clearly said, "…nevertheless crime takes place because Delhi attracts a large number of migrants". Migrants are therefore despondent about their situation and want the government to take more responsibility toward them.

Nevertheless, there are certain governmental policies that go unnoticed. Majority of the migrant population are unaware of the government policies and hence are unable to reap its benefits. And the few who are aware, do not know the procedure to apply. Others complain that even if they apply, the time and effort taken will never be worth the kind of benefits they receive.

4.9 Accessibility of Opportunities

The question which becomes more critical than opportunities is migrant's ability to access such resources. Perhaps a redistribution of resources rather than concentration in the hands of few would enable migrants to build a basis in their native lands. This would also give them a medium of life which would then encourage them to stay in their respective homelands. Since majority of the migrants tend to be from the rural areas, the Indian government has nonetheless made many facilities for the development of the rural belt. One such case in point would be the MGNREGA scheme and support systems. Despite this, the benefits of such schemes very rarely trickle down to the grass-root level. "*Sarkar kya hi karegi!*" said Suresh, a 24-year old rickshaw puller. It is unfortunate that many migrants' faith on "good governance" is no longer present. They feel that they have got themselves trapped in the vicious circle of poverty and inequality and are gripped in it forever. Their voices and demands find no space in the mainstream city.

Hence, a number of migrant population tend to be employed as newspaper distributors, motor mechanic, rickshaw pullers, security guards, plumbers, construction workers, etc. They provide services without which the life of the city would be in jeopardy and this contribution of theirs should be highly appreciated.

4.10 Migrants' Satisfaction Level

Satisfaction seemed a really vague concept to address to the migrants. Given their conditions and daily experiences, the question of satisfaction was a far-sighted dream.

It has been found out that majority of the people were not satisfied with the migration process, as it has not helped them to fulfill their dreams and aspirations in any major way. They believe that by staying back in the villages they would have lived a more satisfying life with a free and fearless mind unlike the city where they always remained dependent on people higher up in the social and economic ladder. Moreover, the cost of living in urban areas is much more as compared to the villages where they were often self-reliant. In addition, the socioeconomic exploitation faced by the migrants is also agonizing.

However, some do feel that migration has helped them in major ways. It has enhanced their self-confidence and self-reliability to some extent. Cities have nevertheless provided them with better benefits. Referring to the limited facilities available in terms of health, education, transport and communication, etc., Neelu, a 32-year old women, mother of two has said "*yaha pe toh zindagi guzarti hain*" (Life still goes on).

5 Conclusion

From this study, it has been found that majority of the migrants in our study area are regretting their decision to migrate to urban centers as they feel that cities which promised of opportunities, have actually deprived them of the basic conditions of well-being. The income gained through their work is not adequate to satisfy their wants, as a result of which many of them resort to detrimental practices like robbery, crime, thievery, etc. This lack of income also prohibits them from sending their wards to educational institutions while many of them do not even see its advantage. Lack of education among their children tends to continue this terrible condition for the successive generations too. Hence there emerges a vicious circle of exploitation and suffering. Moreover, the attitude of the government has done little to help the migrants come out of this suffering. The migrants are not provided with any social security measures by their managers, which further compel them to live in dilapidated conditions. The study also showed that there was a majority of women population migrating to Mukherjee Nagar in search of better income opportunities, as the city of Delhi seemed to promise them of infinite job opportunities. We have found out that giving extra opportunities can help them gain economic stability to some extent, but before they can enroll themselves for any such kind of services, they should be given proper training in various fields. Without proper training, they cannot give their level best in any kind of economic activity which often leads to a question of their commitment level. The managerial commitment level, however, should also be taken into account. But in spite of all the barriers, a few proportion of the migrants have been able to engross themselves in daily jobs like carpentry, plumbers, electricians, construction work, vendors, tailoring, beauty parlor assistant domestic help (in case of women), etc. As a result of this the city continues to be the "pull factor" for innumerous migrants. Thus, our study tries to question the often meted out unequal treatment against the migrants in the name of "development" and see the quality services they provide to the "urbanized" city life. By pointing out their problems and the possible solutions to their deplorable conditions, we hope to bring to the fore the 'subaltern' voice.

Acknowledgement We would like to extend our heartiest gratitude to our professors Dr. Dinaz Mirchandani and Dr. Reema Bhatia from Department of Sociology in Miranda House, Delhi University. They have guided us in the successful completion of the paper. We are also grateful to our informants in the study who have given us useful information and helped us for a better comprehension of their problems. At last, we would like to acknowledge ENSURE 2012, which has given us a platform to raise our concerns for the migrants and help them have an enhanced life in the future.

Appendix

QUESTIONNAIRE:

NAME:
AGE:
SEX:
EDUCATIONAL QUALIFICATION:
OCCUPATION:

MARITAL STATUS:
NO. OF CHILDREN:
INCOME:

1. Cities seem to be a promising factor for opportunities. However, have you been able to avail these opportunities?
2. Is your income proportional to the work done and is that enough to support your family?
3. What are the other services provided to you by your managers?
4. Do you aspire for better educational qualification for your children than you do?
5. Did your entire family move to the city with you? And if not, are you morally satisfied to leave them behind?
6. What do you think are the reasons for increasing number of police cases in your area?
7. What do you have to say about the condition and state of women migrants in your area?
8. Are you aware of any governmental policies for migrants? If yes then are you able to reap its benefits?
9. Would it help if small-scale job opportunities are provided to you in the vicinity of this slum? If yes, what type of opportunities would you like to have?
10. Finally, do you think migration has helped you in any way to fulfill your dreams and aspirations?

References

Census of India (2001)
Human Development Report, Delhi (2010)
Mehra D (2012) Protesting publics in Indian cities: the 2006 sealing drive and Delhi's traders. Economic and Political Weekly, XLVII(30)
timesofindia.indiatimes.com
Webster A (1990) Introduction to the sociology of development (2nd edn). Humanities Press International
www.assocham.org

Role of Urban Local Bodies and Opportunities in Municipal Solid Waste Management

Waikhom Roshan Singh and Ajay S. Kalamdhad

Abstract Due to inefficient organisational set up and inadequate resources, the municipal solid waste management services of the majority of the urban local bodies fall far short to be desired. There is a sense of urgency prevailing all over India for positive changes in the municipal solid waste management system in the light of the Municipal Solid Waste (Management and Handling) Rules, 2000. This paper reviews the role of the urban local bodies in municipal solid waste management vis-à-vis with that envisaged in the aforesaid rules as well as in the municipal by-laws and the opportunities available to perform the role. Various systems for the effective management of the urban waste have been evolved and some put into place, as they are suitable for the urban authorities. The systems are ways to augment their resources and capabilities and at the same time enable them to apply the managerial and technological options available in waste management. The urban local bodies can avail central funding under the Jawaharlal Nehru National Urban Renewal Mission apart from the soft loans offered by many national and bilateral funding agencies. The Government of India propagates outsourcing certain waste management services through Public–Private Partnership and involvement of Non–Governmental Organisations.

Keywords Urban local body · Public–Private Partnership · Funding

W.R. Singh (✉)
Manipur Pollution Control Board, Imphal West DC Office Complex,
Lalmphelpat 795004, India
e-mail: wroshan123@gmail.com

A.S. Kalamdhad
Department of Civil Engineering, Indian Institute of Technology Guwahati,
Guwahati, India
e-mail: kajay@iitg.ernet.in

A.K. Sarma et al. (eds.), *Urban Hydrology, Watershed Management and Socio-Economic Aspects*, Water Science and Technology Library 73,
DOI 10.1007/978-3-319-40195-9_27

1 Introduction

286 million out of 1028 million people of India live in urban areas (2001 census) and generate 46 million tonnes per annum of municipal solid waste (MSW) (Kumar and Gaikwad 2004). It is estimated that the urban population of India is continuously growing by 3–3.5 % per annum and the annual MSW generation in the country is increasing by about 5 % every year. Integrated Solid Waste Management (ISWM) is the selection and application of suitable techniques, technologies and management programs to achieve specific waste management objectives and goals. The goals of ISWM is to achieve environmental and health regulations, economic reliability and social acceptability. Developing and implementing an ISWM plan is essentially a local activity that involves the selection of proper mix of alternatives and technologies to meet the changing local waste management needs, while meeting the legislative mandates.

2 Present Scenario of MSWM in India

Municipal Solid Waste Management (MSWM) is an important obligatory function of the urban local bodies (ULBs). However, this service falls far short of the desired levels resulting in problems of health, sanitation and environmental degradation (Report of the committee constituted by the Hon'ble Supreme Court of India 1999). Solid waste management (SWM) has three basic components, viz., collection, transportation and disposal. It is estimated that ULBs spend from Rs. 500 to Rs. 1500 per tonne on MSWM. 60–70 % of the total cost is spent on collection, 20–30 % on transportation and a minimal 5 % on final disposal. New Delhi, for instance, spends 71 % in collection, 26 % in transportation, and 3 % in final disposal (Kumar 2009).

In many cities, the residents collect their household wastes in plastic buckets and deposit it regularly in community bins located near their houses. In some areas, the waste is stored at their individual household bins. Source segregation at the households is reported in some cities such as Ahmedabad, Ludhiana, Gandhinagar, parts of Mumbai, Chandigarh, Goa, Pondicherry, etc. For the purpose, separate bins for wet and dry wastes are provided. Around 27–28 cities have partially initiated house-to-house/door-to-door collection (DTDC) and a few cities have implemented it wholly. DTDC is reported to be fully covered in Nashik, Chennai, Panjim, Vijayawada, Visakhapatnam, Nagpur, and Pondicherry. DTDC is also starting in megacities such as Delhi, Mumbai, Bangalore, Kolkata and Hyderabad (Kumar et al. 2008). However, waste collection is still inefficient in most of the cities. Residents usually leave their wastes in front of their homes to be picked up by the sweepers. The wastes are often scattered by rag pickers while searching for recyclables, as well as by the stray animals in search for food. Generally, overcrowded low-income settlements do not have MSW collection and disposal services.

The reason is that these settlements are often illegal and the inhabitants are unwilling or unable to pay for MSWM services. They throw away their wastes near or around their houses, which make effective collection and transportation of the wastes very difficult in these areas.

During transportation in vehicles, the waste tends to spill on the road resulting in unhygienic conditions. Nowadays, in many cities the wastes are covered with tarpaulin or any other quality cover during transportation in order to minimize the nuisance. In some cities, modern enclosed vehicles are now used for transportation (Bhide and Shekdar 1998; Reddy and Galab 1998).

Cities usually lack recycling programs, but a large number of rag pickers recover recyclables (paper, cardboard, plastics, metal scraps, etc.) from wastes. It has been estimated that up to 1 million individuals make a living from scavenging activities throughout India. Surprisingly, rag pickers are rare in cities like Surat, Kanpur, Coimbatore, Kochi, Visakhapatnam, Nagpur, Bhopal, Indore, and Panjim (Kumar et al. 2008).

90 % of the MSW generated in India is normally disposed off in open dumps, which are the cheapest option available (Das et al. 1998). Open dumps pose environmental hazards and cause ecological imbalances with respect to land, water and air pollution (Kansal 2002). Some cities such as Nashik, Pune, etc., have adopted sanitary land filling. Leachate collection and treatment, and biogas recovery at landfills are not practiced in most of the cities. Daily earth cover is provided in Mumbai, Kolkata, Chennai, Ahmedabad, Kanpur, Lucknow, Coimbatore, Nashik, Vadodara, Jamshedpur, Allahabad, Amritsar, Rajkot, Simla, Thiruvananthapuram, and Dehradun. In hilly regions, disposal of urban waste is carried out along the valley ridges. Landfill sites have not yet been identified by many municipalities and in several municipalities, the landfill sites have been exhausted and the respective local bodies do not have resources to acquire new land.

3 Role of ULBs

In India, the municipal governments are the ULBs responsible for the implementation of MSWM programs and facilities. The responsibility of ULBs in MSWM is clearly defined in Rule 4 of the Municipal Solid Waste (Management and Handling) Rules, 2000 notified by the Ministry of Environment and Forests (MoEF), New Delhi—"Every municipal authority shall, within the territorial area of the municipality, be responsible for the implementation of the provisions of these rules, and for any infrastructure development for collection, storage, segregation, transportation, processing and disposal of municipal solid wastes". Some of the broad responsibilities of the municipal bodies in MSWM as envisaged in the aforesaid Rules are indicated below:

- To take proper steps for prohibiting littering and dumping of solid wastes anywhere in the city and to facilitate compliance thereof;
- To notify waste collection schedule and the likely method to be adopted for the benefit of public and to ensure collection thereof;
- To encourage the citizens to segregate waste through organizing awareness programmes on the benefits of waste segregation in recycling or reuse and ultimately in disposal;
- To establish and maintain storage facilities in such a manner that they do not create unhygienic and insanitary conditions around;
- To clear collected waste daily and transport them in a hygienic manner to the place of processing or disposal; and
- To adopt suitable technology or combination of such technologies to make use of wastes so as to minimize burden on landfill.

The responsibility of implementation of the Rules including development of required infrastructure lies with the municipal authorities. The Rules requires that they should obtain authorization from the State Pollution Control Boards/Committees for setting up waste processing and disposal facilities and further furnish their annual reports to the Boards/Committees. Even though some positive changes are now observed in the Indian MSWM system, the practice still leaves ample room to be desired. To implement this service efficiently, proper technical, managerial, financial and institutional arrangements are required. Growing costs, shortage of funds, institutional deficiencies, indiscipline among the work force, lack of trained personnel and perhaps political pressure are making the situation worse for the ULBs (Ansani 2006). In the face of the grim situation, the ULBs are called in to perform varied roles in their MSWM services, which are discussed below.

3.1 Keep Solid Waste Management in Priority

Technologies/tools are available to tackle most of the environmental issues of cities, but what is often lacking is the willingness to clearly define the priorities and play respective roles. It is high time that the ULBs keep MSWM on the list of their priorities. Elected representatives as well as municipal authorities must see that MSWM receives the attention it deserves by making serious efforts to adapt to the latest technologies of waste management, treatment and disposal.

3.2 Institutional Strengthening and Capacity Building

Officers and staff involved in MSWM require exposure to modern method of waste management from time to time through intensive training, seminars and field visits. They have to be exposed with appropriate technologies through demonstrations and disseminations at international and national levels. There are reports that the work forces at lower levels are generally adverse to change of their daily practices. The notion needs to be cleared through motivations to do their work better and by infusing pride in their work (Agrawal 2001). Further, there is a need for re-organisation of the ULBs to suit the present demands in MSWM by involving environmental engineers, health professionals, etc.

3.3 Preparation of City Wide Plan for Scientific Waste Management

ULBs prepare action plans in-house or by hiring professionals for the purpose (CMAM 2005). ULBs must initiate development of Action Plans that have to go substantially beyond the general trend of acquisition of a landfill site; purchase of vehicles and community bins; and employing more staff. What is required is a detailed plan that would comprehensively and optimally manage the waste generators and the waste generated in the city. A macro action plan can identify the quantum and characteristics of the wastes generated in the city and the broad strategy to be adopted to manage them. This may be followed by a micro or locality wise plan with details of collection route and timing and also the equipments, manpower and their mode of deployment.

3.4 Regulations and Policies

Vision, goals and strategies laid in the action plan can be achieved through proper monitoring. The Municipal Solid Waste (Management and Handling) Rules, 2000 is a step in this direction. As per the said Rules, all metropolitan cities are required to submit an annual report in Form-II to the Secretary, Urban Development Department. All other cities and towns are required to submit the report to the District Magistrates. Further, a copy of the report is to be furnished to the respective State Pollution Control Board. The system of submission of annual report put in place an effective platform for monitoring the performance of the ULBs. Apart from the central rules, some state governments have made their own Rules to prevent throwing or depositing non-biodegradable garbage in public drains, roads, wetland, water bodies, places open to public view and to regulate the use of non-biodegradable materials. There are few cases of inconsistencies between the central and the local by-laws, which may be avoided. A revised Municipal Solid Waste Management Rules, 2016 is in line which mandates source segregation of waste by the generators and/or collectors.

3.5 Maintaining an Up-to-Date Database

A common mistake is to precede a solid waste management policy without the analysis of appropriate data. ULBs need to maintain an adequate database for the design and implementation of the expensive waste collection service as it makes it possible to improve the quality and coverage of the service at lower costs. Optimisation of collection costs requires accurate current information on parameters such as number of stops per route; amounts of waste collected per stop per route; time requirements; and work output per worker and per vehicle for a crew size. These data should influence the design of collection routes and the rationale for assigning crew size and number of vehicles per given route. Reliance upon a management information system can bring about a reduction in the overall cost of solid waste management. Often, the accuracy of data is unfavorably compromised because of failure to adhere to prescribed protocols. Ultimately, effectiveness of cost reduction measures is a function of quality of the professional skill with which they are formulated and applied (UNEP 2005). In this context, Geoinformatics will help in monitoring the unauthorised activities and also in optimization of transportation routes (Ramchandra and Bachamanda 2007).

3.6 Public–Private Partnership

The role of the municipal government typically involves making a decision as to whether to take care to implement the whole MSWM services by themselves or licensing private partners to provide these services or portion of the services within their jurisdiction. The decision should be consistent with legislation, guidelines, policies, and programs that have been adopted at the National and State levels. Public–Private Partnership (PPP) is an arrangement between a government/statutory entity/government owned entity on one side and a private sector entity on the other, for the provision of public assets and/or public services, through investments being made and/or management being undertaken by the private sector entity, for a specified period of time, where there is well defined allocation of risk between the private sector and the public entity and the private entity receives performance linked payments that conform (or are benchmarked) to specified and pre-determined performance standards, measurable by the public entity or its representative. Experience the world over has shown that private sector participation results in cost savings and improvement in efficiency and effectiveness in service delivery mainly due to financial and managerial autonomy and accountability in private sector operations. Besides, it brings in new investment and better technologies (Asnani 2000).

The need for PPP in MSWM has been recognized, both by the Technical Committee appointed by the Supreme Court and also by the 12th Finance Commission. The later has prescribed that 50 % of the funds flowing to the

Municipal Bodies would have to be spent for SWM activities through PPP. Participation may be based on models such as Build, Own and Operate (BOO); Build, Own, Operate and Transfer (BOOT); and Design, Build, Finance, Operate and Transfer (DBFOT) as per the convenience of the ULBs. The contract for partnership to be entered into between the ULB and the Private sector should be clear with regards to delivery, evaluation, penalty and even reward for the service providers. The Union Finance Minister, in the Budget speech for the year 2011–12 announced, "It is our endeavor to come up with a comprehensive policy that can be used by the Centre and the State Governments in further developing Public-Private Partnerships".

Cities like Bangalore, Chennai, Hyderabad, Ahmedabad, Surat, Delhi, Coimbatore, Guwahati, etc., have entered into PPP for collection, transportation and sanitary filling of MSW. In cities like Bangalore, Hyderabad, Ahmedabad, Kolkata, Bhopal, Nashik, etc., compost plants have been established and commissioned by private agencies. A waste-to-energy plant of capacity 500 TPD of MSW was established at Vijayawada by M/S Shriram Energy Systems Ltd., Hyderabad and has been in operation since December 2003. Another plant of capacity 700 TPD of MSW and with a power generating capacity of 6.6 MW was installed at Gandhamguda near Hyderabad by M/S SELCO International, Ltd in November, 2003. Vermicomposting of MSW has been initiated in five cities, i.e., Hyderabad (7 TPD capacity), Nagpur (30 TPD capacity), Pune (50 TPD capacity), Indore (1.25 TPD capacity), and Pondicherry (5 TPD capacity). A biomethanation plant for the treatment of MSW (300 TPD capacity) has been commissioned at Lucknow to generate electrical energy (Kumar et al. 2008).

3.7 Involving Non-governmental Organisations and Community-Based Organisations

The informal system plays an important role in waste management system and exists as a parallel system to the formal waste management system. The sector includes rag pickers, Itinerant Waste Buyers (IWBs), and small recycling enterprises. Non-Governmental Organisations (NGOs) and Community-Based Organisations (CBOs) can play an important role in incorporating the micro-enterprises and the informal waste recycling groups. In Ahmedabad, Self-Employed Women's Association is involved in organizing women rag pickers. Collection efficiency is reported to be high in the cities and states, where private contractors and NGOs are employed for the collection and transportation of MSW. In addition, the citizen's welfare associations are also entrusted works of collection in some urban areas on specified monthly payments. Involvement of local public is important in preparation of SWM Action Plans for ensuring smooth implementation. Endorsing of the plan from such varied groups shall not only help in bringing innovative ideas but also help in creating a sense of belonging, responsibility and awareness among them to keep the city clean. EXNORA, Swabhimana, Waste Wise, Swachha Bangalore, Shuchi Mitras, etc. are some NGOs that support MSWM. They carry out public grievance meetings to identify problems and convey them to the authorities. Few NGOs have also set up decentralised composting plants in residential areas and for this they also carry out DTDC and educate the public to segregate the waste prior to disposal. They also hold awareness programmes in schools, colleges, public places, etc., to about the positive aspects of waste segregation, non-littering, etc.

3.8 Involvement of Other Stakeholders

ULBs also need to look for new ways to share their traditional responsibilities of MSWM with neighbourhood communities, micro and small enterprises (MSEs) and large private entrepreneurs and industries. Increasingly, the local authority may seek to mobilise the human and financial resources of these actors in order to develop an adequate system of waste services. Apart from the aforementioned, other stake holders in MSWM may be

- City planners—Keeping waste management in mind while developing city plans
- Teachers/academicians—Influencing minds especially of students on the culture of solid waste management and carry out relevant research and development
- Vendors/shop owners—Ensuring that the waste/litter is properly put in a nearby garbage bin
- Hospitals—Implementing the requirements of bio-medical rules
- Politicians—Leading 'Clean City' campaigns and pressurizing the municipal corporation to make 'Clean City' an issue of priority
- Corporations—Ensuring that all employees understand the gravity of the situation by taking serious actions on the cleanliness front within the office/factory premises and spreading the message across the city. They may also provide dustbins outside the office/company premises so that the passers-by do not throw garbage on the road (Joseph 2005).

The MoEF is responsible for the environmental policies at the national level, including SWM. The Ministry has an overview of all the activities of the MSWM sector and makes sure that it is performed well. It reviews the Environmental Impact Assessment carried out by the agencies prior to the construction of a landfill site or a processing plant. The Central Pollution Control Board keeps a check on all the activities that have potential to pollute the environment and this includes the monitoring of the MSWM activities in the country. Likewise, the State Pollution Control Boards (SPCBs) keep a check on all the activities that have the potential to pollute the environment including monitoring of MSWM in their respective states.

3.9 Improving the Financial Condition of ULBs

According to the report of the Technological Advisory Group on SWM, most of the municipalities are financially weak and are, therefore, not able to discharge their obligatory functions satisfactorily. MSWM costs 20–50 % of the total Municipal Expenditure Budget. The report also mentions a survey by the National Institute of Urban Affairs (NIUA) indicating that MSWM services generate negligible or very insignificant revenues. In order to meet the mandatory MSWM Rules, ULBs are required to raise funds through efficient collection of taxes and user service charges. The ULBs cannot shirk from their responsibilities for the reasons of financial limitations. In addition, since the obligatory duty has to go on indefinitely, it has to be a self supporting, or low cost activity. It should operate on business principles and has to be financially viable in the long run. Introducing the concept of payment of service charges by each waste generator against ensured service delivery would lead to direct involvement of the citizens and also help in attaining a self-monitoring system. Once the service charges are linked to the quantity of waste handed over to the public system, the residents would be motivated to reduce the quantum of waste or to segregate the waste before handing over. Service charges are levied within the framework of the by-laws passed by the municipal bodies for MSWM (CMAM 2005).

In the XIth Five Year Plan of India (2007–2012), the total fund requirement for implementation of the Plan target in respect of solid waste management was Rs. 2212 crore. Jawaharlal Nehru Urban Renewal Mission (JNNURM) is a reform-linked urban infrastructure investment project launched by the Government of India. Under the mission, approximately Rs. 100,000 crore was earmarked for the seven year period 2005–2012 for improvement of infrastructure and providing basic services for the poor in urban areas. The Government has identified 65 cities under Urban Infrastructure and Governance component of the JNNURM program. As on April 2009, the Government approved 461 projects at total cost of Rs. 49,422.4 crore under urban infrastructure and governance component of the program out of which Rs. 2186 crore was earmarked for SWM. The States and the ULBs to access the benefits of JNNURM had to complete a total of 22 mandatory and optional reforms, during the seven-year period (2005–12). Many of the reforms were steps to empower ULBs. Some of the reforms are:

1. Implementation of 74th Constitution Amendment Act
    - (i) Election in ULBs
    - (ii) Transfer of 12th Schedule functions to ULBs
    - (iii) Formation of District Planning Committees/Metropolitan Planning Committees
    - (iv) Formation of State Finance Commission and implementation of its recommendations

2. Public Disclosure Law

   Regular disclosure budgets, projects, revenues, financial statements, etc.

3. Formation of Area Sabhas for active participation of community in the budget making process, project implementation monitoring, etc.
4. Integration of City Planning and Delivery Function with ULBs
5. Accounting Reform

   (i) Introduction of accrual-based double entry accounting system
   (ii) Preparation of Annual Balance Sheets.

6. Property Tax Reform

   (i) Introduction of Self Assessment System
   (ii) More than 90 % of properties to be on tax records
   (iii) More than 80 % of tax collection

7. User Charges—at least 100 % collection of operation and maintenance expenditure
8. Administrative and Structural Reforms

   (i) Human Resource Development policy
   (ii) Municipal Cadre

9. Encouraging Public-Private Partnership

There are many funding institutions available where the ULBs can avail soft loans for the components of MSWM services. Some of the funding agencies are Housing and Urban Development Corporation (HUDCO), Industrial Credit and Investment Corporation of India (ICICI), Infrastructure Development Finance Company (IDFC), Indian Renewable Energy Development Agency (IREDA), etc. Bilateral and Multilateral donors such as World Bank, Indo-US Financial Institution Expansion, US Asia Environmental Partnership, JICA, Asian Development Bank (ADB), etc., are also available for providing technical and financial assistance. In addition, through foreign direct investment (FDI) funds may be mobilized to support the sectoral activities.

Another option is the Kyoto Protocol which provides carbon trading between developed and developing countries. If a developing country like India can reduce Green House Gas (GHG) emissions from solid waste by capturing methane, then it is eligible for certified emission reductions (CER) which can be sold in the carbon market. The World Bank has estimated that methane capture from landfills and bio-methanation will be eligible for carbon finance at about Rs. 200 per ton of municipal solid waste. The Protocol also provides another route for carbon financing known as the Clean Development Mechanism (CDM), whereby a developed country invests in solid waste disposal in a developing country and claims some or all of the emission reduction for its country. Developed countries find it cost-effective to either buy carbon credits in the market or invest in CDM projects compared to incurring the cost of reducing greenhouse gas emissions.

Since GHG are global pollutants, it does not matter where the reduction takes place as long as the reduction targets for each country are met.

4 Conclusions

The ULBs have varied roles in MSWM including policy-making to legitimise and support the roles of local communities, NGOs, CBOs and MSEs. They also support for participation of private sectors and other stakeholders in the MSWM process in accordance with the policy and guidelines at the national and regional level. It is something of a challenge for the local authority to adjust its operational procedures to reliable co-ordination with new partners. Nevertheless, they need to be well equipped and trained to perform their responsibilities and in the process find answers.

References

Agrawal GD (2001) Sustainable waste management-objectives, targets and policies for India. J IAEM 28:79–83

Ansani PU (2006) Solid Waste Management. In: India Infrastructure Report, 2006

Bhide AD, Shekdar AV (1998) Solid waste management in Indian urban centers. Int Solid Waste Assoc Times (ISWA) (1):26–28. 466

City Managers Association Maharasthra (2005) Compendium of best practices in Municipal Waste Management

Das D, Srinivasu M, Bandyopadhyay M (1998) Solid state acidification of vegetable waste. Indian J Environ Health 40(4):333–342

Joseph K (2005) Stakeholder participation for sustainable waste management. Habitat Int 30:863–871

Kansal A (2002) Solid waste management strategies for India. Indian J Environ Prot 22(4):444–448

Kumar S, Gaikwad SA (2004) Municipal solid waste management in Indian Urban Centres: an approach for betterment. In: Gupta KR (ed) Urban Development Debates in the New Millennium. Atlantic Publishers & Distributors, New Delhi, pp 100–111

Kumar S, Bhattacharyya JK, Vaidya AN, Chakrabarti T, Devotta S, Akolkar AB (2008) Assessment of the status of municipal solid waste management in metro cities, state capitals, class I cities, and class II towns in India: an insight. Waste Manag 29(2009):883–895

Ramchandra TV, Bachamanda S (2007) Environmental audit of municipal solid waste management. Int J Environ Technol Mange 7(3/4)

Reddy S, Galab S (1998) An integrated economic and environmental assessment of solid waste management in India—The Case of Hyderabad, India

Report of the Committee constituted by the Hon'ble Supreme Court of India (1999) Solid Waste Management in Class 1 Cities

UNEP (2005) Solid waste management

Economic Evaluation of Transportation Project: A Case Study of Ferry System for IIT Guwahati

Abegaonkar Amit Anantrao, A. Mani Venkat Sai Kumar and A.K. Maurya

Abstract Transportation projects hold a major share of the economic investments made in India. Evaluation of alternate strategies is necessary to optimize gain from investments in terms of facilities and costs. Various methods like NPV, IRR can be used for feasibility studies of projects. The study consists of measuring the direct–indirect costs and overall benefits. For the economic evaluation of the proposed project, we have used a cost-effective approach comparing the costs and the outcomes from the different means of transportation. In IIT Guwahati campus, the major means of public transportation is a Bus service provided by Green Valley Travels Private Ltd. The rise in campus population is pushing this service to an extreme level causing discomfort to passengers. Located on the north side of river Brahmaputra, IITG Campus is 20 km away from the main Guwahati city with Saraighat Bridge being the only connecting link by road. Saraighat Bridge is the first rail-cum-road bridge constructed over the River Brahmaputra in Guwahati in 1962 by the then Prime Minister Jawaharlal Nehru. It is the only bridge over the river Brahmaputra for about 100 km upstream and downstream. Daily huge amount of highway traffic is forced to take Saraighat Bridge in order to enter city and northeast States, resulting in slow moving long queues, almost choking the vehicular flow. This traffic jam on the way to city is responsible for the further increment in travel time. It takes about 75–90 min to reach city due to high level of congestion. The alternate ways to reach city are either by Auto-rickshaw or by Ferry. The existing Govt. Ferry Transport hardly takes 20 min to reach city, but there is no proper facility available to reach Ferry-Ghat. Our work elaborates the socioeconomic evaluation of establishing Ferry Transportation system from the IITG campus to Guwahati city. After a thorough survey of the costs and benefits

A.A. Anantrao (✉) · A.M.V. Sai Kumar · A.K. Maurya
Department of Civil Engineering, Indian Institute of Technology Guwahati, Guwahati, India
e-mail: abegaonkar@iitg.ac.in

A.M.V. Sai Kumar
e-mail: mani.venkat@iitg.ac.in

A.K. Maurya
e-mail: maurya@iitg.ac.in

A.K. Sarma et al. (eds.), *Urban Hydrology, Watershed Management and Socio-Economic Aspects*, Water Science and Technology Library 73,
DOI 10.1007/978-3-319-40195-9_28

involved, final Cost benefit analysis of the proposed ferry system show a high internal rate of return. The project appears to be beneficial both in terms of cost-saving and time-saving. Practical implementation of the proposed project will promote overall socioeconomic growth.

Keywords Economic evaluation techniques · Transportation projects · Case study · Net present value · Internal rate of return · IIT Guwahati

1 Introduction

The life cycle of project begins with the initial conception of the project and continues through planning, design, procurement, construction, start-up, operation, and maintenance. It ends with the disposal of a facility when it is no longer productive or useful. In the planning stage, it is important to evaluate facilities rationally with regard to both the economic feasibility of individual projects and the relative net benefits of alternative and mutually exclusive projects. Methods of economic evaluation include the net present value method, the equivalent uniform annual value method, the benefit-cost ratio method, and the internal rate of return method, which will be discussed in the following sections. Facility investment evaluation includes time preference for consumption, opportunity cost, minimum attractive rate of return, cash flows over the planning horizon and profit measures. Factors affecting cash flows are depreciation and tax effects, price level changes, and treatment of risk and uncertainty.

A systematic approach for economic evaluation of facilities consists of the following major steps:

1. Generate a set of projects or purchases for investment consideration.
2. Establish the planning horizon for economic analysis.
3. Estimate the cash flow profile for each project.
4. Specify the minimum attractive rate of return (MARR).
5. Establish the criterion for accepting or rejecting a proposal, or for selecting the best among a group of mutually exclusive proposals, on the basis of the objective of the investment.
6. Perform sensitivity or uncertainty analysis.
7. Accept or reject a proposal on the basis of the established criterion.

The basic principle in assessing the economic costs and benefits of new facility investments is to find the aggregate of individual changes in the welfare of all parties affected by the proposed projects. The changes in welfare are generally measured in monetary terms, but there are exceptions, since some effects cannot be measured directly by cash receipts and disbursements. Examples include the value of human lives saved through safety improvements or the cost of environmental degradation. The difficulties in estimating future costs and benefits lie not only in the uncertainties and reliability of measurement, but also on the social costs and benefits generated as side effects.

Furthermore, proceeds and expenditures related to financial transactions, such as interest and subsidies must also be considered by private firms and public agencies. Another difference between public and private facility owners is in their relative liability for taxes. Public entities are often exempt from taxes of various kinds, whereas private facility owners incur a variety of income, property, and excise taxes. However, these private tax liabilities can be offset, at least in part, by tax deductions of various kinds.

2 Existing Evaluation Techniques

A profit measure is defined as an indicator of the desirability of a project from the standpoint of a decision maker. A profit measure may or may not be used as the basis for project selection. Since various profit measures are used by decision makers for different purposes, the advantages and restrictions for using these profit measures should be fully understood.

There are several profit measures that are commonly used by decision makers in both private corporations and public agencies. Each of these measures is intended to be an indicator of profit or net benefit for a project under consideration. Some of these measures indicate the size of the profit at a specific point in time; others give the rate of return per period when the capital is in use or when reinvestments of the early profits are also included. If a decision maker understands clearly the meaning of the various profit measures for a given project, there is no reason why one cannot use all of them for the restrictive purposes for which they are appropriate. With the availability of computer-based analysis and commercial software, it takes only a few seconds to compute these profit measures. However, it is important to define these measures precisely.

2.1 Net Future Value and Net Present Value

When an organization makes an investment, the decision maker looks forward to the gain over a planning horizon, against what might be gained if the money were invested elsewhere. A minimum attractive rate of return (MARR) is adopted to reflect this opportunity cost of capital. The MARR is used for compounding the estimated cash flows to the end of the planning horizon, or for discounting the cash flow to the present. The profitability is measured by the net future value (NFV) which is the net return at the end of the planning horizon above what might have been gained by investing elsewhere at the MARR. The net present value (NPV) of the estimated cash flows over the planning horizon is the discounted value of the NFV to the present. A positive NPV for a project indicates the present value of the net gain corresponding to the project cash flows.

If there is no budget constraint, then all independent projects having net present values greater than or equal to zero are acceptable. For mutually exclusive proposals, a proposal should be selected if it has the maximum nonnegative net present value among all proposals.

The strength of calculating NPV is that we are recognizing the value of a rupee today is greater than the value of a rupee received a year from now. That's the time value of money concept. NPV rule tells us to accept all investments where the NPV is greater than zero. However, the measure doesn't tell us when a positive NPV is achieved. Does it happen in 5 years or 15? Another limitation of the NPV approach is that the model assumes that capital is abundant; that is there is no capital rationing. If resources are scarce, then the analyst has to look carefully at not just the NPV for each project they are evaluating, but also the size of the investment itself.

2.2 *Equivalent Uniform Annual Net Worth*

The equivalent uniform annual net worth (EUAW) is a constant stream of benefits less costs at equally spaced time periods over the intended planning horizon of a project. It is a measure of the net return of a project on an annualized basis. The use of EUAW alone presupposes that the discounted benefits of all potential projects over the planning horizon are identical and therefore only the discounted costs of various projects need be considered. Therefore, the EUAC is an indicator of the negative attribute of a project which should be minimized.

The equal uniform annual net worth method has an advantage of comparing alternatives with different life times. Alternatives are compared based on their equivalent annual cash flow.

2.3 *Benefit-Cost Ratio*

The benefit-cost ratio (BCR), defined as the ratio of discounted benefits to the discounted costs at the same point in time, is a profitability index based on discounted benefits per unit of discounted costs of a project. It is sometimes referred to as the savings-to-investment ratio (SIR) when the benefits are derived from the reduction of undesirable effects. Its use also requires the choice of a planning horizon and a MARR. Since some savings may be interpreted as a negative cost to be deducted from the denominator or as a positive benefit to be added to the numerator of the ratio, the BCR, or SIR is not an absolute numerical measure. However, if the ratio of the present value of benefit to the present value of cost exceeds one, the project is profitable irrespective of different interpretations of such benefits or costs.

BCR Analysis is a powerful, widely used, and relatively easy tool for deciding whether to go with the project. Since some savings may be interpreted as a negative cost to be deducted from the denominator or as a positive benefit to be added to the numerator of the ratio, the BCR or SIR is not an absolute numerical measure. This method again does not indicate time after which positive returns start, so it is required to work out the time it will take for the benefits to repay the costs.

2.4 *Internal Rate of Return*

The internal rate of return (IRR) is defined as the discount rate, which sets the net present value of a series of cash flows over the planning horizon equal to zero. It is used as a profit measure since it has been identified as the "marginal efficiency of capital" or the "rate of return over cost." The IRR gives the return of an investment when the capital is in use as if the investment consists of a single outlay at the beginning and generates a stream of net benefits afterwards. For cash flows with two or more sign reversals of the cash flows in any period, there may exist multiple values of IRR; in such cases, the multiple values are subject to various interpretations.

The term *internal rate of return method* has been used by different analysts to mean somewhat different procedures for economic evaluation. The method is often misunderstood and misused, and its popularity among analysts in the private sector is undeserved even when the method is defined and interpreted in the most favorable light. The method is usually applied by comparing the MARR to the internal rate of return value(s) for a project or a set of projects. The IRR rule tells us to accept projects where the IRR is greater than the opportunity cost of capital. But if this discount rate changes each year, then it's impossible to make this comparison.

A major difficulty in applying the internal rate of return method for economic evaluation is the possible existence of multiple values of IRR when there are two or more changes of sign in the cash flow profile. When that happens, the method is generally not applicable either in determining the acceptance of independent projects or for selection of the best among a group of mutually exclusive proposals unless a set of well-defined decision rules are introduced for incremental analysis. In any case, no advantage is gained by using this method since the procedure is cumbersome even if the method is correctly applied. This method is not recommended for use either in accepting independent projects or in selecting the best among mutually exclusive proposals.

2.5 Adjusted Internal Rate of Return

IRR does not take into consideration the reinvestment opportunities related to the timing and intensity of the outlays and returns at the intermediate points over the planning horizon. If the financing and reinvestment policies are incorporated into the evaluation of a project, an adjusted internal rate of return (AIRR) which reflects such policies may be a useful indicator of profitability under restricted circumstances. Because of the complexity of financing and reinvestment policies used by an organization over the life of a project, the AIRR seldom can reflect the reality of actual cash flows. However, it offers an approximate value of the yield on an investment for which two or more sign reversals in the cash flows would result in multiple values of IRR. The adjusted internal rate of return is usually calculated as the internal rate of return on the project cash flow modified so that all costs are discounted to the present and all benefits are compounded to the end of the planning horizon.

2.6 Return on Investment

When an accountant reports income in each year of a multi-year project, the stream of cash flows must be broken up into annual rates of return for those years. The return on investment (ROI) as used by accountants usually means the accountant's rate of return for each year of the project duration based on the ratio of the income (revenue less depreciation) for each year and the undepreciated asset value (investment) for that same year. Hence, the ROI is different from year to year, with a very low value at the early years and a high value in the later years of the project. ROI has become a central financial metric for asset purchase decisions for computer systems, factory machines, or service vehicles, for example, and approval and funding decisions for projects and programs of all kinds, such as marketing programs, recruiting programs, and training programs, and more traditional investment decisions, such as the management of stock portfolios or the use of venture capital.

2.7 Payback Period

The payback period (PBP) refers to the length of time within which the benefits received from an investment can repay the costs incurred during the time in question while ignoring the remaining time periods in the planning horizon. Even the discounted payback period indicating the "capital recovery period" does not reflect the magnitude or direction of the cash flows in the remaining periods. However, if a project is found to be profitable by other measures, the payback period can be used as a secondary measure of the financing requirements for a project. The payback period is what generally known as a break-even point.

IRR does not take into account the variation in time-period for returns on investment and it does not justify the quick early returns to slower returns. Hence, IRR hinders in the utilization of capital to the maximum period of the project, as it favors for the early returns for higher NPVs. The Profitability Index (PI), computed as the (PV of future cash flows)/(PV Initial investment), is probably a better measure as it allows you to identify the relationship of investment to payoff of a proposed project. Hence we should adhere to the following route when assessing investment returns: NPV, then PI, and finally only IRR. Also, in the case of mutually exclusive projects IRR and NPV may lead to different decisions. The decision as shown by IRR may not be correct. This can be brought to light using the Incremental Cash flow between two projects.

3 Problem Statement

Guwahati is a major city in eastern India, often considered as the gateway to the northeast region of the country and is the largest city within the region. Dispur, the capital of the Indian state of Assam is situated within the city. It is a major commercial and educational center of eastern India and is home to world class institutions, such as the Indian Institute of Technology Guwahati. The city is also an important hub for transportation in the north-east region. Guwahati possesses many places of interests with its lively urban activities, ancient temples, numerous scenic natural features, and the recreational activities they provide. Guwahati is also situated at the center of an attractive region (within 200 km radius) with natural parks, wildlife sanctuaries, hill stations of different types, and a colorful cultural landscape.

Located on the north side of river Brahmaputra; IITG Campus is around 20 km away from the main Guwahati city with Saraighat bridge being the only connecting link by road. Saraighat Bridge is the first rail-cum-road bridge constructed over the River Brahmaputra in Guwahati. It was opened to traffic in 1962 by the then Prime Minister Jawaharlal Nehru. It is the only bridge over the river Brahmaputra for about 100 km upstream and downstream, becoming a backbone of the city's economic development. Daily huge amount highway traffic is forced to take Saraighat Bridge to enter northeastern states.

Most of the campus staff and students rely on the private bus system hired by IITG administration for their daily back and forth travel to the city. The existing bus system is quite convenient and became an important part in day-to-day life of campus residents. However, the recent tremendous increase in the congestion over the Saraighat Bridge is affecting the travel time significantly. In addition, the continuous increase in the campus population is resulting in extreme utilization of the bus system, reducing the comfort level of passengers. Alternately, students and faculty travel by their own private vehicles or hire an auto (Fig. 1).

Fig. 1 Route for the proposed ferry system for IITG

The other available alternate travel mode is the Govt. Ferry System which is around 3 km southeast of the campus. The Ferry system is time efficient as it takes very less time (15 min) compared to the IITG bus system (90 min), but a proper access to the Ferry Ghat from IITG campus is not available. The unpleasant surroundings at the River Ghat, lack of better seating space and safety measures on the ferry is a big letdown for IITG residents from using this mode. All these drawbacks lead to propose an alternate time and cost efficient travel facility without compromising the comfort level of the campus residents.

3.1 *Case Study: Ferry System for IIT Guwahati*

The concept of proposed travel facility is based upon

- Easy Accessibility
- Optimum Travel time
- Minimum Travel cost

These three factors led to the proposal of the self-owned ferry system for the IITG with the modified route from that of existing Govt. Ferry system.

This ferry system will not only reduce the travel time of the passenger but will also be cost efficient. Ferry will be accessible from the KV Gate of IITG Campus and will lead to the Fancy Bazar ferry Ghat. Thus, this being a new establishment; complete economic analysis of the project needs to be done before practically implementing the ferry system.

3.2 *Economic Analysis of Proposed Ferry System*

Amongst several methods available for the economic evaluation of the projects; the IRR is used taking the following factors into consideration.

(a) This method calculates the rate of return considering the net benefit value to be zero. The direct benefit in terms of money would not be a major factor for IITG against the other positive aspects of the project.
(b) The rate of return shows the actual value of return, which can directly be compared to the interest rate charged by the bank and thus the decision for feasibility of the project can be drawn.

3.2.1 Cost Analysis

The economic evaluation of the project considers the following costs.

1. Fixed Construction cost
2. Operational cost
3. Maintenance cost
4. Other Benefits (user, environmental, etc.)

The fixed construction cost involves the purchasing cost of a couple of ferries and the cost of constructing a platform to dock and aboard the ferry at both the stations. The market study and various surveys Table 1 done at the existing Govt. ferry system says the total fixed cost will not exceed fifty-five lakh. The lifetime of a single ferry can be comfortably assumed to be more than 20 years.

The operational cost includes fuel cost along with the logistics cost which includes drivers and conductor's salaries, Govt. taxes, etc. The fuel cost per day is determined by considering the total daily run with a maximum of ten trips each way. Thus, taking the mileage of three Km per liter of diesel and a run of 6 km per trip; the fuel cost totals to sixty thousand per month. Adding the approximate value of logistics and the salaries of workers; the total monthly maintenance cost turns out to be one lakh and twenty thousand. A sum of two lakh per year is further added as maintenance cost to the total cost. Thus, the total annual operational and maintenance cost will sum up to the approximate value of sixteen lakh per annum. The benefit obtained from charging 10 rupees per head per trip with a ferry capacity of fifty will total to around Rs. 40 lakh per annum.

Table 1 Estimation of fixed construction cost

S. no.	Item	Cost	No of items	Total
1	Ferry	12.5	2	25
2	Docking platforms	10.0	2	20
3	Tender (5 years), Govt. clearance and taxes	2.5	4	10
Total (Rs. in Lakh)				55

Table 2 Estimation of operational and maintenance cost

S. no.	Items	Cost per item	Number	Total cost per month
1	Fuel cost	Rs. 50 per liter	40 × 30 days	60,000
2	Logistic	Rs. 8000 PM	6	48,000
3	Maintenance	Rs. 15,000 PM	1	15,000
Total				123,000

Table 3 Total average variable cost per year

S. no.	Item	Cost (per year)
1	Operational and maintenance cost	Rs. 1,600,000
2	Feeder bus in city and campus	Rs. 1,500,000
Total		Rs. 3,100,000

3.2.2 Cost Estimation of Feeder Bus Service

For the convenience and comfort of the campus residents; an additional feeder bus needs to be provided which will pick and drop the passengers from the KV Gate to the respective residences. Considering an equal number of to and fro trips as that of ferry; the intra campus feeder bus system will approximately cost half a lakh per month on rent basis. On the other side a couple of buses can be arranged to ply between Pan Bazar and Jalukbari via Fancy Bazar, a distance of 9.5 km. This facility would cost around 1.25 lakh per month considering a bus rent to be Rs. 30,000 per month and operational cost to be Rs. 12 per km. Thus adding the cost of feeder bus system to operational cost of ferry, the total expenditure per year is Rs. 31 lakh per year (Tables 2 and 3).

3.2.3 Calculation of IRR

Based upon the IRR concept, equating the net present benefit to zero and taking the return period of twenty years,

$$\mathrm{NPV}_0 = (B_0 - C_0) + \frac{(B_1 - C_1)}{(1+r)^1} + \frac{(B_2 - C_2)}{(1+r)^2} + \cdots + \frac{(B_t - C_t)}{(1+r)^t} + \cdots + \frac{(B_n - C_n)}{(1+r)^n}.$$

The r value obtained is 15.4 % considering feeder buses and ferry together for a period of 20 years. When ferry service alone is considered, resulting r value is 43.6 %; which clearly indicates the benefits seen in this project in terms of direct return value. As per the Govt. guidelines, any transportation project with return value more than 12 % is feasible and can be practically implemented.

Table 4 Rent paid to Green valley bus service in year 2011

Months	Total payable	Tax (13.5 %)	Total
Jan'11	706,813	184,763	891,576
Feb'11	679,849	162,310	842,159
Mar'11	708,123	181,811	889,934
Apr'11	774,568	196,567	971,135
May'11	771,123	196,567	967,690
Jun'11	755,772	177,066	932,838
July'11	772,826	196,567	969,393
Aug'11	782,909	196,567	979,476
Sep'11	772,631	196,567	969,198
Oct'11	725,637	196,567	922,204
Nov'11	788,589	196,567	985,156
Total (in rupees)			10,320,759

3.3 Comparison Existing and Proposed

Distance between IIT campus and Panbazar is about 20 km which takes around 90 min by bus whereas traveling by ferry takes 20 min till Fancy Bazar. Table 4 gives the amount of rent IIT Guwahati pays to Green Valley (Establishment cell, Administration IIT Guwahati) during the last 11 months all in Rupees.

On an average, the total payable amount to Green Valley per month is around 9.3 lakh rupees. The salary for drivers and conductors is handled by green valley. The total amount of rent IIT Guwahati pays to Green Valley is approximately Rs. 103 lakhs in the year 2011. Whereas the total cost involved in initiating ferry services is around 50 lakh and from then onwards is a positive income of 9 lakh per year. A lot of fuel can be cut down as it is the shortest route between city and campus.

4 Conclusion

It is always important to evaluate facilities rationally with regard to both the economic feasibility of individual projects and the relative net benefits of alternative and mutually exclusive projects. Methods of economic evaluation include the net present value method, the equivalent uniform annual value method, the benefit-cost ratio method, and the internal rate of return method. For the proposed project of initiating a ferry service from IIT Guwahati to city, IRR method is considered to economically evaluate the project. The internal rate of return for a horizon of 20 years is determined to be 15.4 %, which makes it economically feasible to initiate the project.

The project is not just economically profitable. The most significant benefit out of this service would be the reduction in travel time. Travel time is cut down by

75 % from 90 to 20 min. Guwahati is a city of great scenic beauty and this beauty can be appreciated very well if one could cross the river by ferry rather than traveling through congested and highly polluted roadways. Environmentally it is beneficial by cutting down a lot of fuel consumption compared to bus service. Because of comfort and lesser travel time, many students who own private vehicles would tend to take the ferry. This might reduce the number of road accidents of campus residents that happen in the city. Faculty members, mostly avoid traveling by existing ferry due to many reasons like unpleasant surroundings, sometimes it is overloaded and it appears unsafe to travel. In the proposed project ferries will be equipped with good lighting, lifebuoys, lifejackets, and fire extinguishers. Drivers and conductors would be recruited who are good swimmers and they would be trained to act accordingly in emergency situations.

The project can be extended further by conducting economic evaluation of ferry and bus systems together. Surely fraction of the existing bus fleet can be cut down to an optimum number of buses that could be kept operational along with the ferry system.

Acknowledgments I would like to express my sincere gratitude to my supervisor Prof. Dr. A.K. Maurya for his valuable guidance, support. I would like to thank the establishment cell and accounts section IIT Guwahati and all the people responsible for successful completion of the paper.

References

http://pmbook.ce.cmu.edu/06_Economic_Evaluation_of_Facility_Investments.html, Jan 2012
http://www.moneyzine.com/Investing/Investing/Evaluating-Cash-Flow-Results/, Jan 2012

Author Index

A.K. Sarma et al. (eds.), *Urban Hydrology, Watershed Management and Socio-Economic Aspects*, Water Science and Technology Library 73,
DOI 10.1007/978-3-319-40195-9

Subject Index

A.K. Sarma et al. (eds.), *Urban Hydrology, Watershed Management and Socio-Economic Aspects*, Water Science and Technology Library 73,
DOI 10.1007/978-3-319-40195-9

FSC
www.fsc.org
MIX
Papier | Fördert
gute Waldnutzung
FSC® C083411

Kolibri 360 GmbH
Ferdinand-Jühlke-Straße 7
99095 Erfurt, Deutschland
produktsicherheit@kolibri360.de